ENVIRONMENTAL REMEDIATION TECHNOLOGIES, REGULATIONS AND SAFETY

LANDFILLS AND RECYCLING CENTERS

PROCESSING SYSTEMS, IMPACT ON THE ENVIRONMENT AND ADVERSE HEALTH EFFECTS

ENVIRONMENTAL REMEDIATION TECHNOLOGIES, REGULATIONS AND SAFETY

Additional books in this series can be found on Nova's website
under the Series tab.

Additional e-books in this series can be found on Nova's website
under the e-book tab.

ENVIRONMENTAL REMEDIATION TECHNOLOGIES, REGULATIONS AND SAFETY

LANDFILLS AND RECYCLING CENTERS

PROCESSING SYSTEMS, IMPACT ON THE ENVIRONMENT AND ADVERSE HEALTH EFFECTS

CARLA H. JACKSON
EDITOR

nova publishers
New York

Library of Congress Cataloging-in-Publication Data

Landfills and recycling centers : processing systems, impact on the environment and adverse health effects / editor, Carla H. Jackson.
pages cm
Includes index.
ISBN 978-1-63463-792-3 (hardcover)
1. Sanitary landfills--Environmental aspects. 2. Sanitary landfills--Health aspects. I. Jackson, Carla H.
TD795.7.L34845 2015
628.4'4564--dc23

2014047569

Published by Nova Science Publishers, Inc. † New York

CONTENTS

Preface vii

Chapter 1 Long-Term Perturbations in the Soils, Waters and Plant Communities of Two Closed Landfills under a Dry-Mediterranean Climate 1
A. J. Hernández, Mª. J. Gutiérrez-Ginés and J. Pastor

Chapter 2 Surface Geophysical Methods in the Assessment of Environmental Impacts of Landfills: An Overview 25
Elsa Cristina Ramalho, Manuel João Senos Matias and Rui Miguel Marques Moura

Chapter 3 Municipal Landfill Leachates: Water Contaminants or Source of Nutrients in Agricultural Soils? 53
Marco Antonio Nogueira, Cristiane Alcantara dos Santos, Sandra Márcia Cesário Pereira da Silva, Daniel Bini, Luciana Cadioli Panchoni, Vanessa Lázaro Mangieri and Grazielle Favero Valério

Chapter 4 Integrated Technologies Based on the Use of Activated Carbon and Radiation to Remove Contaminants Present in Landfill Leachates 69
R. Ocampo-Perez, J. Rivera-Utrilla, M. M. Abdel daiem and M. Sánchez-Polo

Chapter 5 Phytocapping of Municipal Landfills 111
Kartik Venkatraman, Nanjappa Ashwath and Ninghu Su

Chapter 6 Health and Ecological Risk of Methane and Volatile Organic Compounds Emission from Solid Waste Disposal: A Case Study in Thailand 147
Sansarith Thianpopirug, Sudarat Kucamau, Wilai Chiemchaisri and Chart Chiemchaisri

Chapter 7 Models for Evaluating Prevention of Waste on a Macroeconomic Level **167**
Predrag Korica and Andreja Žgajnar Gotvajn

Chapter 8 From Dumping to Sustainable Landfilling: The Concept of Aerobic Landfills **199**
Gabriela Kalčíková and Andreja Žgajnar Gotvajn

Index **223**

PREFACE

Landfills have been targeted by geophysical methods in order to investigate their environmental impacts. In fact, landfills have been the classic way to deposit domestic and industrial waste and have generated a large range of negative environmental impacts in groundwater and soils. These problems often persist even after the effective use of the landfills and subsequent recovery processes. Owing to their characteristics, landfills are difficult to access and because of the general lack of accurate information regarding the shape, nature of the refuse, history and development of the landfill, non-invasive, non-destructive methods and sometimes autonomous data acquisition devices must be used to monitor impacts and to investigate and prevent groundwater and soil contamination. This book discusses processing systems, environmental impacts and adverse health effects of landfills and other recycling centers.

Chapter 1 – In the Mediterranean, scarce soil on the slopes of sealed solid waste landfills and frequent erosive processes have left exposed many materials that are consequently mobilized along with leachates. This chapter focuses on two typical sites used for disposal: mixed urban/industrial waste landfills and natural depressions close to streams and marshy areas. These settings all pose risks to the groundwater and to the birds and herbivores (wild and domestic) that drink from wetlands and consume their plants when surface water evaporates. They should thus be a priority in any geochemical assessment of the health of ecosystems affected by waste disposal sites. Results obtained so far indicate the importance of monitoring compounds related to ecosystems' health since intense environmental impacts are still produced. Essentially, we need to evaluate the kinds of pollutant most easily transferred to soils, waters and the trophic networks of the affected ecosystems. Elements used as micronutrients (Zn, Cu, Fe, Mn and B), anions indicative of high salinity and heavy metals are frequent problems. Organic compounds need to be particularly considered given the high levels detected in water samples, including high concentrations of phthalates. The latter have generated much debate concerning their possible repercussions on health and have been described as weak to not so weak pollutants. Equally revealing has been the study of heavy metal and organic pollutants in the landfills' cover soils with a view to an effective revegetation that will promote succesion processes. Finally, the evolution of plant communities has been monitored for 20 years. Natural revegetation has been observed in one of the landfills, indicating a recovery process that is helping to minimize the transfer of pollutants to leachates. In contrast, contstant anthropogenic pressure on the other landfill is degrading the ecosystem, compromising its biodiversity.

Chapter 2 – The fast development of field, data processing equipment and software has allowed the use of geophysical methods to an ever increasing range of applications. Hence, nowadays it is much easier to conduct massive field surveys combining different methods, to obtain more accurate and denser data, so that complex modeling and interpretation at limited costs can be carried out. Landfills have been targeted by geophysical methods in order to investigate their environmental impacts. In fact, landfills have been the classic way to deposit domestic and industrial waste and have generated a large range of negative environmental impacts in groundwater and soils. These problems often persist even after the effective use of the landfills and subsequent recovery processes. Owing to their characteristics, landfills are difficult to access and because of the general lack of accurate information regarding the shape, nature of the refuse, history and development of the landfill, non-invasive, non-destructive methods and sometimes autonomous data acquisition devices must be used to monitor impacts and to investigate and prevent groundwater and soil contamination. Geophysical methods can be applied to investigate a wide range of aspects related with the assessment of the environmental impact of landfills. Problems such as geometry definition, geological settings, contamination plume location and monitoring, investigation of internal structure and refuse zoning, determination of fluid flow direction and paths or the determination of sealing conditions and leakage may be more successfully evaluated if a carefully chosen geophysical survey is part of any investigation program. Because of the nature and complexity of the problems to investigate, only multidisciplinary approaches, involving geophysics, hydrochemical, hydrogeological and geological information, can provide meaningful results for a thorough assessment of the landfills impact on the environment. This work intends to demonstrate the application of geophysical methods in the investigation of the environmental impacts, as described above, of industrial and domestic landfills during their life time and after closure. Thus, several examples will be discussed illustrating the use of 2D, 3D and time lapse resistivity, electromagnetic, ground probing radar, self-potential, magnetic, gravity surveys and airborne thermal mapping. Most of the geophysical data will be presented and shortly discussed together with information from boreholes, geology, hydrogeology and hydrochemical data. As it will be shown, it is clear that only a judicious combination of methods and information from different nature can provide tools for the diagnosis and assessment of the impact of landfills in the environment, for the investigation of the best engineering solutions to remediate them and for the possible recovery of refuse with economic interest.

Chapter 3 – Municipal landfill leachates are an environmental concern worldwide due to its polluting potential, especially for surface and ground water. Despite available technology for installation of municipal landfills and adequate treatment of leachate, many landfills are improperly established and the formed leachate has no adequate treatment and destination, especially in developing countries. However, even after treated, the resulting residues or the effluent contains nutrients like nitrogen (N), phosphorus (P) and potassium (K), which may cause eutrophication of stream waters. These nutrients are essential for plant growth and thus leachates free from contaminants like heavy metals, organic chemicals or pathogens above safe levels, could be used as source of nutrients in agricultural soils. If a decision is taken for use of landfill leachates in agricultural soils, monitoring of soil chemical, physical and biological properties are essential for a safer destination. In this chapter we discuss on aspects related to leachate formation, its environmental risks, potential use as nutrient source in agricultural soils, and procedures for monitoring the effects of landfill leachates on soil functions to prevent degradation.

Chapter 4 – Landfills remain a prime concern of waste management systems throughout the world and have been identified as one of the main threats to groundwater resources. The two most serious environmental effects of landfills are related to the leachate formed during decomposition from the percolation of rainwater through the waste layer and to the toxic byproducts generated by chemical and biological processes during waste leaching. Once landfill leachate is released into the environment, it can pollute soils, surface water, groundwater, and wildlife, due to its toxicity. Various organic pollutantsand chemical species have been detectedin the composition oflandfill leachates, including:i)plasticizers and organic estersaddedto polymersin orderto improve the propertiesof the final product, ii) antioxidants orstabilizers used to improve the durabilityof materials,andiii) herbicides/pesticidesor other widely-used anti-pest compounds. This chapter is concerned with conventional and novel technologies for the elimination of pollutants present in landfill leachates stand out: (a) Photocatalysis in the presence of activated carbon; (b) Photolysis with the participation of $HO^{\bullet}$ y $SO_4^{\bullet}$-radicals; (c) Application of gamma radiation to generate oxidative and reductive species; and (d) adsorption on activated carbon. Thus, the objective of this chapter is to summarize the main technologies applied to the elimination of contaminants present in landfill leachates. The chapter will be divided in four main topics. Adsorption/bioadsorption of phthalic acid, an organic micropollutant present in landfill leachates, on activated carbons. Comparative study of the photo-degradation of bisphenol A by $HO^{\bullet}$, $SO_4^{\bullet}$- and $CO_3^{\bullet}$-/$HCO_3^{\bullet}$ radicals in aqueous phase; Role of activated carbon in the photocatalytic degradation of 2, 4-dichlorophenoxyacetic acid by the UV/TiO_2/activated carbon system. Treatment of water contaminated with diphenolic acid by gamma radiation in the presence of different compounds.

Chapter 5 – Landfill capping is a part of the post closure management protocol to minimise environmental problems such as leachate generation from landfills. The costs of conventional caps (e.g. clay caps, HDPE) and remediation of landfills have escalated over the last decade due to increase in the cost of capping material and with the introduction of stringent environmental laws. Research is therefore required to identify alternative capping systems that will not only minimise the costs, but also demonstrate increased environmental performance and community benefits.An alternative landfill capping technique known as 'Phytocapping' (establishment of plants on a layer of soil placed over the waste) was trialled at Lakes Creek Landfill, Rockhampton, Australia. This particular system checks on the seepage as well as allows preferential amount of water to promote biodegradation. In this technique, trees were used as 'bio-pumps" and soil cover as "water storage". The trees and soil together help minimise percolation of water into the landfill. Phytocaps also know as Evapotranspiration (ET) covers have been very successful in the US. This study is the first of its kind in Australia to evaluate the performance of 21 different tree species over two soil thicknesses (1400 mm and 700 mm). This paper reviews the critical points of the current knowledge and study conducted on phytocapping till date. The paper also reports the progress done on phytocapping till date; identifies the missing links and recommends further research that needs to be done to make this system more sustainable.

Chapter 6 – The objective of this research was to evaluate the emission and distribution of methane and volatile organic compounds (VOCs) from solid waste disposal site in Thailand. Gaseous emissions from solid waste piles with and without soil cover were determined. Highest average methane emission rate was found to be 53.51 $g/m^2.d$ from solid waste pile with cover soil during rainy season. This rate was significant higher than 20.87 $g/m^2.d$ under dry season and 5.23 $g/m^2.d$and3.99 $g/m^2.d$ from open waste pile under rainy and dry conditions respectively. The emission of BTEX compounds was found highest from the waste pile with cover soil under dry condition at $7.11x10^{-2}$, $9.76x10^{-2}$, $4.8x10^{-2}$ and$7.62x10^{-2}$ $g/m^2.d$. The source of VOCs was identified in fresh waste samples. The dispersion of methane and BTEX compounds in ambient air was modeled. The toxicity of BTEX compounds to workers and nearby communities was at low level as determined by hazard index.However, thecarcinogenicity risk from benzene to those people were found to be $2.45x10^{-5}$ and $2.85x10^{-5}$, respectively. Geno-toxicity study using plant comet assay (Pothos) suggested that there was a relationship between DNA damage after 96 hours exposure and VOC emission rate from solid waste disposal area.

Chapter 7 – The aim of this work is to give review of methods for measuring prevention of waste used by authors in their attempts to assess the impacts which prevention of waste activities can have on the environment, economy and society. The most evident problem with the measuring of waste prevention on a macroeconomical level is that without the clear measurements prevention activities cannot be chosen, proscribed, implemented or evaluated. Proscribing as target just the reduction of amounts of produced waste without the consideration of other variables that influence the generation of waste wouldn't give a right description of the results of waste prevention activities. Mostly used principles for the evaluation of prevention of waste include models for the conducting of life-cycle assessment (LCA), monitoring approaches of the certain locality (measuring the produced waste before and after the implemented activity), and econometric models. These models still don't give the clear answer to the question on how to measure prevention of waste; however, they make the effects of implemented measures of waste prevention more clear to decision makers so that problems like transferring of an environmental pressure to other side can be avoided. General approach in the European Union for measuring of prevention of waste on a macroeconomic level is to observe the connections which economic variables (e.g. GDP, domestic material consumption etc.) with the amounts of produced waste. In this paper, by using the statistical data available, connections between the generation of waste and economic variables are evaluated in other to assess their potential to be used as waste prevention indicators on a macroeconomic level.

Chapter 8 – Landfilling has been for decades the most often used method for waste management. However, the growing awareness of the detrimental environmental effects of current waste processing led to important movement towards sustainable waste management practices in Europe. The concept includes clear hierarchy of waste management, where the main priority of waste management is waste reduction, reuse, recycling and composting, energy recovery and the last favorable option is landfilling. However, a reduction of municipal waste has been minimal over past years and landfilling is still the dominating waste management option in many countries of the European Union. Because of that, municipal waste will be still disposed in landfills and therefore, there is still the need for landfilling. Landfilling of municipal waste under aerobic conditions is one of the most promising methods for sustainable municipal waste disposal. Hence, the main aim of our study was to compare different ways of landfilling during past years. Special attention was focused on the concept of aerobic landfilling. As a case study, biodegradation of residual municipal waste in aerobic landfill was investigated. The biodegradation proceeded in a pilot-scale aerobic landfill bioreactor and as a result landfill leachate was generated. Due to the fast and intensive biodegradation, leachate pollution considerably decreased and after several months, all investigated parameters of leachates were below legislative limits. Furthermore, leachate became non-toxic to many organisms and it is expected to cause minimal adverse effect to the environment. Aerobic landfilling thus meet requirements of sustainability concepts and it seems to be an appropriate option for modern landfilling in the framework of the sustainable waste management.

In: Landfills and Recycling Centers
Editor: Carla H. Jackson

ISBN: 978-1-63463-792-3

Chapter 1

LONG-TERM PERTURBATIONS IN THE SOILS, WATERS AND PLANT COMMUNITIES OF TWO CLOSED LANDFILLS UNDER A DRY-MEDITERRANEAN CLIMATE

A. J. Hernández[1], Mª. J. Gutiérrez-Ginés[2,*] and J. Pastor [3]

[1] Department of Life Sciences, University of Alcalá, Spain
[2] Dept. Soil and Physical Science, Faculty of Agriculture and Life Science, Lincoln University, New Zealand
[3] Dept of Biogeochemistry and Microbiology Ecology, National Museum of Natural Sciences, CSIC, Spain

ABSTRACT

In the Mediterranean, scarce soil on the slopes of sealed solid waste landfills and frequent erosive processes have left exposed many materials that are consequently mobilized along with leachates. This chapter focuses on two typical sites used for disposal: mixed urban/industrial waste landfills and natural depressions close to streams and marshy areas. These settings all pose risks to the groundwater and to the birds and herbivores (wild and domestic) that drink from wetlands and consume their plants when surface water evaporates. They should thus be a priority in any geochemical assessment of the health of ecosystems affected by waste disposal sites. Results obtained so far indicate the importance of monitoring compounds related to ecosystems' health since intense environmental impacts are still produced. Essentially, we need to evaluate the kinds of pollutant most easily transferred to soils, waters and the trophic networks of the affected ecosystems. Elements used as micronutrients (Zn, Cu, Fe, Mn and B), anions indicative of high salinity and heavy metals are frequent problems. Organic compounds need to be particularly considered given the high levels detected in water samples,

[*]Corresponding autor: MªJ Gutiérrez-Ginés. Dept. Soil and Physical Science, Faculty of Agriculture and Life Science P O Box 85084. Lincoln University. Lincoln 7647. Christchurch, New Zealand. email: MariaJesus.Gutierrez@lincoln.ac.nz.

including high concentrations of phthalates. The latter have generated much debate concerning their possible repercussions on health and have been described as weak to not so weak pollutants. Equally revealing has been the study of heavy metal and organic pollutants in the landfills' cover soils with a view to an effective revegetation that will promote succesion processes. Finally, the evolution of plant communities has been monitored for 20 years. Natural revegetation has been observed in one of the landfills, indicating a recovery process that is helping to minimize the transfer of pollutants to leachates. In contrast, contstant anthropogenic pressure on the other landfill is degrading the ecosystem, compromising its biodiversity.

Keywords: Polluted soils, heavy metals, anions, organic pollutants, community ecotoxicology

1. Introduction

In some of the more recently developed countries, there is growing concern over the state of former waste disposal sites whose use was discontinued in the 1970-90's. Industrial growth and the rise in population over this time period generated vast amounts of household and industrial waste that was simply buried in landfills without any type of lining, compacting, sorting or processing. Most of these uncontrolled landfills contained mixed urban and industrial waste, and over the years they have released considerable amounts of toxic substances affecting the atmosphere, water resources and soils. Today's interest in recovering these areas is dependent upon a good diagnosis of each site (Hernández and Pastor, 2008; Hernández et al., 2012; Pastor and Hernández, 2012). Moreover, common European policy on this issue is becoming increasingly strict and even being applied to countries seeking development help.

As rainwater percolates through a refuse tip, it extracts, dissolves, and washes out a wide range of organic and inorganic constituents. In effect, leachate production is recognized as one of the greatest problems associated with old landfills. Such liquid waste causes considerable pollution problems by contacting the surrounding soil and ground or surface waters. Old landfills are thus considered major pollution hazards unless precautionary measures are implemented (Baccini et al., 1987). In an investigation of the effects of landfill leachates on soils, Calace et al. (2001) noted that the adsorption of the leachate's constituents led to enhanced heavy metal uptake.

Few investigations on the impacts of landfills have examined those of arid or semiarid environments (Al-Yaqout and Hamoda, 2003; Illera et al., 2000; Tatsi and Zouboulis, 2002).

In the present study we tried to apply our acquired knowledge in geochemically assessing landfill ecosystems' health (Hernández et al., 2009; Pastor and Hernández, 2012; Pastor et al., 2014) to the disused waste disposal site scenario. Our working protocol first involves an adequate assessment of a set of chemical properties of the soils and waters of the affected ecosystems that can be linked to both the environment and plant communities, and by extension to animal and human health (Hernández and Pastor, 2008; Hernández et al., 2012).

2. GENERAL IMPACTS OF LANDFILLS IN THE MEDITERRANEAN AREA

2.1. Study Scenarios

The most outstanding impacts are linked to pollution by both surface and underground leachates produced by rain infiltration and surface runoff, mainly affecting ecosystems in the principal discharge areas (Adarve et al., 1998; Cyr et al., 1987; Stegman, 1982) and surrounding ecosystems (Hernández et al., 1998; Pastor et al., 1993a). It is fairly common to find wet areas grazed by domestic and wild animals as well as cereal crops at leachate discharge sites, such that some plant species may accumulate pollutants which may be passed on to the trophic networks of other ecosystems. In addition, in many cases surface leachates reach streams that act as vehicles for the transport of pollutants outside the landfill system.

The surface area occupied by each landfill has varied since their initial sealing due to subsequent waste dumping. These areas are generally 50,000 to 70,000 m^2. They may all be classed as "uncontrolled" given that since their initial sealing no efforts have been made to avoid the outflow of leachates from the landfills.

The sealing material is a 20-50 cm layer of soil and rubble from the surroundings that do not protect the landfill from the penetration of rain water. Cover soils show different phytoedaphic properties according to their sorce rocks.

Landfill slopes are generally over 15 m high and are sometimes superimposed due to the further deposition of waste after their initial sealing. Slopes are usually marked many being over 40%. These features affect both the plant colonization of slopes and surface runoff discharge areas. Even when a landfill consists of only one enourmous slope, its runoff adopts a fan-like pattern thus differently affecting the biodiversity of the discharge area. There is also much variation in edaphic factors, as often occurs on the slopes of a single landfill (see Table 1). Surface water salt concentrations also vary according to the yearly rainfall regime (see Table 2). These issues are discussed in more detail in Gutiérrez-Ginés et al. (2014), Pastor and Hernández (2012) and Pastor et al. (2014).

Most of the edaphic variables examined are generally higher in the urban/industrial waste landfills than in the soils of the surrounding reference ecosystems. Of concern are their appreaciable amounts of heavy metals (Zn, Cu, Pb and Ni), and lower levels of organic matter and total N.

Landfill plant cover is determined by the presence in the sealing soil of a seed bank and propagules of different species in the surrounding areas. According to the close relationship detected among floral diversity, soil fauna and nutrient recycling, the lower diversity of nematodes observed in the landfill soil covers (Table 3) is hindering the evolution of these emerging systems. Nutrition deficiencies are among the main limiting factors for the development of an adequate plant cover.

2.2. Methodological Aspects of the Study

The characteristics of each landfill define their environmental impacts. Thus the starting point for any study is landfill characterization including chemical analyses to guide the ecotoxicological bioassays required for each particular case.

Table 1. Hg concentrations (mg/Kg) detected in soil samples collected from landfill slopes in Torrejón (Madrid) 28 years after initial sealing

Sample	Hg	Sample	Hg
1	0.022	13	1.366
2	0.020	14	0.049
3	0.866	15	0.131
4	0.051	16	0.032
5	0.036	17	0.051
6	0.023	18	0.029
7	0.091	19	0.048
8	1.149	20	0.067
9	1.156	21	0.117
10	0.242	22	0.029
11	2.590	23	0.042
12	0.263		

Table 2. Composition of water samples obtained from the leachate discharge area (wetland) of one closed landfill in Community of Madrid (Spain). Ion concentrations given in mg/L and electrical conductivity (EC) in µS/cm. NO_2^- and PO_4^{3-} were not detected

Parameters	Samples		
	1	2	3
pH	7.55	8.1	7.99
F^-	-	3.8	-
Cl^-	201	177	90
NO_3^-	8.4	28.5	32.8
SO_4^{2-}	1683	2730	2126
Na	120	114	77
EC	3220	4110	3690
Fe	0.124	0.068	0.052
Cu	0.02	0.014	0.016
B	0.376	0.324	0.379
Ca	509	622	668
Mg	143	355	263
K	583	268	82

Table 3. Biotic indices based on soil nematofauna determined in landfill versus reference ecosystem soils

Index	Landfill	Reference ecosystem
Shannon *H'*	1.02	2.14
Shannon transformed *e**H'*	3.38	9.05
Simpson *D*	1.91	3.68
Pollution *MI-2*	0.10	0.40

The first question to address is thus the flow directions of deep and surface leachates and relationships with the environment's surface water. This question includes diffentiation between distance pollution or diffuse contamination of ecosystems outside the immediate environment (aquatic ecosystems-surface water courses such as rivers and streams- and land ecosystems, including pollution effects on animals that can disrupt the structure of an ecosystem). Another issue to consider is the composition of leachates and their main flow directions in relation to discharge zones since these are linked to the so-called spatial-temporal pollution of ecosystems in the landfill environment. The different water table levels in discharge zones affect the punctate pollution of stable ecosystems of the surroundings (mostly wetlands and pastures).

The ideal tools for assessing environmental health are living organisms. This idea is currently being applied to environmental impact assessment and highlights the important role of plants and animals in biological monitoring (Koehler, 1996; Pankhurst, 1994; Spellerberg, 1991; Verhoef and van Gestel (1995). Any ecotoxicological evaluation (Hernández et al., 2012) should consider the concepts of risk analysis (Hernández et al., 2009). In addition, as when the health of a person is assessed, the factors examined at the population or community level need to be compared with those considered as references in settings free from pollution.

An extensive protocol including several chronological steps and complementary methods will lead to a reliable diagnosis and realistic prognosis in each case. Besides an ecosystem analysis of a landfill in its own environment, we need to center on populations that may pose risks for the transfer of pollutants to other components of the system through the use of biomarkers and biochemical variables. For instance, monitoring plant cover can be an adequate biomarker of toxicity for a grass community as may be serum and urine biochemicals in rats treated with surface leachates from the landfill under study (Pastor et al., 1993b).

3. Objectives

In the study by Hernández and Pastor (2008), we describe the general features of municipal solid waste (MSW) landfills in the central region of Spain and the methods used to assess their impacts on soils and plants communities. As an extension of this work, we are presently addressing the question of whether after many years these landfills continue to have significant impacts. In this chapter, we examine the changes produced in several factors both in affected waters and the plant communities that have spontaneously colonized these systems in two of Madrid's former waste disposal sites.

4. Material and Methods

4.1. Study Sites

Two former waste disposal sites in the Madrid province (central Spain) were selected. The operation of the two landfills was uncontrolled and they lack currently recommended protection and drainage layers or leachate collectors.

Móstoles. This household waste disposal site on an arkosic substrate was sealed in 1986. Its discharge ecosytems are a wetland and stream. Subsequent uses of the landfill have been rainfed cereal cultivation in the higher areas and itinerant sheep grazing in the remaining zones. The natural vegetation of the Móstoles landfill would have been holm oak woods but we have only been able to find shrublands of thyme and broom. The zone occupied by the landfill was used in the past for crops and pastures sloping towards the stream.

Torrejón. The municipal waste landfill of Torrejón lies on a limestone and marl substrate and was first sealed in 1982 and then resealed in 1990. Most of the tip lies on an old river terrace in a highly permeable zone between two streams such that its location is very inadequate and carries a high risk of pollution. Besides domestic waste, this landfill contains a large amount of industrial waste and construction rubble.

An initial characterization of these landfills may be found in Pastor et al. (1993a; 1993c), and is reviewed by Pastor and Hernández (2012).

4.2. Soil Sampling and Analysis

The soils and waters of discharge areas were sampled twice. The first sampling year, 1991, was the 5th and 9th year since the first sealing of the two landfills, respectively. In that year, the land use of these sites had not changed since their closure. Twenty years later in 2010, a second sampling effort was made at both sites. Results of these two sampling time points are hereafter referred to as 1st year results and 2nd year results. In the second sampling year, the Torrejón landfill had received continuous wastes of all kinds, whereas the Móstoles landfill was unmodified since its closure. Both sites have been frequently grazed by sheep.

The discarge areas as well as the thin soil layer that covers the waste materials were monitored during both the first and second sampling years. Soil sampling in each landfill and period was stratified as discharge zones, slopes and flat areas. Samples were collected at random in every area of the landfill. A spade was used to obtain an average sample of topsoil (0-12 cm) in each zone. In total, 71 soil samples were collected. They were dried at room temperature and passed through a 2 mm sieve. Samples were subsequently used for physico-chemical determinations (MAPA 1982, Pastor and Hernández, 2012; Pastor et al., 2014). The following variables were monitored in soils: pH, COD, nitrates, nitrites, sulphates, chlorides, phosphates and metal concentrations. Kjeldahl nitrogen content was determined after preliminary acid sample digestion. The concentrations of anions were determined by ion-chromatography (using a Dionex ion chromatograph).

Available elements were determined according to the method of Lakanen and Ervio (1971) based on plasma emission spectrophotometry (ICP-OES). Pseudototal metal contents were determined by ICP-OES on ground soil samples acid treated with a 4:1 mixture of HNO3 and HClO4. Existing organic pollutants were also determined in the soil and water samples collected in the second sampling year using standardized EU techniques when available or otherwise using EPA procedures. The organic pollutants determined in soils were: total hydrocarbons by infrared spectrophotometry (UNE 77307); chlorinated insecticides and polychlorinated biphenyls by gas chromatography (ISO 10382); PAHs by gas chromatography (ISO 18287); and phenols by gas chromatography (U.S. E.P.A 3550B, U.S. E.P.A 3650B and U.S. E.P.A 8401). Total Hg levels were determined using an Advanced

Mercury Analyser (AMA-254, LECO Company, Czeck Republic) according to the procedure described by Millán et al. (2006).

4.3. Water Sampling and Analysis

We also collected water samples from the affected discharge areas: Peñaca stream in Mostoles and Pelayo stream in Torrejón. These samples were stored cold in the dark until the time of analysis. Before analysis, they were filtered through a 7-9 μmfilter. pH, nutrients, EC, COD, and anions (Cl^-, NO_3^-, SO_4^{2-}) were determined in water samples according to the methods cited in the soil section. In 2nd year water samples, organic compounds were qualitatively detected by gas chromatography/mass spectrometry, according to US EPA method 525.2.

4.4. Data Analysis

Statistical tests (ANOVA) were performed on log (X+1) transformed data using SPSS v.18 software.

4.5. Plant Communities

A plant species inventory was made in both sampling years. Sampling areas were $1m^2$. Observed species were catalogued in phytosociological classes according to the Mediterranean climate vegetation (Rivas-Martínez et al., 2002).

5. Results and Discussion

The landfills were sealed by covering the solid waste with a thin layer of soil usually obtained from marginal agricultural zones cleared to construct houses or industrial estates. Thus, an important starting point of our study was to evaluate these source zones and compare the impacts of the landfills. In a preliminary comparative study, we lacked data for some of the soil variables (Pastor et al., 1993a; 1993c) as well as any prior data for the area reported by other researchers. We gradually incorporated new chemical variables and landfills to our study (Hernández et al., 1998) and started examining the effects of leachates, especially those containing heavy metals and salts, on plants and animals (Adarve et al., 1998; Hernández et al., 1999; Pastor et al., 1993c; Urcelai et al., 2000).

During the course of our research we have become aware that the variables considered in the present study are the most appropriate in terms of relating aspects of geochemistry to the environmental health of systems affected by waste landfills in a dry Mediterranean climate. These variables have also been successfully used in the evaluation of landfill impacts and risks in other environments (Akinbile, 2012).

5.1. Comparing the Chemical Properties of Surface Waters Around the Household Waste Landfills

When evaluating the effects of landfills, the different authors have generally focused on the study of leachates. Emerging data on the composition of leachates from waste landfills and their effects on surface and ground waters have been an important focus of both the international (Chu et al., 1994; Fatta et al., 1999; Hogland et al., 2004; Kjeldsen, 1993) and Spanish (Adarve et al., 1994) literature. Since these landfills are not provided with leachate collection systems, here we examined spring water compositions in the discharge area of the two landfills presumably affected by landfill leachates: streams that flow next to the Móstoles landifll and the stream over which the Torrejón landfill was created.

Variables such as the sulphate/chloride ratio and the behavior of pH, reflect the degree of leachate stabilization and thus, could roughly correspond to leachate compositions of differently aged samples. A decrease in this ratio has been found for leachate generated from landfills (Pohland et al., 1985; Tatsi and Zouboulis, 2002). According to Al-Yaqout and Hamoda (2003), in arid regions the generation of leachate from MSW landfills has long been neglected on the assumption that minimal leachate could be formed in the absence of precipitation. Leachates collected from both active and old (closed) landfills where co-disposal of MSW and other solid and liquid wastes occurred were severely contaminated with organic compounds, heavy metals and salts.

Table 4 shows the water composition of the Peñaca stream as it enters and exits the Móstoles landfill, as well as in its middle section. In the zone where the stream enters the landfill, Mg, K, Na, Mn and B contents, EC and anion levels were greatest in 1991, while the water pH and COD were clearly higher in the 2nd sampling year.

In the middle of the landfill, the composition of the stream water was similar to that revealed in the previous analyses. Both in the first and second sampling years, highest values of the variables examined were detected in the stream as it exits the disposal site. The high pH of the stream water sampled the second year warrants further investigation. Although salinity has shown a general decrease in recent years compared with the 1st year results, it continues to be much higher in the stream outflow reach than at the head, which indicates an obvious effect of the landfill, even two decades after its sealing.

In Table 5, which shows the results of an analysis conducted on the waters of the Pelayo stream at the foot of the Torrejón MSW landfil, similar behavior to the previous case was observed including higher EC, chloride and Na levels in the 1st year than 2nd year, and higher COD values in the 2nd year.

In general, salinity of affected waters of the two landfill sites has decreased over the past two decades (lower EC, Na and Cl). COD and pH increased in the 2nd year compared with the 1st year. In contrast, Chu et al. (1994) reported increasing chloride concentrations with increasing leachate age. However, our results are consistent with those of Kalcikova et al. (2012), except in the case of COD, which decreased in the cited work. Kalcikova et al. (2012) discovered that leachates of old sealed parts of a landfill showed very low levels of pollutants and salinity, and phytotoxicity tests performed on these leachates indicated no toxicity. Alkibile (2012) reported similar results related to an increasing distance from the landfill.

Table 4. Comparing the pH and means and standard deviations (3 replicates) of nutrients, metals, B (mg/L), EC (μS/cm), COD (mg/l O_2) and main anions (mg/L) recorded in the first sampling year (4 years after sealing) and second sampling year (23 years after sealing) in water samples taken in early spring from three reaches of the Peñaca stream flowing around the Móstoles household waste landfill. *: significant at the 95% level <0.05: significant at the 99% level <0.01***: significant at the 99.9% level <0.001**

Variable	Inflow reach			Middle reach			Outflow reach		
	1st year	2nd year	Sig.	1st year	2nd year	Sig.	1st year	2nd year	Sig.
pH	6.9 ± 0.1	8.1 ± 0.1	***	6.9 ± 0.2	8.3 ± 0.11	***	7.3 ± 0.1	8.7 ± 0.2	***
Ca	43 ± 7.5	38 ± 2.1		235 ± 25	33 ± 1.2	***	250 ± 30	110 ± 3.5	***
Mg	10 ± 0.2	5.8 ± 0.7	***	35 ± 5.2	3.9 ± 0.3	***	39.7± 0.6	25 ± 1.3	***
K	15 ±2. 0	7.7 ± 0.3	***	15.4 ± 5.6	7.8 ± 0.2	*	90 ± 2.3	61 ± 1.9	***
Na	166 ± 14	29 ± 1.5	***	178 ± 17	27 ± 2.2	***	465 ± 15	210 ± 3.8	***
Fe	0.41 ± 0.32	0.18 ± 0.01		0.05 ± 0.05	0.03 ± 0.06		0.15 ± 0.11	0.36 ± 0.02	*
Mn	0.2 0± 0.06	0.12 ± 0.01	***	2.05 ± 1.15	0.06 ± 0.01	*	1.9 ± 0.52	1.97 ± 1.1	
B	0.55 ± 0.05	0.10±0.05	***	0.15± 0.05	0.42 ± 0.51		0.35 ± 0.05	0.23 ± 0.8	
Zn	0.02 ± 0.02	0.02 ± 0.01		0.04 ± 0.01	0.03 ± 0.04		0.05 ± 0.01	0.03 ± 0.01	
Cu	0.05 ± 0.05	0.0 4± 0.05		0.0 ± 0.0	0.04 ± 0.04		0.0 ± 0.0	0.0 ± 0.0	
EC	945 ± 75	216 ± 12	***	1935 ± 85	344 ± 12	***	4175 ± 75	2492 ± 18	***
COD	10 ± 0.05	66 ± 5.3	***	12 ± 2.3	98 ± 3.2	***	21 ± 2.2	207 ± 2.1	***
Cl	120 ± 14	17.5 ± 0.5	***	222 ± 118	18 ± 0.7	**	699 ± 8.3	569 ± 8.3	***
NO3	26 ± 9.2	0.18 ± 0.02	***	4.5 ± 2.4	0.06 ± 0.02	**	19 ± 0.1	4.5 ± 0.45	***
SO4	114 ± 5.5	25 ± 1.1	***	302 ± 44	28 ± 1.6	***	307 ± 57	61 ± 4.2	***

Table 5. Comparing the pH and means and standard deviations (3 replicates) of nutrients, metals, B (mg/L), EC (μS/cm), COD (mg/L O_2) and main anions (mg/L) recorded in the first and second sampling year in water samples taken in early spring from the Pelayo stream at the foot of the Torrejón municipal waste landfill. R: reliably significant at the 90% level <0.1*: significant at the 95% level <0.05: significant at the 99% level <0.01***: significant at the 99.9% level <0.001**

Variable	1st year	2nd year	Sig.
pH	7.8 ± 0.11	7.5 ± 0.1	R
Ca	700 ± 159	630 ± 40	
Mg	610 ± 430	246 ± 14	
K	9.2 ± 6.1	42 ± 1.1	**
Na	1643 ± 1183	217 ± 14	R
Fe	0.13 ± 0.06	0.03 ± 0.06	R
Mn	0.33 ± 0.49	0.81 ± 0.02	
Zn	0.03 ± 0.06	0.06 ± 0.00	
Cu	0.10 ± 0.0	0.20 ± 0.24	
B	1.6 ± 0.44	0.44 ± 0.02	**
EC	14060 ± 7705	2260 ± 111	*
COD	16 ± 7.7	158 ± 7.1	**
Cl	3119 ± 2324	302 ± 7.8	R
NO_3	72 ± 52	52 ± 2.7	
SO_4	2641 ± 1225	2107 ± 42	

During the years of monitoring the waters of the two wetlands, we have noted that birds, rabbits, hares and other small wild animals, as well as sheep, drink this water and over long periods consume plants growing on these wetlands. The high salinity and presence of metals in the soils and waters pose a clear risk for the health of these animals.

As landfill age increases, the consequent increase in pH has led to a certain decrease in metal solubility. Moreover, the lower metal concentration in stabilized leachates is mainly due to adsorption and precipitation reactions. Heavy metals in waste have for years been disposed of in dumps and landfills. It is an established fact that organic matter in landfill leachates binds heavy metals strongly (Christensen et al., 2001; Slacka et al., 2005).

At the time of writing (2012), the wetland at the Torrejón waste disposal site continues to receive vast amounts of rubble, inert materials and uncontrolled industrial waste and this, in large measure, is reducing the size of this ecosystem and enormously disturbing it due to the presence of extremely harmful organic pollutants and heavy metals.

Table 6 lists the organic compounds present in waters affected by their proximity to the waste landfills. Numerous detected organic compounds were presumably originally natural and not expected to cause pollution. Only those compounds reported to be potentially toxic are shown in the table. Especially noticeable is the presence of phthalates in our water samples. Phthalates are colourless and odourless liquid organic compounds used as plasticizers, but also used as solvents in perfumes, pesticides, waxes, adhesives, modeling clay, and paint pigments. The effects of phthlates on health and the environment have been intensely investigated. A summary of the characteristics, toxicity and management of these compounds can be found in the EPA Phthalates Action Plan (US EPA, 2012). The compound diethyltoluamide is an insect repellent that does not kill insects. Case reports of possible adverse health effects have described neurological effects (seizures, encephalopathy, tremor, slurred speech, coma, and rarely, death) in children and adults most commonly following the ingestion of DEET or skin applications not consistent with label directions (US EPA, 2007). Di 2-ethylhexyl adipate is a liquid of low volatility, widely used as a plasticizer. This compound has been reported to produce liver adenomas and carcinomas in mice (IARC, 2000).

In a characterization study, Paxeus (2000) examined leachates from three municipal landfills in western Sweden in terms of their contents of individual organic compounds. Matejczyk et al. (2011) examined the risks related to 22 landfill leachates and detected several risk factors: faecal and other epidemiogenic microorganisms, heavy metals and organic compounds. However, they observed no relationship betweeen pollutant concentration and toxicity as assessed by tests on bacteria, protozoans, rotifers and two crustaceans. Based on our results those of others, we do not share the opinion of Cheng and Chu (2007; 2011) regarding the use of landfill leachates to irrigate crop lands. Despite their high nutrient concentrations, we consider that the risks of their use for crops are much higher than their possible benefits.

5.2. Analyzing the Soils in Landfill Discharge Zones in the First and Second Decades after Their Initial Sealing

Table 7 provides the means and standard deviations recorded for the variables pH, OM, N, nutrients, pseudototal and available Zn, electrical conductivity and the main anions in soil samples taken from the main discharge zones of the two landfills. These landfills were selected to represent the real scenarios of many landfills in Mediterranean countries, which share the features of siliceous and calcareous soils, an arkosic and limestone-marl substrate and mixed household/industrial waste contents.

Table 6. List of the organic compounds detected in water samples obtained from the streams and wetlands of the landfill discharge areas

Móstoles (Peñaca Stream)	Torrejón (Pelayo Stream)
Di 2-ethylhexyl adipate	Di 2-ethylhexyl adipate
Diethyltoluamide	Diethyl phthalate
Diethyl phthalate	Diisobutyl phthalate
Butyl isobutyl phthalate	Butyl isobutyl phthalate
Diisobutyl phthalate	Dibutyl phthalate
Dibutyl phthalate	Di 2-ethylhexyl phthalate
Di 2-ethylhexyl phthalate	

Table 7. Means and standard deviations of soil pH, OM (%), N (%), nutrients (mg/100g), Zn (mg/kg), EC (μs/cm) and anions (mg/100g) recorded in the Móstoles household waste landfill. R: reliably significant at the 90% level <0.1, *: significant at the 95% level <0.05, **: significant at the 99% level <0.01, *: significant at the 99.9% level <0.001**

	Móstoles			Torrejón		
Variable	1st year	2nd year	Sig.	1st year	2nd year	Sig.
No	21	15		7	6	
pH	5.6 ± 1.7	5.5 ± 0.9		7.2 ± 0.22	7.6 ± 0.13	**
OM	0.81 ± 0.51	1.9 ± 1.0	***	7.4 ± 4.1	5.8 ± 2.4	
N	0.05 ± 0.03	0.10 ± 0.05	***	0.40 ± 0.21	0.49 ± 0.40	
Ca	127 ± 72	110 ± 52		956 ± 407	631 ± 201	
Mg	12 ± 6.3	11 ± 4.1		123 ± 107	81 ± 96	
K	16 ± 7.7	27 ± 14	**	49 ± 19	68 ± 22	
Na	13 ± 25	2.9 ± 3.8	*	78 ± 91	7.6 ± 9.1	*
P_2O_5	8.6 ± 5.8	9.9 ± 4.4		65 ± 27	43 ± 17	
Zn T	38 ± 13	24 ± 8.2	***	84 ± 76	276 ± 185	*
Zn B	0.30±0.15	0.11 ± 0.14	***	2.1 ± 3.1	8.0 ± 6.0	*
EC	5965 ± 15776	270 ± 416	***	7044 ± 3716	1779 ± 1028	**
Cl^-	31 ± 9.2	27 ± 44	**	209 ± 208	44 ± 30	
SO_4	35 ± 54	310 ± 747		1385 ± 2052	3225 ± 2626	R
NO_3	4.5 ± 4.4	37 ± 53	**	7.8 ± 5.1	75 ± 100	**
PO4	0.0 ± 0.0	13 ± 8.6	***	0.0 ± 0.0	1.2 ± 1.7	R
F	0.0 ± 0.0	0.62 ± 0.4	***	0.0 ± 0.0	1.7± 1.4	**

Móstoles and Torrejón showed similar behavior in terms of the decreases produced in EC, Na, Cl and increases in sulphates, nitrates and fluorine from the 1st to the 2nd sampling year. These results are similar to those reported by Adarve et al. (1994) and not surprising. Thus, in a landfill sealed with a soil cover, initially chlorides and Na will abide in the soils of the discharge areas, since cooking salt is present in food condiments and Cl occurs in books and magazines, which in those days were not separated from the rest of the waste; both elements are of course readily soluble. Subsequent to this, other anions linked more to the decomposition of organic matter appear such that higher levels of these were recorded in the soils sampled in the 2nd year. The increased Zn concentrations of the Torrejón soils reflects the two kinds of received waste, mixed urban and industrial, as well as its continuous use receiving wastes even after its initial sealing.

Table 8 provides the latest results obtained for the presence of hazardous pollutants (heavy metals and organic compounds) in these former waste disposal sites, which are today much closer to inhabited areas. Our observations include high heavy metal levels at several sampling points especially in the Torrejón landfill, as well as the presence of particularly harmful organic pollutants in this landfill. Zn and Cu in this order were the most abundant heavy metals present. Many of the organic compounds found in soils in high concentration are those addressed in the Stockholm Convention, and were probably discarded since they were considered dangerous and therefore useless, as reviewed by Weber et al. (2011). Few literature reports exist of soil pollution caused by landfills, and available data are usually based on leachate effects. As far as we are aware, there are no references regarding soil organic pollutants.

Table 8. Heavy metals and organic pollutants (maximum values, mg/kg) detected in the soils of the two landfills in the second sampling year. Ref. Reference level above which a soil is considered polluted

Pollutant		Landfill	Ref.
	Móstoles	Torrejón	
Zn	38	1073	140
Cu	8.8	3271	36
Pb	9.4	624	50
Cr	2.8	169	100
Ni	3.1	114	35
Cd	0.0	41	0.8
Phenols		Phenol Cresols Pentachlorophenol	
Insecticides	Lindane	Lindane p,p'-DDE	
Hydrocarbons		190	50
PCBs		4.14	0.01
PAHs		Several PAHs	

Taken together, our results indicate that the two former waste disposal sites continue to have a high environmental impact and reveal a need for the constant follow up of the chemical features of the soils and waters that can be most easily transferred to the different substrates in which the landfills are found and the trophic networks of the affected ecosystems. First and foremost, elements that are micronutrients for living organisms (Zn, Cu and B) should be monitored both in the soils and waters of the landfills' surroundings, as should chlorides, sulphates and nitrates, as indicators of the high salinity produced.

5.3. Plant Community Ecotoxicology

To address the ecotoxicology of the landfill communities, species richness was used as an indicator of ecological integrity; reduced species richness being one of five general indicators of "ecosystem distress syndrome" (Rapport et al., 1985). Species richness is a simple count of the number of different species within a local habitat or region (alpha diversity). Because assessments of anthropogenic disturbances are generally site specific, alpha diversity is the measure most relevant to ecotoxicologists. Basic approaches in biological monitoring begin with a species list for the study site or sampling unit (Clements and Newman, 2002). This was done here by defining several zones across the length and breadth of two of the landfills and recording the main species observed.

Community-level biological monitoring is based on the assumption that the composition and organization of plant communities reflect local environmental conditions and respond to anthropogenic alterations of those conditions. Relative few studies have examined community-level responses (Clements and Newman, 2002). According to Cairns and Pratt (1993), physical and chemical measures of contaminant effects dominated the field of pollution assessment until the 1970s. The historic emphasis on abiotic measures has gradually been replaced by an understanding that biological indicators of ecological integrity are equally important. We examined changes produced in plant community composition at the polluted landfill sites. These changes are the result of differences in sensitivity among plant populations as well as interactions among populations.

The presence of pollutants prevents or at least delays the natural recolonisation of waste landfills (Hernández et al., 2008). Toxicity appears in an uneven manner depending on the type of residue buried at each point of the landfill. Moreover, leachate discharge zones, which run through the landfill's interior, act as pollution channels and help spread pollutants to other zones.

The plant communities identified were ruderals that barely progress in the succession including annual or pioneer species that need little or even poor soil to thrive. The species comprising the different associations start their development in spring and are substituted by others as temperatures rise.

Table 9 shows the 16 phytosociological classes to which the species detected belong along with a summary of their main characteristics according to Rivas-Martinez et al. (2002).

Classes 9, 12 and 20 correspond to communities found in damp or transiently wet areas to which a few of the species identified belong. Classes 34, 37, 38, 39, 40 and 41 represent ruderal plant associationsanthropisized to a greater or lesser extent. Classes 38, 39, 39A, 39e and 41 are "moderately impacted" ecosystems and classes 34, 37, 39B, 39f and 40 "severely impacted" ecosystems.

Communities of biannual large thistles belong to the class *ARTEMISIETEA VULGARIS* and occupy dry, arid environments on disturbed soils or rubble.

Among the annual communities linked to the presence of ground water, we should mention the hemlocks of the class *GALIO-URTICETEA*.

Table 9. Phytosociological classes and mean characteristics of the main species found at the landfills

Phytosociological class	Main characteristics
9. ISOETO-NANOJUNCETEA Br.-Bl. &Tüxen ex Westhoff, Dijk&Passchier 1946	Pionner annual and dwarf perennial ephemeral isoetid communities on periodically flooded bare soils
12. PHRAGMITO-MAGNOCARICETAE Klika in Klika & V. Novák 1941	Swampy, fenny, lacustrine and riverine helophyte communities dominated by perennial graminoids, sedges, forbs and herbs of fresh and brackish waters
20. JUNCETEA MARITIMI Br.-Bl. in Br.-Bl. Roussine & Nègre 1952	Perennial grasslands growing on coastal and inland temporary wet or inundated salt marshes
34. ARTEMISIETEA VULGARIS Lohmeyer, Preising&Tuxen ex von Rochow 1951	Perennial and tall biennial forbs, grasses and thistle pioneer ruderal and nitrophilous sunny communities growing on rich soils.
37. PEGANO-SALSOLETEA Br.-Bl. & O. Bolós 1958	Nitrophilous or halo-nitrophilous dwarf scrub communities, including anthropogenic alloctonous shrubby vegetation
38. POLYGONO-POETEA ANNUAE Rivas Martinez 1975	Annual pioneer ephemeral and exceptionally small creeping perennial nitrophilousanthropozoogenic heavy trodden communities of urban and rural paths
39. STELLARIETEA MEDIAE Tüxen, Lohmeyer & Preising ex von Rochow 1951. 39A. Stellarieneamediae 39B. Chenopodio-Stellarienea Rivas Goday 1956 39e. Thero-Brometalia (Rivas Goday&Rivas Martinez ex Esteve 1973) O. Bolós 1975 39f. Sisymbrietaliaofficinalis J. Tüxen in Lohmeyer & et al. 1962 em. Rivas-Martinez, Fernándz-González & Loidi 1991	Annual ephemeral weeds, ruderal, nitrophilous and semi-nitrophilous communities -Cultivated field weed communities -Ruderal, nitrophilous and seminitrophilous communities. -Subnitrophilous Mediterranean annual ephemeral grassland-like spring blooming communities. - Nitrophilous and temperate annual ephemeral grassland-like. Path, roadside and rural, often trampled communities
40. GALIO-URTICETEA Passarge ex Kopecký	Perennial hemycriptophyte and climbing tall herbs of nitrified wood fringes and other semi-shaded anthropogenic biotope communities.
41. CARDAMINO HIRSUTAE-GERANIETEA PURPUREI (Rivas-Martínez, Fernández-González & Loidi 1999) Rivas-Martínez, Fernández-González & Loidiclassis nova, stat. Nov.	Annual spring and summer ephemeral internal and external shrub fringes slightly nitrified semi-shaded communities, growing on rich organic nutrient soils.
43. TRIFOLIO-GERANIETEA Müller 1962	Semi-shaded perennial herb communities of scarce moisture, external fringe woodlands. Calcareous or mesoeutrophic rich soils in temperate submediterranean central Iberian territories.
50. TUBERARIETEA GUTTATAE (Br.-Bl. in Br.-Bl., Roussine & Nègre 1952) Rivas Goday& Rivas Martínez 1963 nom. mut. Propos.	Therophytic grasslands. Pioneer spring and early summer ephemeral plant acidophilous or calcifugous communities, dominated by non nitrophilous annual short herbs and grasses, but localized only in dry or initial soils, mostly in submediterranean or step territories.

Phytosociological class	Main characteristics
51. FESTUCO-BROMETEA Br.-Bl. &Tüxen ex Br.-Bl. 1949	Perennial xerophytic and mesophytic grasslands. Anthropogenic grazed baso-neutrophilous or slightly acidophilousmesophytic or slightly xerophytic nutrient rich-pastures largely covered by perennial grasses.
54. POETEA BULBOSAE Rivas Goday& Rivas Martínez in Rivas-Martinez 1978	Western Mediterranean oceanic thermo- to supramediterranean upper semiarid to humid pastures, grazed and manured, dominated by dwarf perennial grasses and other nutritious prostrate chamaephytes.....
56. LYGEO-STIPETEA Rivas-Martinez 1978 nom. Conserv. Propos.	Mediterranean perennial basophilousxerophytic tall bunchy dense or short open grasslands.
57. STIPO GIGANTAE-AGROSTIETEA CASTELLANAE Rivas-Martínez, Fernández-González &Loidi 1999	Silicicolous perennial grasslands rich in endemics, serial of Quercusrotundifolia and other Quercus natural potential forest communities.
59. MOLINIO-ARRHENATHERETEA Tüxen 1937	Mesophile to wet often manured meadows and pasture communities on deep and moist soils, widely spread by grazed and anthropic activities

The annual community representing the most incipient pioneer state is a member of the class *STELLARIETEA MEDIAE*, which colonizes the most recently deposited or mobilized soils. The annual communities growing in more compacted soils belong to this same class (Order *Sisymbretaliaofficinalis*); these are dense graminoid communities and are the most extended in the landfills. In places where we locally detected greater amounts of nitrogenated or phosphorus compounds, nitrophile communities of the subclass *Chenopodio-Sterallieneа*appearedas dense grasslands. In fertile lowlands, appeared some species of the classes *FESTUCO-BROMETEA* and *MOLINIO-ARRHENATHERETEA*. These vary depending on moisture needs.

Among the scores of measures used to assess the effects of contaminants, reduced species richness is probably the most consistent and least controversial. Most measures are still based on the simple assumption that the absence of pollution-sensitive taxa and presence of pollution-tolerant taxa is indicative of degradation (Clements and Newman, 2002). Table 10 lists the species growing at the Móstoles landfill in the 1st and 2nd sampling years and Table 11 those observed in the Torrejón landfill in these same years.

In Tables 10 and 11, we may see that the surface of all the landfills is a mosaic. Four years after sealing, the soil of the Móstoles landfill is no longer bare but sustains the growth of 95 well-represented species. Of these, 38 species that appear in the group of reference species are those found in adjacent pastures. Another group of 37 species was ascribed to the group "moderately impacted" (the landfill also bordered with land left fallow) including ruderals that are nitrophillic, but less so that those comprising the group "severely impacted"; most of these species appear in the class *STELLARIETEA MEDIAE*, in orders and alliances of less marked anthropogenic nature. In this "severely impacted" group, we see species from intensely disturbed places belonging to this class.

Data of 2nd sampling year in Mostoles landfill indicate the soil has been colonized increasing the diversity of species in the three impact groups. The landfill is today grazed by sheep. In the 2nd year the ecosystem showed signs of recovery and in the high areas of the landfills slopes, 49 species present in the reference pastures predominate. The species of "moderately impacted" sites were more frequently found in the middle zone, while the low slopes (discharge zones) show the predominance of 29 species typical of "severely impacted" sites.

Table 10. Main species observed at the Móstoles landfill in the 1st and 2nd sampling years. Species observed only in the 1st sampling year are *underlined*; species observed only in the 2nd year appear in bold

Reference	Class	Moderately impacted	Class	Severely impacted	Class
Juncusbufonius L.	9	*Plantagocoronopus* L.	38	*Carduustenuiflorus* Curtis	34
Pulicariapaludosa Link	9	<u>*Polygonumaviculare* L.</u>	<u>38</u>	***Carlinacorymbosa*L.**	**34**
***Phragmites australis* (Cav.)Trin ex Steud.**	**12**	*Spergularia rubra* (L.) J. et C. Presl	38	*Carthamuslanatus* L.	34
***Elymuspungens*(Pers.) Melderis**	**20**	***Bromushordeaceus* L.**	**39**	*Cichoriumintybus*L.	34
Brachypodiumdystachion (L.) Beauv	43	*Bromustectorum* L.	39	*Convolvulus arvensis* L.	34
***Anthoxanthumaristatum*Boiss.**	**50**	*Capsella bursa-pastoris* (L.) Medik	39	***Dittrichiaviscosa* (L.) Greuter**	**34**
***Astragaluspelecinus*(L.)Barneby**	**50**	*Cerastiumglomeratum*Thuill	39	*Echiumvulgare*L.	34
*Filagolutescens*Jord.	50	***Chamamelummixtum* (L.) All.**	**39**	*Eryngiumcampestre* L.	34
Hymenocarposcornicina(L.) Vis	50	***Chenopodium album* L.**	**39**	*Lactucaserriola* L.	34
Hymenocarposlotoides(L.) Vis	50	***Echiumplantagineum* L.**	**39**	*Marrubiumvulgare* L.	34
Jasionemontana L.	50	*Filagopyramidata*L.	39	***Onopordumillyricum* L.**	**34**
Lathyrusangulatus L.	50	***Medicagolupulina*L.**	**39**	***Picnomonacarna*(L.) Cass**	**34**
Leontodontaraxacoides (Vill.) Mérat	50	***Papaverrhoeas* L.**	**39**	*Scolymushispanicus*L.	34
***Micropyrumtenellum* (L.) Link**	**50**	*Polypogonmonspeliensis* (L.) Desf.	39	***Seneciojacobaea* (L.)**	**34**
Ornithopuscompressus L.	50	***Raphanusraphanistrum* L.**	**39**	*Silybummarianum* (L.) Gaertn.	34
Petroragiananteuilli (Burnat)P.W. & Heywood	50	***Setariaviridis*(L.) P. Beauv.**	**39**	***Verbascumsinuatum*L.**	**34**
Psilurusincurvus(Gouan) Schinz&Thell	50	*Sonchusasper* (L.) Hill.	39	*Artemisia herba-alba* Asso	37
*Silenecolorata*Poiret	50	***Tribulusterrestris* L.**	**39**	*Cnicusbenedictus* L.	39B
*Silenenocturna*L.	50	*Viciabenghalensis* L.	39	***Conyzabonariensis* (L.) Cronq.**	**39B**
Trifoliumarvense L.	50	***Xanthium strumarium*L.**	**39**	<u>*Malvaparviflora* L.</u>	<u>39B</u>
*Trifoliumcampestre*Schreber	50	*Torilisnodosa*(L.) Gaertn.	41	***Malvanicaeensis*All.**	**39B**
Trifolium striatum L.	50	*Lamiumamplexicaule* L.	39A	***Xanthium spinosum*L.**	**39B**
*Trifoliumsylvaticum*Gèrard in Loissel	50	<u>*Papaverrhoeas* L.</u>	<u>39A</u>	*Anacyclusclavatus* (Desf.) Pers.	39f
***Trisetumovatum* (Cav.) Pers.**	**50**	<u>*Raphanusraphanistrum* L.</u>	<u>39A</u>	*Crepisvesicaria* L.	39f
***Vulpiabromoides*(L.) S.F. Gray**	**50**	*Aegilopsgeniculata*Roth	39e	*Diplotaxisvirgata*(Cav.) DC.	39f
Vulpiamyuros(L.) C. C. Gmel.	50	*Aegilopstriuncialis* L.	39e	*Hirchsfeldiaincana*(L)Lagr. Fos	39f
*Ononisspinosa*L.	51	*Alyssum granatense*Boiss& Reuter	39e	*Hordeumleporinum* L.	39f
***Poabulbosa*L.**	**54**	***Astragalushamosus* L.**	**39e**	***Malvasylvestris*L.**	**39f**
***Trifoliumgemellum*Pourret**	**54**	*Avena barbata*Pott ex Link	39e	*Medicagopolymorpha* L.	39f

Reference	Class	Moderately impacted	Class	Severely impacted	Class
ex Willd					
*Trifoliumglomeratum*L.	54	*Avenasterilis*L.	39e	*Plantagolagopus* L.	39f
*Trifoliumscabrum*L.	54	*Bartsiatrixago*L	39e		
*Trifoliumtomentosum*L.	54	*Bromusdiandrus*Roth.	39e		
Andryalaintegrifolia L.	56	*Bromushordeaceus* L.	39e		
***Andryalalaxiflora* (Salzm.) DC.**	**56**	*Bromusmadritensis* L.	39e		
Convolvulus althaeoides L.	56	*Bromusrubens* L.	39e		
Daucuscarota L.	56	*Corynephorusfasciculatus* Boiss& Reuter	39e		
*Agrostiscastellana*Boiss& Reuter	57	*Echiumplantagineum*L.	39e		
*Trifoliumcernuum*Brot.	57	*Loliumrigidum*Gaudin	39e		
*Trifoliumretusum*L.	57	*Lupinusangustifolius* L.	39e		
***Crepiscapillaris*(L.) Wallr.**	**59**	***Medicagoorbicularis*(L.) Bartal**	**39e**		
Cynodondactylon (L.) Pers.	59	***Melilotussulcatus*Desf.**	**39e**		
Cynosurusechinatus L.	59	***Plantagoafra* L.**	**39e**		
*Dactylisglomerata*L.	59	*Silenegallica* L.	39e		
Elymuspungens (Pers.) Melderis	59	*Silenenocturna* L.	39e		
***Equisetum ramosissimum*Desf.**	**59**	*Taeniatherum caput-medusae* (L.) Nevski	39e		
*Holcuslanatus*L.	59	*Trifoliumangustifolium* L.	39e		
Hypochaerisradicata L.	59	*Trifoliumcherleri* L.	39e		
***Loliumperenne* L.**	**59**	*Trifoliumhirtum* All.	39e		
Poatrivialis L.	59	*Trisetumpaniceum* (Lam.) Pers.	39e		
Scirpoidesholoschoenus(L.) Sojak.	59	*Vulpiaciliata*Dumart	39e		
Seneciojacobaea L.	59				
***Trifoliumpratense* L.**	**59**				
***Trifoliumrepens* L.**	**59**				

Table 11. Main species observed at the Torrejón municipal landfill in the 1st and 2nd sampling years. Species observed only in the 1st sampling year are underlined; species observed only in the 2nd year appear in bold

Reference	Class	Moderately impacted	Class	Severely impacted	Class
*Juncus**bufonius*L.	9	*Polygonumaviculare*L.	38	***Carduusbourgeanus*Boiss&Reut**	**34**
Phragmites australis (Cav.) Trin.exSteud.	12	*Spergularia rubra* (L.) J. Presl & K. Presl.	38	*Carduuspycnocephalus* L.	34
***Sonchus*crassifolius P. ex Willd**	**20**	*Bromushordeaceus*L.	39	*Carduustenuiflorus* Curtis	34
*Viciavillosa*Roth	43	*Bromussterilis*L.	39	***Carlynacorymbosa*L.**	**34**
*Andryalaintegrifolia*L.	50	*Bromustectorum*L.	39	*Carthamuslanatus*L.	34
Coronillascorpioides(L.) W.D. J.Koch	50	*Calendula arvensis*L.	39	*Chondrillajuncea*L.	34
Galiumparisiense L.	50	*Cerastiumglomeratum*Thuill.	39	*Cichoriumintybus*L.	34

Table 11. (Continued)

Reference	Class	Moderately impacted	Class	Severely impacted	Class
*Leontodontaraxacoides*subsp. *longirostris* Finch & P. D. Sell	50	***Erucasativa*(Mill) Thellin Hegi.**	**39**	*Convolvulus arvensis*L.	34
Medicago minima (L.) L.	50	***Erucavesicaria*L.**	**39**	***Daucusmaximus*Desf.**	**34**
*Silenenocturna*L.	50	*Filagopyramidata*L.	39	***Echiumasperrimum* Lam.**	**34**
Tolpisbarbata (L.) Gaertn.	50	*Lophochloacristata* (L.) Hyl	39	*Eryngiumcampestre*L.	34
*Trifoliumcampestre*L	50	*Silene vulgaris* (Moench) Garcke	39	*Foeniculumvulgare*Miller	34
Xeranthemuminapertum(L.) Willd.	50	*Sinapisarvensis*L.	39	*Lactucaserriola*L.	34
***Brachypodiumphoenicoides*(L.) Roem. &Schult**	**51**	*Sonchusoleraceus*L.	39	*Marrubiumvulgare*L.	34
Mantisalcasalmantica(L.) Briq. &Cavill.	51	*Anagallisarvensis* L.	39A	***Melilotusalbus*Medik.**	**34**
Medicagosativa L.	51	***Diplotaxiserucoidis*(L.) DC.**	**39A**	*Melilotusofficinalis*(L.) Pall.	34
*Ononisspinosa*L.	51	***Fumariaofficinalis*L.**	**39A**	*Onopordumillyricum*L.	34
*Astragalusstella*Gouan.	54	*Fumariaparviflora*Lam.	39A	***Onopordumnervosum*Boiss.**	**34**
*Trifoliumglomeratum*L.	54	***Glauciumcorniculatum* (L.) J.H. Rudolph**	**39A**	***Phlomisherba-venti*L.**	**34**
*Trifoliumscabrum*L.	54	***Papaverhybridum*L.**	**39A**	*Reseda lutea*L.	34
*Trifoliumtomentosum*L.	54	*Papaverrhoeas*L.	39A	***Scolymushispanicus*L.**	**34**
*Carexdivisa*Huds.	59	*Raphanusraphanistrum*L.	39A	*Silybummarianum*(L.) Gaertner	34
***Cirsiummonspessulanum* L. (Hill)**	**59**	*Sherardiaarvensis*L.	39A	***Verbascumsinuatum*L.**	**34**
Cynodondactylon(L.) Pers.	59	*Aegilopsgeniculata*Roth.	39e	*Salsolavermiculata*L.	37
*Dactylisglomerata*L.	59	*Aegilopstriuncialis*L.	39e	***Conium maculatum* L.**	**40**
*Loliumperenne*L.	59	*Astragalushamosus*L.	39e	*Galiumaparine*L.	40
Scirpoidesholoschoenus(L.) Soják	59	*Avena barbata*Pott ex Link	39e	***Rumexcrispus*L.**	**40**
***Trifoliumlappaceum* L.**	**59**	*Avenasterilis*L.	39e	*Cnicusbenedictus*L.	39B
*Trifoliumpratense*L.	59	*Bartsiatrixago* L.	39e	*Malvaneglecta*Wallr.	39B
*Trifoliumrepens*L.	59	*Biscutellaauriculata*L.	39e	***Malvaparviflora*L.**	**39B**
Trifoliumresupinatum L.	59	*Bromusdiandrus* Roth	39e	***Sisymbriumirio* L.**	**39B**
		*Bromusmadritensis*L.	39e	*Urticaurens* L.	39B
		*Bromusrubens*L.	39e	*Anacyclusclavatus* (Desf.) Pers.	39f
		*Centaureamelitensis*L.	39e	*Crepis vesicaria* L. ssp. *taraxacifolia* (Thuill.) Thell.	39f
		Hedypnois cretica (L.) Dum-Courset	39e	*Diplotaxisvirgata*(Cav.) DC.	39f
		*Loliumrigidum*Gaudin	39e	*Erodiumciconium*(L.) L`Her	39f
		Medicago orbicularis (L.) Bartal.	39e	*Erucavesicaria*L.	39f
		Medicagorigidula(L.) All.	39e	*Geranium molle* L.	39f
		*Melilotussulcatus*Desf.	39e	*Hirchsfeldiaincana* (L.) Lagr. - Foss.	39f
		*Scorzoneralaciniata*L.	39e	*Hordeumleporinum*L.	39f
		Taeniatherum caput-medusae(L.) Nevski	39e	*Malvasylvestris*L.	39f
		*Trifoliumangustifolium*L.	39e	*Medicagopolymorpha* L.	39f

Reference	Class	Moderately impacted	Class	Severely impacted	Class
		*Trigonellamonspeliaca*L.	39e	*Plantagolagopus*L.	39f
		*Trigonellapolyceratia*L.	39e		
		Trisetumpaniceum(Lam.) Pers.	39e		
		*Vulpiaciliata*Dumort	39e		
		*Asperugoprocumbens*L.	41		
		Torilisleptophylla(L.) Rchb. f.	41		
		Torilisnodosa (L.) Gaertn.	41		

The greater freshness of soils in Torrejón landfill has prompted the somewhat higher presence of species from the initial years of sealing than at the Mostolés landfill. 28 species belong to the reference communities; 41.6 % are species found in “moderately impacted” ecosystems and 30.7% are characteristic of “severely impacted” ecosystems.

At the Torrejón landfill in the 2nd sampling year, both natural plant communities and “moderately impacted” ecosystems have diminished and “severely impacted” ecosystems have expanded owing to the repeated dumping of rubble, rubbish and industrial residues, land use as zones of grazing and walkways and the almost complete lack of natural communities in surrounding areas. The most frequent observation at polluted sites is the loss of sensitive species and their replacement by tolerant species (Clements and Newman, 2002) and this was noted at these waste disposal sites.

Studying the spontaenous colonization by native plant communities is fundamental for phytoremediation research and ecological restoration (Hernández and Pastor, 2008; Nagendran et al., 2006). The objectives of this strategy are i) to discover the natural patterns of succesion in such degraded soils and ii) to select suitable plant species able to deal with soil pollutants.

Conclusion

The present chapter examines the main indicators of physico-chemical pollution in soil, leachate and water samples obtained from two waste landfills and their repercussions on the plant communities that grow at these sites. The data presented indicate that landfill age has a significant effect on its characteristics and composition. In the Mediterranean region, although the dry climate conditions promote the stabilization of landfilled refuse, there may be seasonal fluctuations in the quality and quantity of the leachate produced.

Our results reveal the heterogeneity of inorganic pollutants in the soils of the discharge zones affected by these former waste disposal sites. The scarce amount of cover soil and intense weather events in the dry area of this study, have left exposed and promoted the mobilization of many polluting substances and leachates. All this is aggravated by the sites chosen for waste disposal since these included natural depressions close to streams and wells, and even marshy areas. These settings have determined a greater risk of pollutants reaching the groundwater. A further risk arises from the forage plants growing in the affected soils because under a Mediterranean climate when surface water evaporates in spring these plants are consumed by herbivores.

In general, highest contents of inorganic contaminants in the surface waters sampled were detected in streams as they flowed out of the zones affected by the landfills.

Organic compounds occurring in former waste dumps need to be more closely monitored. In effect, after adapting the necessary techniques for their determination in soil samples, we have now started to apply these to our last sampling effort in the Madrid region.

Our observations over the past 20 years indicate that inorganic elements used by living organisms as micronutrients, salts and heavy metals all have impacts on the health of ecosystems and seriously affect plant communities. These communities become less biodiverse, more stable, and succession processes slow down (annual and biannual biotypes are abundant, whereas perennial biotypes are less frequent). Our analysis of organic compounds both in water and soil samples indicates further risks for these ecosystems.

ACKNOWLEDGMENTS

This study was funded by Project CTM2008-04827/TECNO of the Spanish MICINN and the EIADES Programme of the Community of Madrid. MJGG is supported by Fundación Ramón Areces.

REFERENCES

Adarve, M.J., Hernández, A.J. & Rebollo, L.F. (1994) La contaminación de las aguas subterráneas por lixiviados de un vertedero de residuos sólidos urbanos localizado en Torrejón de Ardoz (Madrid). In: *Agua y Medio Ambiente*. (pp. 162-170) Madrid, Spain: TIASA.

Adarve, M.J., Hernández, A.J., Gil, A. & Pastor, J. (1998). B, Zn, Fe and Mn content in four grassland species exposed to landfill leachates. *Journal of Environmental Quality, 27*, 1286-1293.

Akinbile, C.O. (2012) Environmental Impact of Landfill on Groundwater Quality and Agricultural Soils in Nigeria. *Soil & Water Research, 7*, 18-26.

Al-Yaqout, A.F. &Hamoda, M.F. (2003). Evaluation of landfill leachate in arid climate, a case study.*Environment International, 29*, 593-600.

Baccini, P., Henseler, G., Figi, R. &Belevi, H. (1987) Water and element balances of municipal solid waste landfills. *Waste Management & Research, 5*, 483–499.

Cairns Jr, J. & Pratt, J.R. (1993).A history of biological monitoring using benthic macroinvertebrates. In: D.M. Rosenberg, & V.H. Resh, (Eds.), *Freshwater Biomonitoring and Benthic Macroinvertebrates* (pp. 10-27) New York,USA: Chapman and Hall.

Calace, N., Massimiani, A., Petronio, B.M. &Pietroletti, M. (2001). Municipal landfill leachate–soil interactions: a kinetic approach. *Chemosphere, 44*, 1025-1031.

Cheng, C.Y. & Chu, L.M. (2007).Phytotoxicity data safeguard the performance of the recipient plants in leachate irrigation. *Environmental Pollutution, 145(1)*, 195-202.

Cheng, C.Y. & Chu, L.M. (2011). Fate and distribution of nitrogen in soil and plants irrigated with landfill leachate. *Waste Management, 31(6)*, 1239-1249.

Christensen, T.H., Kjeldsen, P., Bjerg, P., Jensen, D.L., Christensen, J.B. &Baun, A. (2001). Biogeochemistry of landfill leachate plumes. *Applied Geochemistry, 16*, 659–718.

Chu, L.M., Cheung, K.C. & Wong, M.H. (1994).Variations in the Chemical Properties of Landfill Leachate.*Journal of Environmental Management, 18*, 105-117.

Clements, W.H. & Newman, M.C. (2002).*Community Ecotoxicology*.Hierarchical Ecotoxicology Series. Chichester, West Sussex: John Wiley & Sons Ltd.

Cyr, F., Mehr, M.C. & Mallet, V. N. (1987) Leaching of chemical contaminants from municipal landfill sites. *Bulletin of Environmental Contamination and Toxicology, 38*, 775-782.

Fatta, D., Papadopoulos, A. & Loizidou, M. (1999). A study of the landfill leachate and its impact on the groundwater quality of the greater area.*Environmental Geochemistry & Health, 21*, 175-190.

Gutiérrez-Ginés, M.J., Hernández, A.J., Pérez-Leblic, M.I., Pastor, J. &Vangronsveld, J. (2014). Phytoremediation of soils co-contaminated by organic compounds and heavy metals: bioassays with *Lupinusluteus* L and associated endophytic bacteria. *Journal of Environmental Management, 143,* 197-207.

Hernández, A.J. & Pastor, J. (2008). Validated Approaches to Restoring the Health of Ecosystems Affected by Soil Pollution. In: J. B. Dominguez (Ed.), *Soil Contamination Research Trends* (pp. 51-72). Hauppauge, NY: Nova Science Publishers Inc.

Hernández, A.J., Adarve, M.J. & Pastor, J. (1998). Some impacts of urban waste landfills on Mediterranean soils. *Land Degradation and Develoment, 9,* 21-33.

Hernández, A.J., Adarve, M.J., Gil, A. & Pastor, J. (1999). Soil salination from landfill leachates: effects on the macronutrient content and plant growth of four grassland species. *Chemosphere, 38*, 1693-1711.

Hernández, A. J., Bartolomé, C., Álvarez, J. & Pastor, J. (2008). Comunidades vegetales de antiguos vertederos sellados pastados por ovinos: caracterización botánica y suelos. In: P. Fdez. Rebollo et al. (Eds.), *Pastos, clave en la gestión de los territorios*. (pp. 143-149). Sevilla, Spain: Consejería de Agricultura y Pesca, Junta de Andalucía.

Hernández, A. J., Gutierrez, M. J. & Pastor, J. (2009). Ecology and health in risk analysis of polluted soils. In: C.A. Brebbia (ed), *Environmental health Risk* (pp.257-268) Southampton, UK: WIT Press.

Hernández, A. J., Pérez-Leblic, M.I., Bartolomé, C., Rodríguez, J., Álvarez, J. & Pastor, J. (2012).Ecotoxicological diagnosis of a sealed municipal landfill.*Journal of Environmental Management, 95*, S50-S54.

Hogland, W., Marques, M. &Nimmermark, S. (2004). Landfill mining and waste characterization: a strategy for remediation of contaminated areas. *Journal of Material Cycles and Waste Management, 6*, 119-124.

IARC. (2000). Di(2-ethylhexyl) adipate (Group 3). International Agency for Research on Cancer (IARC) - Summaries & Evaluations, 77, 149. Available from: http://www.inchem.org/documents/iarc/vol77/77-02.html (date of last check: 02/27/2013)

Illera, V., Walter, I., Souza, P. & Cala, V. (2000). Short-term effects of biosolid and municipal solid waste applications on heavy metals distribution in a degraded soil under a semi-arid environment.*Science of Total Environment, 255*, 29-44.

ISO 10382. Soil Quality - Determination of organochlorine pesticides and polychlorinated biphenyls - Gas chromatographic method with electron capture detection.

ISO 18287. Soil quality - Determination of polycyclic aromatic hydrocarbons (PAH) - Gas chromatographic method with mass spectrometric detection (GC-MS)

Kalčíková, G., Zagorc-Končan, J., Zupančič, M. &ŽgajnarGotvajn, A. (2012).Variation of landfill leachate phytotoxicity due to landfill ageing, *Journal of Chemical Technology and Biotechnology, 87 (9)*, 1349-1353.

Kjeldsen, P. (1993). Groundwater pollution source characterization of an old landfill.*Journal of Hydrology, 142*, 349-371.

Koeler, H.H. (1996) Soils animals and bioindication. In: N.M. Van Straaelen& A. Krivolustky (Eds.) *Bioindicator Systems for Soil Pollution* (pp. 179-188) Dordrecht, The Netherlands: Kluwer Academic Publishers.

Lakanen, E. &Ervio, R. (1971).A comparison of eight extractants for the determination of plant available micronutrients in soils.*Acta AgriculturicaFennica, 123*, 223-232.

MAPA, Ministerio de Agricultura, Pesca y Alimentación, España. (1982) *Métodos Oficiales de Análisis de Suelos y Aguas.*

Matejczyk, M., Płaza, G.A., Nałęcz-Jawecki, G., Ulfig, K. &Markowska-Szczupak, A. (2011).Estimation of the environmental risk posed by landfills using chemical, microbiological and ecotoxicological testing of leachates.*Chemosphere, 82(7)*, 1017-1023.

Millán, R., Gamarra, R., Schmid, T., Sierra, M.J., Quejido, A.J., Sánchez, D.M., Cardona, A.I., Fernández, M. & Vera R. (2006).Mercury content in vegetation and soils of the Almadén mining area (Spain).*Science of The Total Environment, 368(1)*, 79-87.

Nagendran, R., Selvam, A., Joseph, K. &Chiemchaisri, C. (2006). Phytoremediation and rehabilitation of municipal solid waste landfills and dumpsites: A brief review. *Waste Management, 26(12)*, 1357-1369.

Pankhurst, C. E. (1994). Biological indicators of soil health and sustainable productivity. In: D. J. Greeland& I. Szabloes (Eds.) *Soil Resilence and Sustainable Land Use* (pp. 331-351). Oxon, UK: CAB International.

Pastor, J. & Hernández, A.J. (2012). Heavy metals, salts and organic residues in old solid urban waste landfills and surface waters in their discharge areas: Determinants for restoring their impact. *Journal of Environmental Management, 95*, S42-S49.

Pastor, J., Hernández, A.J., Adarve, M.J. &Urcelai, A (1993a). Chemical characteristics of sedimentary soils in the Mediterranean environment: a comparison of undisturbed and disturbed soils. *Applied Geochemistry, 2*, 195-198.

Pastor, J., Alia, M., Hernandez, A.J., Adarve, M.J., Urcelai, A. & Anton, F.A. (1993b). Ecotoxicological studies on effects of landfill leachates on plants and animals in Central Spain. *Science of The Total Environment, 140*, 127-134.

Pastor, J., Urcelai, A., Oliver, S. & Hernández, A.J. (1993c).Impact of Municipal Waste on Mediterranean Dry Environments.*Geomicrobiology Journal, 11*, 247-260.

Pastor, J., Gutiérrez-Ginés, M.J., Bartolomé, C. & Hernández, A.J. (2014). The complexity of polluted soils of sealed landfills: case study in Mediterranean environment. In: M.C. Hernández-Soriano (Ed.) *Environmental Risk assessment of Soil Pollution*, (pp. 199-223). Croatia: InTech.

Paxeus, N. (2000). Organic compounds in municipal landfill leachates. *Water Science & Technology, 42*, 323-233.

Pohland, F.G., Harper, S.R., Chang, K.C., Dertien, J.T &Chian, E.S.K. (1985) Leachate generation and control oflandfill disposal sites. *Water Pollution Research Journal of Canada, 20*, 10–25.

Rapport, D.J., Regier, H.A. & Hutchinson, T.C. (1985).Ecosystem behaviour under stress.*The American Naturalist, 125,* 617-640.

Rivas-Martinez, S., Diaz, T.E., Fernández-Gonzalez, F., Izco, J., Loidi, J., Lousa, M. & Penas, A. (2002). Vascular Plant Communities of Spain and Portugal.Addenda to the syntaxonomical checklist of 2001.*ItineraGeobotanica, 15(1-2)*, 5-922.

Slacka, R.J., Gronowb, J.R. &Voulvoulisa, N. (2005). Household hazardous waste in municipal landfills: contaminants in leachate. *Science of The Total Environment, 337*, 119-137.

Spellerber, I. F. (1991).*Monitoring Ecological Change* (2nd edition). Cambridge, UK: Cambridge Universty Press.

Stegman, R. (1982). The pollution potential of sanitary landfill.In: *Effects of waste disposal on groundwater and surface water*(pp125-135). INMS Publications, 139.

Tatsi, A.A. &Zouboulis, A.I. (2002).A field investigation of the quantity and quality of leachate from a municipal solid waste landfill in a Mediterranean climate (Thessaloniki, Greece).*Advances in Environmental Research, 6*, 207-219.

UNE 77307. Calidad del suelo. Determinación del contenido de aceite mineral. Método de espectrometría infrarroja y método de cromatografía de gases.

Urcelai, A., Hernandez, A.J. & Pastor, J. (2000). Biotic indices based on soil nematode communities for assessing soil quality in terrestrial ecosystems. *Science of the Total Environment, 247*, 253-261.

U.S. EPA. 2007. Diethyltoluamide (DEET) Chemical Summary. U.S. EPA, Toxicity and Exposure Assessment for Children's Health. Available from: http://www.epa.gov/teach/chem_summ/DEET_summary.pdf (date of download: 02/27/2013)

U.S. EPA.2012. Phthalates Action Plan. Revised 03/14/2012. U.S. Environmental Protection Agency. Available from: http://www.epa.gov/oppt/existingchemicals/pubs/actionplans/phthalates_actionplan_revised_2012-03-14.pdf. (date of download: 02/27/2013)

U.S. Environmental Protection Agency. Test Methods for Evaluating Solid Waste. Physical/Chemical Methods.Method 3550B.Ultrasonic Extraction.

U.S. Environmental Protection Agency. Test Methods for Evaluating Solid Waste. Physical/Chemical Methods.Method 3650B.Acid-Base Partition Cleanup.

U.S. Environmental Protection Agency. Test Methods for Evaluating Solid Waste. Physical/Chemical Methods.Method 8401.Phenols by Gas Chromatography.

Verhoef, H.A. & van Gestel, A.M. (1995) Methods to Assess the Effects of Chemicals on Soils. In: R.A. Linfhurst& P.H. Bourdeau (Eds.) *SCOPE 53. Methods to Assess the Effects of Chemicals On Ecosystems* (pp. 223 – 257). New York, USA: John Wiley & Sons Ltd.

Weber, R., Watson, A., Forter, M. &Oliaei, F. (2011).Persistent organic pollutants and landfills - a review of past experiences and future challenges.*Waste Management Research, 29*, 107-121.

In: Landfills and Recycling Centers
Editor: Carla H. Jackson

ISBN: 978-1-63463-792-3

Chapter 2

SURFACE GEOPHYSICAL METHODS IN THE ASSESSMENT OF ENVIRONMENTAL IMPACTS OF LANDFILLS: AN OVERVIEW

Elsa Cristina Ramalho[1]*, Manuel João Senos Matias[2] and Rui Miguel Marques Moura[3]
[1]Laboratório Nacional de Energia e Geologia
Estrada da Portela, Bairro do Zambujal, Amadora, Portugal
[2]University of Aveiro, Campo Universitário de Santiago, Aveiro, Portugal
[3]University of Oporto, Faculty of Sciences, Porto, Portugal

ABSTRACT

The fast development of field, data processing equipment and software has allowed the use of geophysical methods to an ever increasing range of applications. Hence, nowadays it is much easier to conduct massive field surveys combining different methods, to obtain more accurate and denser data, so that complex modeling and interpretation at limited costs can be carried out.

Landfills have been targeted by geophysical methods in order to investigate their environmental impacts. In fact, landfills have been the classic way to deposit domestic and industrial waste and have generated a large range of negative environmental impacts in groundwater and soils. These problems often persist even after the effective use of the landfills and subsequent recovery processes. Owing to their characteristics, landfills are difficult to access and because of the general lack of accurate information regarding the shape, nature of the refuse, history and development of the landfill, non-invasive, non-destructive methods and sometimes autonomous data acquisition devices must be used to monitor impacts and to investigate and prevent groundwater and soil contamination.

Geophysical methods can be applied to investigate a wide range of aspects related with the assessment of the environmental impact of landfills. Problems such as geometry definition, geological settings, contamination plume location and monitoring, investigation of internal structure and refuse zoning, determination of fluid flow direction and paths or the determination of sealing conditions and leakage may be more

* Email: elsa.ramalho@lneg.pt.

successfully evaluated if a carefully chosen geophysical survey is part of any investigation program.

Because of the nature and complexity of the problems to investigate, only multidisciplinary approaches, involving geophysics, hydrochemical, hydrogeological and geological information, can provide meaningful results for a thorough assessment of the landfills impact on the environment.

This work intends to demonstrate the application of geophysical methods in the investigation of the environmental impacts, as described above, of industrial and domestic landfills during their life time and after closure. Thus, several examples will be discussed illustrating the use of 2D, 3D and time lapse resistivity, electromagnetic, ground probing radar, self-potential, magnetic, gravity surveys and airborne thermal mapping. Most of the geophysical data will be presented and shortly discussed together with information from boreholes, geology, hydrogeology and hydrochemical data.

As it will be shown, it is clear that only a judicious combination of methods and information from different nature can provide tools for the diagnosis and assessment of the impact of landfills in the environment, for the investigation of the best engineering solutions to remediate them and for the possible recovery of refuse with economic interest.

1. Introduction

It has long been acknowledged that traditional Geophysical Methods, with a history of success in the Mining and Oil & Gas Industries, can be adapted to Near Surface problems. As Near Surface Geophysics developed, it was proved to provide better and competitive means to characterize shallow geology and man-made infrastructure. Furthermore, the increasing public opinion awareness on Environment issues attracted the Near Surface scientific and engineering community to the study and use Geophysical Methods in the investigation of those issues. Hence, in the last decades, environmental applications have lead practitioners to push forward, adapt, develop and often engineer techniques as Near Surface Geophysical Methods offer fast, economic and non-invasive tools, that can even "see" through insulating covers (as is the case of the sealed landfills) and are able to deliver accurate 3 and 4D images of the areas under investigation.

Therefore, the practice of Near Surface Geophysical Methods has contributed with solutions to Environmental problems, such as location of groundwater contamination plumes (Benson et al., 1983; Greenhouse et al., 1989; Matias et al., 1994), organic contamination (Benson et al., 1991; Sauck et al., 1998; Sauck, 2000; Atekwana et al., 2000), contamination monitoring and time evolution (Brewster et al., 1995; Matias et al., 2006), landfill studies (Lanz et al., 1997; Roberts et al., 1990; Biandini et al., 2004; Bernstone et al., 2000; Soupios et al., 2007; etc.). In general, multi-method strategies, thus integrated surveys, are adopted so that joint interpretations are sought as an attempt to reduce non uniqueness problems.

The diagnosis and the assessment of soil and groundwater environmental problems in and around landfills are often investigated with geophysical methods to complement the existing spatial and in depth information. In fact, their non-destructive and non-invasive characteristics, the low costs involved (Pereira and Lima, 2007) as well as the possibility of a fast and dense geographical coverage, both laterally and in depth (3D and 4D) make them almost indispensable tools in the investigation of these problems. The potentialities of surface geophysical methods increase significantly when borehole information is available to calibrate and correlate results. However, the information to obtain varies both with the characteristics of the geological environment and with the geophysical methods used. These methods allow the appraisal of the physical properties of various geological formations, a preliminary fluid characterization and the general characteristics of landfills and their content. The knowledge and understanding of these features give important clues and often enable to determine the origin of the environmental problems, playing and important role in their resolution.

The effectiveness of the geophysical methods increases with the increasing contrast between the geological media and the landfill itself (Reynolds, 1997). Therefore, their use in environmental situations related with landfills is sustained on the fact that physical properties contrasts in these cases are usually larger than in natural and non-disturbed environments. Nowadays, geophysical data can be modeled and correlated with analytical data and, therefore, it is also possible to extend the available information far beyond the usual confined soil and/or groundwater analysis to larger areas where no information is available.

Furthermore, it is often difficult if not impossible to sample the earth in landfill areas by direct means, and therefore indirect and non-invasive methods must be sought. Thus, Geophysical Methods allow non-invasive data acquisition in depth, conserving local conditions and, in particular, hydrogeological features. Hence, aquifers mixture and consequent contamination are avoided (Senos Matias, 2001) so that exploration operations preserve in-situ characteristics around landfills.

In conclusion, the prevention of environmental risks originated by landfills operations, the assessment and identification of problems related with these infrastructures, the investigation of effective mitigation actions can profit with the use of selected geophysical methods in different geological environments, in a large variety of situations, such as difficult access and hazardous conditions, which could hardly be investigated otherwise.

The use of different surface geophysical methods, such as resistivity, seismic, magnetic or radiometric methods, either individually or in combination, in conjunction with geological and hydrogeological studies is a proven basis to the study of the environmental impact of landfills. Some objectives of the use of surface geophysical methods in landfill studies are shown in Figure 1.

As it was mentioned previously, geophysical methods can cover large areas. Therefore, the use of an initial fast and expeditious geophysical method, enabling a fast coverage of the entire area, leads to a preliminary diagnostic stage. This earlier stage provides important information about the most adequate physical properties to investigate and allows the definition of sub-areas to prospect using a denser mesh or recommends the use of other methods with higher lateral or vertical resolution (Senos Matias, 2001). Modeling and interpretation of geophysical data is always associated with a certain degree of uncertainty. As a matter of fact, anomalies may be erroneously interpreted due to extreme fluid characteristics or to the heterogeneity of the geological environments. A good geological knowledge of the area is therefore very important as well as the use of more than one geophysical method, since each method responds to different properties (Telford et al., 1995). Hence, it is good practice to complement the obtained data with local geological information or to use more than one geophysical method whenever possible (Senos Matias, 2001).

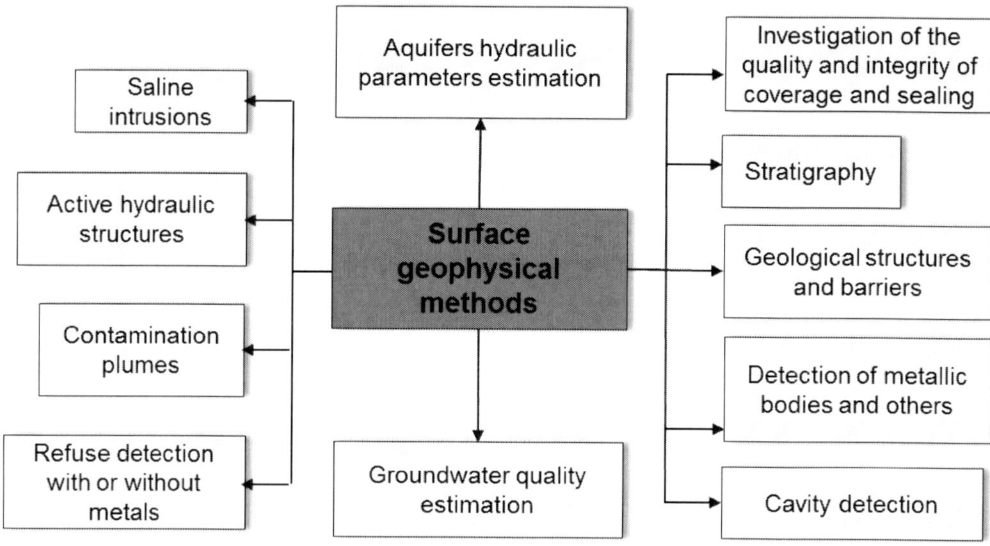

Figure 1. Objectives of the use of geophysical methods in landfill studies.

The application of geophysical methods must be designed in accordance with the objectives of the specific case in hands. Geoelectrical methods (resistivity with the various vertical electrical soundings (VES) arrays, self-potential (SP) and electromagnetics) are effective in porous sedimentary zones, where the resistivity variations obtained in depth are usually related with lithology variations and/or groundwater conductivity. On the other hand, in crystalline environments, they are usually used to investigate groundwater flow and the identification of geological structures that allow groundwater circulation such as weathered areas, fracturing zones or the Hydrostatic Level (HL) depth.

These methods can also be used to check the integrity of landfills sealing. Ground Probing Radar (GPR) can be used in stratigraphic investigations, in the location of metallic bodies and in the detection and delineation of contamination plumes.

Magnetic methods play an important role in the location of metallic refuse. Seismic Refraction surveys are particularly important in stratigraphic studies and in groundwater level estimates. Gravimetry can be used to delineate landfills, especially in crystalline environments. Radiometry is very important when investigating radioactive leakage and radioactive waste.

The use of surface geophysical methods applied in landfills is summarized in Table 1. It must be noticed that in the last column is added information on the application of geophysical logging, out of the scope of the present work, but nevertheless important in these investigations.

Table 1. Use of several geophysical methods applied in landfills in the situations referred in Figure 1 (adapted from Senos Matias, 2001; Martinho, 2003)

Application	Refection seismics	Refraction seismics	Gravimetry	Magnetics	SP	Resistivity/ /IP	Electroma-gnetics	Georadar	Radiometry	Geophysical logging
Geological and hydrogeological mapping	1	1	0	0	1	2	2	2	2	0
Geological structures and barriers	2	2	1	2	1	2	2	2	1	2
Active hydraulic structures	1	1	0	0	2	2	2	2	2	2
Contamination plumes	0	0	0	0	0	2	2	2	1	2
Waste containing metals	0	0	0	2	0	2	2	2	0	2
Waste without metals	0	0	0	0	1	2	2	2	0	2
Metallic objects	0	0	0	2	0	2	2	2	0	1
Stratigraphy	2	2	1	0	0	2	2	2	0	2
Cavities	2	2	1	0	0	2	2	2	0	2
Groundwater quality estimation	0	0	0	0	0	2	2	2	0	2
Hydraulic parameters estimation	1	1	0	0	0	2	2	2	0	2

Use intensity: 0 – weak; 1 – regular; 2 – strong.

This work does not intend to review the general concepts on geophysics. For this purpose, Telford el al. (1995), Reynolds (1997) or Sharma (1997) can be consulted. Instead, case-studies in Portugal are presented and selected in a way to provide a comprehensive vision of the use of geophysical methods in landfill problems. Therefore, landfills located in different geological environments have been considered and distinct environmental problems were selected.

2. Geophysics in Geological Studies Prior to the Implementation of Landfills

Knowledge of local geology is very important in the assessment of the environmental impacts of a landfill. Ideally, such knowledge should be available before the deposition of any refuse in order to engineer the best conditions for waste disposal. Often this is not the case, as landfills have been in use long before the understanding of their negative impacts.

Therefore, unfortunately even after years of operation or even after closure, geological conditions and hydrogeology remain unknown and, hence, it is still necessary to obtain information on local settings in order to minimize groundwater contamination and to design appropriate remediation plans. In any case, geophysics can provide paramount information on these matters and, frequently, geophysical methods have been used in the different phases described above.

Herein two examples will be presented. The first consists on a GPR survey to investigate geology and hydrogeology conditions in a location where a controlled landfill was built (Ílhavo Landfill). The second consists on a joint seismic refraction and resistivity survey to investigate the surroundings of a "chemical waste pond" that has been used for the deposition of chemical waste in the Estarreja Chemical Complex.

The Ílhavo Landfill – Geology and Hydrogeology Settings

Several geophysical surveys were carried out in the old Ílhavo Landfill (Tareco et al., 2010) and in the site where a new controlled landfill was constructed. Local near surface geology consists on Quaternary deposits up to dozens of meters of thickness. The bottom these deposits consist mainly in gravel horizons and coarse sands that are separated from shallower aeolian sands by organic mud layers at depths that vary between 9 and 11 m. The Quaternary sands are very porous and permeable whilst the mud layers act as a natural barrier between the top units and the lower ones, thus preventing groundwater circulation. The controlled landfill was projected to operate above these mud layers.

A GPR survey was carried out to investigate local geological and hydrogeological conditions. This survey used constant space 100 MHz antenna. Data processing included: drift removal, time zero adjustment, dewow, filtering and gain. Depth conversion used velocity values estimated from local CMP gathers. One of the profiles is shown in Figure 2, with data from a borehole imposed on the GPR radargram.

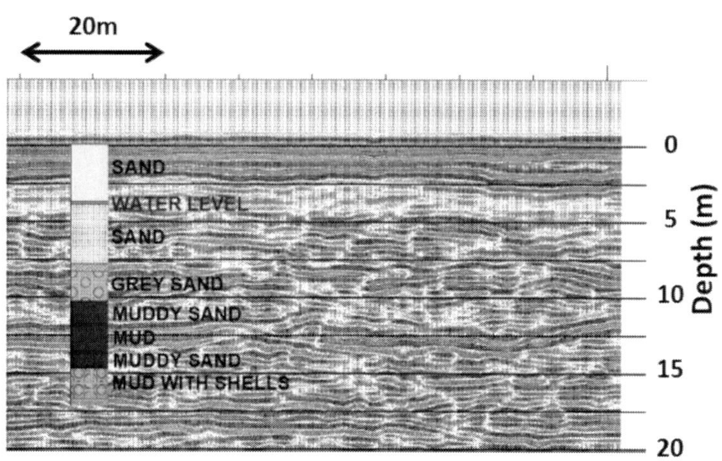

Figure 2. GPR radargram in the Ílhavo Landfill site.

The radargram shows clearly a near horizontal sequence of Quaternary deposits, with a strong reflection between 11 and 13 m depth that corresponds to the mud layers. There is a close agreement between borehole data and the radargram. The groundwater level is also marked, although the radargram seems to locate it about 1 m deeper than direct measurements in nearby boreholes. The mud layer separates the upper free aquifer from a lower confined or semiconfined aquifer, that must be preserved when constructing the landfill. It is clear that the conservation of this lower aquifer requires the integrity of the detected mudlayer.

Thus, GPR radargrams in the area allowed the definition of local geology and hydrogeology conditions, their variability and gave the necessary information to the optimum location and depths of monitoring boreholes.

The Estarreja "Chemical Mud Pond" Study – Geology and Contamination

The Estarreja Chemical Complex, located 30 km North of Aveiro, has been operating for more than sixty years. During this time, a large amount of chemical muds was acumulated in a "pond" – PL in Figure 3 – that has been contaminating groundwater in its vicinity. Remediation procedures need an understanding of local geology and hydrogeology that was not available when deposition operations started.

Locally, there is a crystaline bedrock consisting on schists and greywakes covered by Cretaceous formations and Quaternary sand deposits on top. It is not known the extent of the Cretaceous formations and the nature of the Quaternary sands. These are coarser towards the bottom and, in places, there are mud layers separating coarser and deeper sediments from shallower and finer aeolian sands. It is known that groundwater flow is, in broad terms, from the East towards West.

Bearing in mind the existing information a combined seismic refraction and resistivity survey was planned and carried out to the W of the PL (Figure 3), to investigate both the extent of contamination areas and local geology (Senos Matias, 2001).

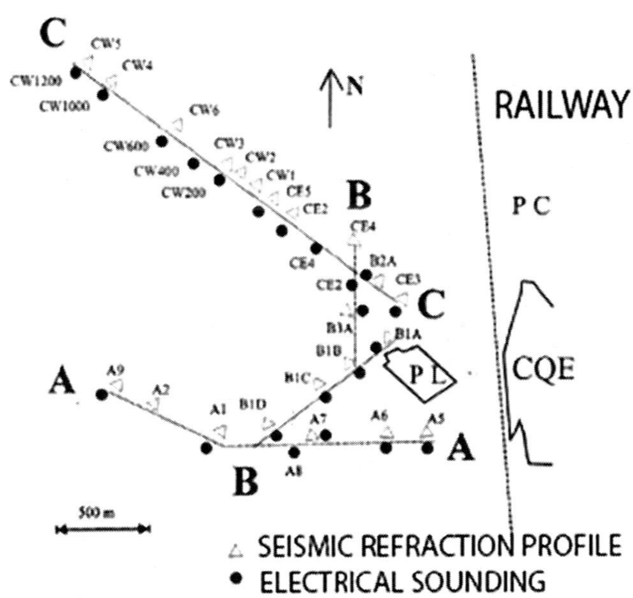

Figure 3. Location of the Geophysical survey (combined seismic refraction and electrical resistivity) in the Estarreja Chemical Complex west of the "chemical pond" (PL).

The seismic refraction survey consisted of profiles AA, BB and CC and the resistivity survey consisted of a set of Wenner array soundings located on the same profiles. The interpretation is shown in Figure 4.

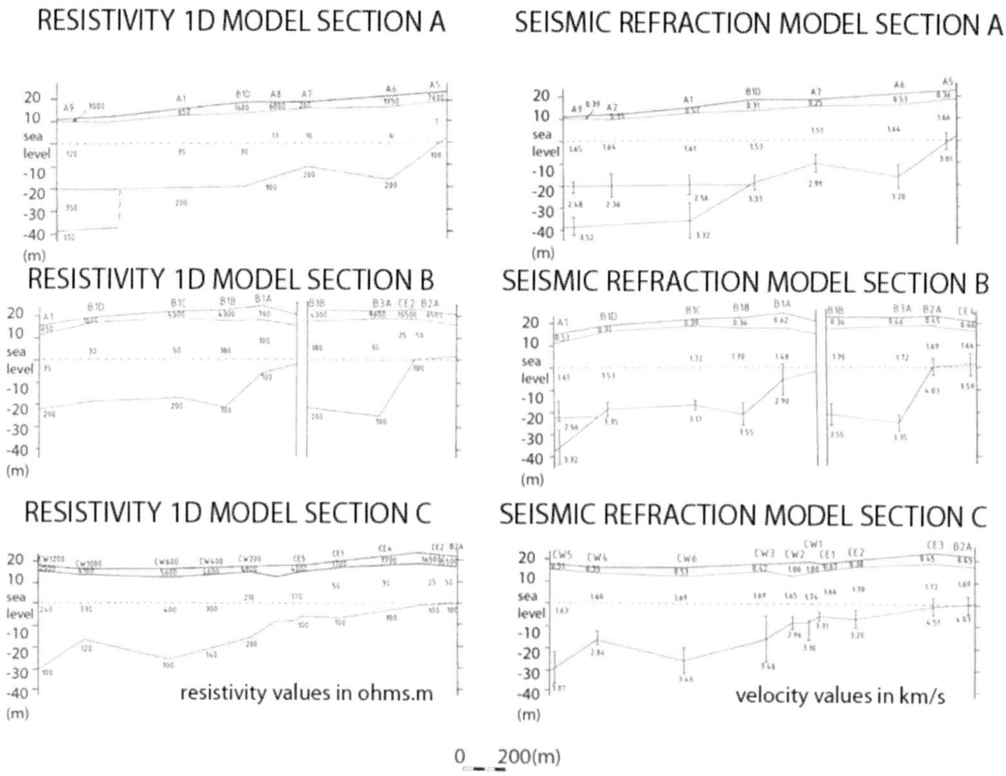

Figure 4. Interpretation of the geophysical survey in the Estarreja Chemical Complex: (left) electrical resistivity 1D models; (right) seismic refraction models.

The proposed models for seismic refraction data and for the resistivity survey provide very similar interpretation and the assignement of geological unities to geophysical layers is based on information from boreholes and wells in the area. The bedrock, composed by schists and greywakes, is deeper towards W and shows velocities above 3 km/s as well as higher resistivities. The sedimentary cover is estimated to be thick (10 to more than 20 metres) and contamination potential is large. Cretaceous formations, with velocity about 2.5 km/s, are interpreted to the W on both A and C profiles.

The second layer on the seismic profiles shows velocity values in the range of 1.5 to 1.7 km/s, that are typical of saturated formations. Hence the top of this unity must correspond to the HL. Above this layer, dry aeolian sands are interpreted, seismic velocities ranging from 0.35 to 0.5 km/s and electrical resistivies higher than 1000 ohm.m.

The Geophysical project was conducted long after the PL became operational and inspection of the resistivity models reveals a lateral variation of resistivity values for the second saturated layer. Thus the lower resistivity values, between 6 and 35 ohm.m, must correspond to a heavily contaminated aquifer, that is, from sounding A5 to A1 in the resistivity section A, from sounding B1C to A1 and B3A to B2A in section B and, finally, from sounding B2A to CE1 in section C. Higher resistivity values are modelled westwards, further away from the pond, and should correpond to non contaminated media and to the Cretaceous formations.

Therefore, this case demonstrated that the combined use of electrical resistivity and seismic refraction methods contributed to the modeling of local geology and hydrogeology and delineated the boundries of groundwater contamination both in depth and lateraly.

3. Leaks and Infiltrations in Landfills

The protection of landfills neighboring areas is very important to prevent groundwater and soil contamination. This objective can be achieved with proper engineering techniques and monitoring of the area around the landfill. When a new landfill is planned, it is possible to install a grid of electrodes at its base before engineering operations start. This grid of electrodes can be later used to monitor fluid leaks by measuring electrical conductivities (White and Barker, 1997). Usually, this procedure is not possible, and thus leaks and infiltrations can only be investigated by indirect methods, such 2D or 3D electrical resistivity and GPR.

GPR and 2D Resistivity in Ílhavo Landfill to Identify Leaks and Infiltrations

GPR and 2D resistivity data are shown on Figure 5 and clearly demonstrate their advantages in the investigation of leaking through the bottom of a landfill (Matias and Hermozilha, 2006). Figure 5 (top) shows a radargram carried out over the Ílhavo sealed landfill. As it can be observed, the bottom of the landfill is marked by the white solid thicker line and, from the radargram only there are no signs of leaking. However, the 2D electrical resistivity data (Figure 5, bottom), show clear signs of leaking through the bottom of the landfill. The model shows an interrupted resistive horizon at the larger depths, that is, areas with electrical resistivities ranging between about 1000 and 5000 ohm.m, at the bottom of the model, are interrupted by blue/green conductive areas, positions 100 and 160 meters, where leakage must be occurring. The top right of the resistivity data, thus at very shallow depth, also shows a conductive region near position 180. This area is in clear contrast with the resistivity character of the top resistive layer. Hence, it is also possible that there is also a failure in the sealing of the landfill. It must also be added that comparison between the figure on the top (GPR) and that on the bottom (resistivity model) shows a good consistency in the interpreted depth to the bottom of the landfill as depicted in the very bottom of Figure 5.

The GPR time slices on Figure 6 refers to data from Ílhavo Landfill (Tareco et al., 2010). Data were processed using velocity values of 0.08 m/ns, as estimated from local data analysis. This landfill was sealed in the end of the 20th century but there is evidence of surface water infiltration through the top insulator of the landfill. The time slices reveal several interesting features. The area contoured with a solid thin white line corresponds to the infiltration area on top of the landfill as revealed by on-site inspection. The infiltration is confirmed to penetrate deep in the landfill as it is depicted in the time slices from lower depths and in the electrical resistivity data from Figure 5. The other areas from Figure 6 (dot-dashed and dashed lines) correspond to inhomogeneities inside the landfill that will be discussed later.

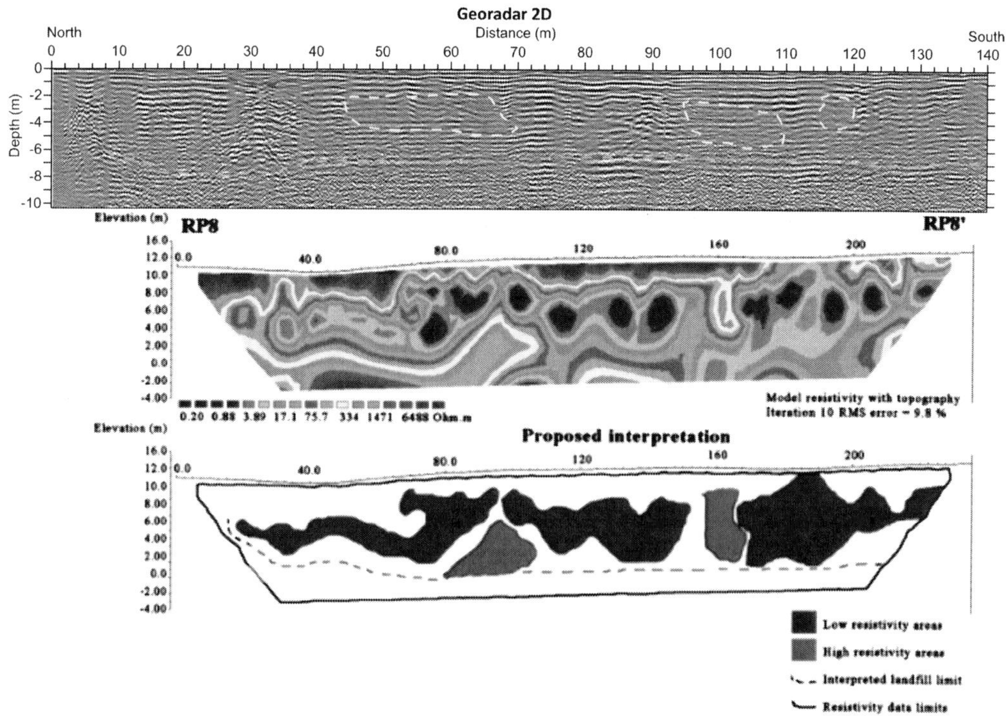

Figure 5. Ílhavo Landfill: Georadar 2D (top) – the solid bold white line marks the bottom of the landfill; 2D resistivity model (middle) and interpretation (bottom).

4. Landfills Materials and Leachates

Landfill materials identification and leachates characterization is often needed after landfill closure operations. Physical properties of the materials inside a landfill vary significantly with the nature of refuse, waste decomposition processes and with the presence and characteristics of groundwater and leachates (Sharma, 1997). Therefore, electrical resistivity may be used to identify eventual layering and lateral variations inside a landfill.

Either in natural media or in landfills, variations in groundwater conductivity are due to natural changes in its total mineralization, or to the introduction of external factors in its composition usually from anthropogenic origin. Hence, it is possible to estimate the geoelectrical characteristics of a contamination plume, and consequently the electrical resistivity of the leachates and undisturbed fluids in background zones. It is also possible to distinguish several types of layers and, using Archie's Law (Archie, 1942; Carpenter et al., 2009), to estimate groundwater conductivity and indirectly, its contamination degree. If borehole information is available, data can be crossed-checked and interpretation improved. Several Geophysical Methods can be used in the investigation of the materials inside a landfill. However, the choice of a method or a combination of methods depends on the specificities and objectives of each case. The application of electrical methods will be demonstrated in the Gaeiras Landfill case study, whilst gravimetry, GPR and airborne thermal photography will be used in the Ílhavo Landfill project. Finally magnetic and SP methods will also be demonstrated in the Gaeiras Landfill.

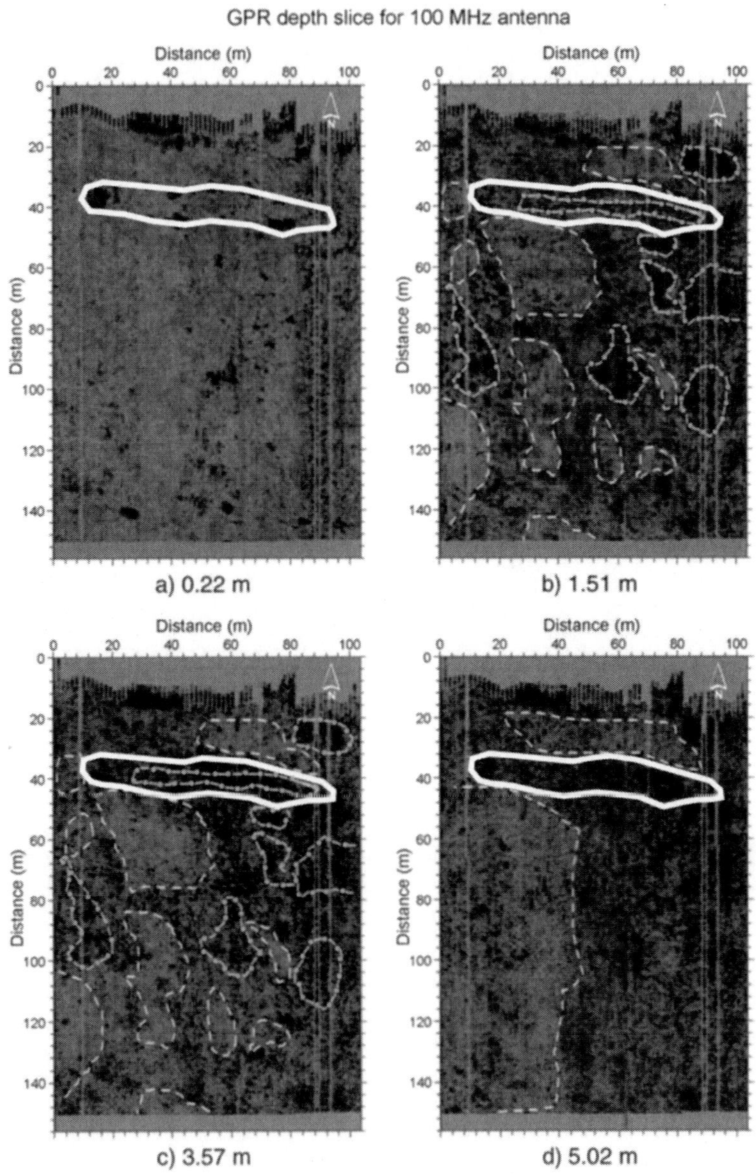

Figure 6. GPR time slices from Ílhavo Landfill, processed using velocity values of 0.08 8 m/ηs (Tareco et al., 2010). White solid lines correspond to the infiltration area on top of the landfill as revealed by on-site inspection. Dot-dashed and dashed lines correspond to inhomogeneities inside the landfill.

Defining Landfill Layering in the Gaeiras Landfill

The Gaeiras Landfill case study shows how geophysical methods were successfully applied to identify the origin and consequences of unexpected leachates that were produced in a sealed landfill. The landfill is located on a 9% streamline slope and was used for domestic and industrial wastes disposal for 17 years. Owing to the complex physical and chemical processes inside the sealed landfill and due to the lack of information prior to the sealing engineering process, 10 m^3/day leachates unexpectedly poured off the protecting geotextile.

Due to the sealed state of the landfill, several geophysical methods were used over its surface and a 81 points regular grid was the basis for the conducted surveys (Figure 7). The survey consisted on electrical (VES with Schlumberger array), frequency domain electromagnetics (FDEM with Geonics EM34) that allowed to estimate of the bedrock depth, thickness of the geoelectrical layers and leachates characteristics. Although waste hydraulic characteristics were difficult to define, due to the waste heterogeneous distribution inside the sealed landfill, it is admitted that permeability in the landfill was higher than that in the geological environment. So the application of Archie's Law (Archie, 1949) allowed porosity estimates ranging from 12 to 44%.

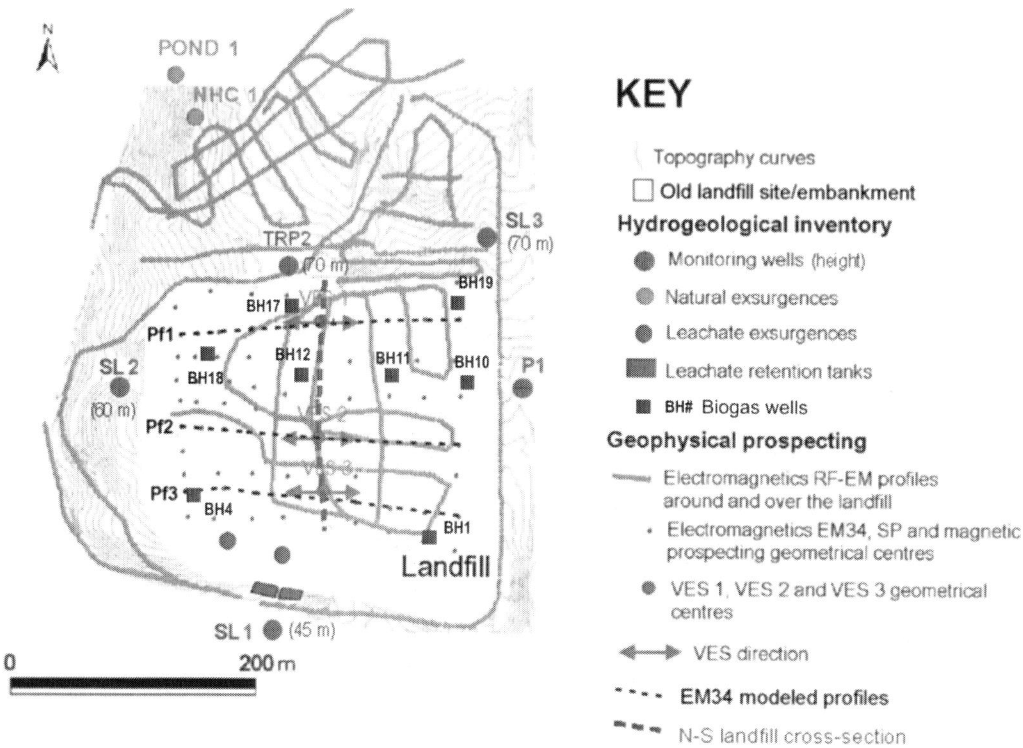

Figure 7. Geophysical surveys conducted over and around the sealed Gaeiras landfill. Electromagnetic RF–EM profiles location around and over the sealed landfill and location of the 81 EM34 points, SP and magnetic surveys located over the landfill. VES 1, 2 and 3 locations are indicated. Hydrogeological inventory (SL1, SL2, SL3 and P1 are the monitoring wells drilled after the landfill sealing; TRP2 is a monitoring well that was not sampled). Height above mean sea level of the monitoring well heads (m) is indicated in brackets. Slope is about 9%. Electrical conductivity EM34 modeled profiles Pf1, Pf2 and Pf3 are also showed (Ramalho et al., 2102). N-S landfill cross-section represented in Figure 8 is also showed. BH# are the biogas wells.

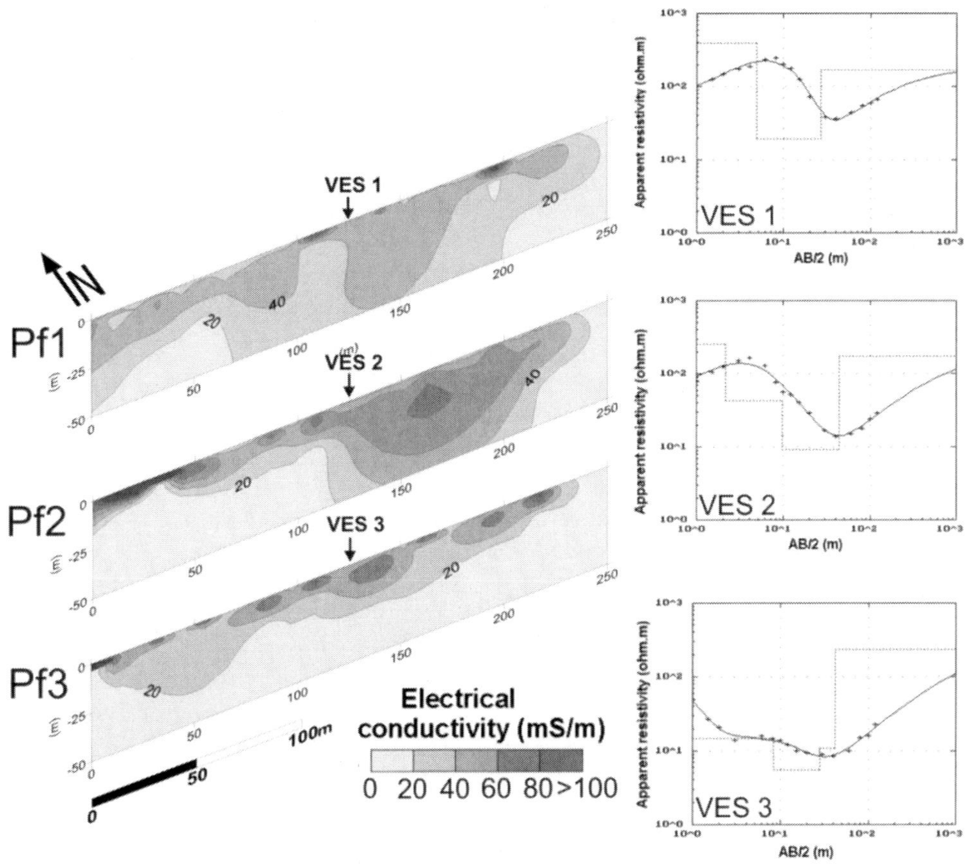

Figure 8. (Left) FDEM Geonics EM34 data electrical conductivity inverse modeling with a software developed by Monteiro Santos (2004); (Right) 1D electrical resistivity inverted models of VES 1, 2 and 3.

A conceptual schematic model was constructed for the N-S profile (Figure 7) from 1D VES modeling. These data are in broad agreement with the inverted FDEM survey data 2D models that were modeled with software developed by Monteiro Santos (2004) (Figure 8, left). The FDEM models Pf1, Pf2 and Pf3 show zones corresponding to leachates accumulations with very high electrical conductivity and confirmed by the 1D electrical inversion of VES (Figure 8, right).

In the N-S 3 layer conceptual model (Figure 9) based on VES and FDEM inverted data from Figure 8, an upper dryer and more resistive layer with thickness varying from 1 to 10 m and resistivity of about 150 ohm.m (layer ρ_1), should correspond to the sealing materials and dry waste. The layer ρ_2, probably composed by waste materials and leachates is located immediately below the previous layer. This second layer shows increasing thickness towards the S reaching values up to 40 m and electrical resistivities between 10 and 20 ohm.m. These findings imply that excavations were conducted in original topography, possibly in the streamline alluvium where the landfill was implanted. Finally, the sandy silts, composing the bedrock, must correspond to the deepest layer ρ_3, with resistivity values of about 180 ohm.m.

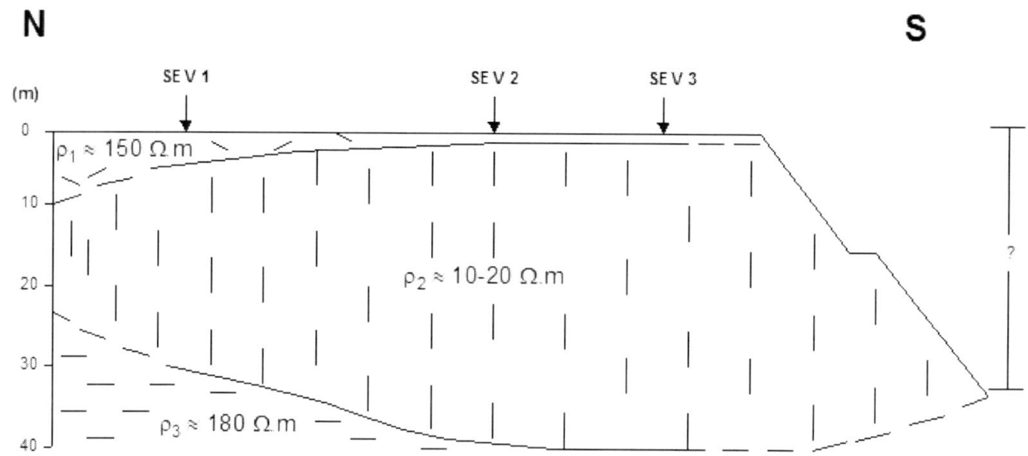

Figure 9. Schematic N-S landfill cross-section resistivity 1D model showing the clayey bedrock depth after inverting VES 1, VES 2 and VES 3 results.

Samples from local boreholes and from the biogas wells (Costa et al., 2002) allowed the calculation of some physical and chemical parameters (Table 2).

Table 2. *In-situ* parameters determined for the leachates in biogas wells (BW#) and boreholes from the hydrogeological inventory (Costa et al., 2002)

Point	Length (m)	Water Table Depth (m)	Temperature (°C)	pH	Electrical Conductivity (mS/cm)	TDS (g/L)
SL1	20	3.20	18.6	7.12	4.050	3.430
SL3	17	3.16	18.6	7.38	0.832	0.599
P1	unknown	0.97	23.3	7.46	0.995	0.599
BW1	15.10	7.00	28.3	7.08	>20	>10
BW4	14.60	6.40	32.2	7.66	>20	>10
BW10	11.90	6.80	26.8	7.28	>20	>10
BW11	9.40	3.80	35.7	7.70	>20	>10
BW12	12.90	4.90	23.5	7.43	>20	>10
BW17	11.40	4.90	19.7	6.36	1.37	0.68
BW18	10.70	5.50	37.2	7.62	>20	>10
BW19	11.21	3.31	22.9	7.22	16.87	8.41

As expected from the interpretation of the overall information, both the electrical conductivity and the TSD from piezometer SL1 located near the Southern edge of the embankment shows the highest groundwater electrical conductivity values.

Leachates and Waste Distribution inside Gaeiras Landfill

Sometimes environmental problems are caused by an abnormal flow inside landfills. In these cases, Spontaneous Potential (SP) can be applied over the sealed landfills to allow the identification, distribution and characterization of preferential leachate pathways. This is the case of the Gaeiras Landfill and the data in Table 2 are consistent with the SP map (Figure 10, top) (SP stations are located in Figure 7), indicating preferential fluid circulation inside the landfill (arrowed lines), which was in agreement with the patterns of shallowest apparent electrical conductivity values mapped from FDEM (Ramalho et al., 2012)

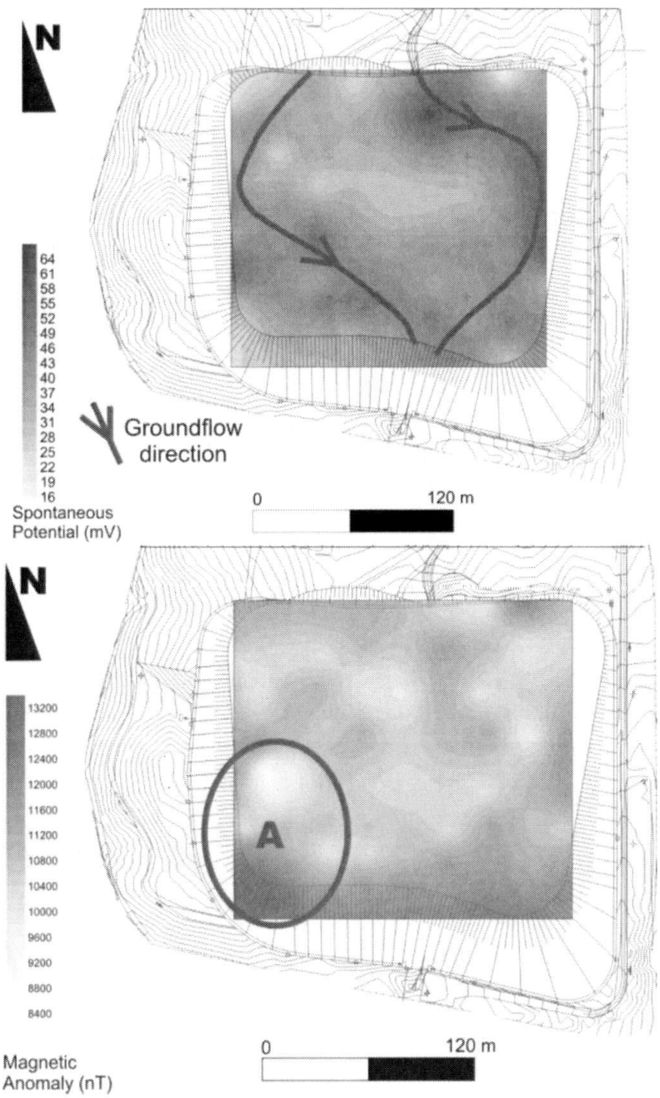

Figure 10. (top): Spontaneous Potential map (mV) over the Gaeiras Landfill. The obtained results show that higher SP values – positive anomalies – are due to preferential leachate flowpaths with very high electrical conductivitiy, depicted by the arrowed line. (below): Intensity of Magnetic Anomaly map of the area over the Gaeiras Landfill (nT). Area A is a zone with a magnetic anomaly, attributed to metallic waste.

Magnetic methods were also applied to locate deposition areas of metallic refuse inside the sealed landfill. The magnetic survey (stations are also located in Figure 7) showed high magnetic anomaly values, with a heterogeneous distribution. However, the dipole A locates an abnormal accumulation of metallic objects, mostly in the SW zone of the embankment (Figure 10, below).

The Ringing Effect in Ílhavo Landfill

The Ílhavo Landfill was used to test several Near Surface Geophysical methods in landfill studies as mentioned before. On Figure 11, some geophysical data from GPR, relevant to the investigation of the internal structure of the landfill are depicted and used to identify the occurrence of objects inside the landfill.

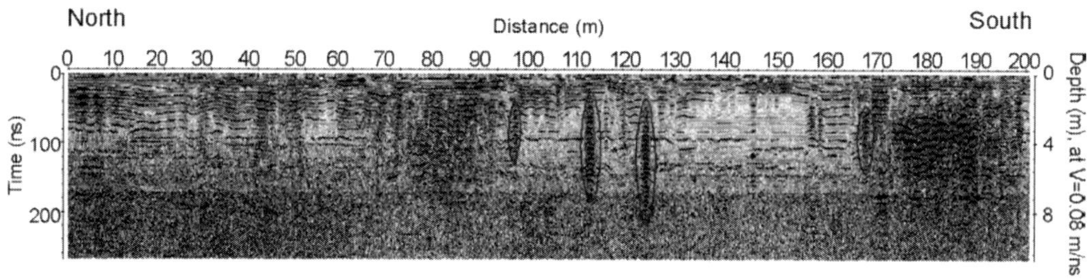

Figure 11. Internal structure of the Ílhavo Landfill using GPR data to identify the occurrence of objects inside the landfill.

The radargram reveals the so called ringing effect identified by the enclosed blue envelops. These effects do not start at time zero and, therefore, must be explained by the occurrence of objects inside the landfill. Such events have been observed when metal targets are in the subsurface (Tareco et al., 2010; GSSI, 2006).

Further information on the nature of the waste in Ílhavo Landfill can be obtained from GPR data. Thus, earlier in this work, Figure 5 was introduced to demonstrate that GPR can be used to locate infiltration in landfills. This figure also shows areas enclosed by dot-dashed and dashed lines. The dot-dashed areas correspond to high energy areas whilst the dashed areas correspond to low energy zones where attenuation is higher and, thus, refuse fluid content should prevail.

Gravity Studies in Ílhavo Landfill

Gravity methods are not common in landfill studies, but some work has already been experimented (Silva et al., 2008; Galland and Stierman, 2005; Mantl'ik et al., 2009). Due to the fact that in a landfill, refuse varies from residential waste to industrial waste, it is possible that gravity can be used to distinguish between different types of waste, contributing to the "waste zoning" of a landfill by mapping different bulk densities of a landfill site. However, some very important aspects need consideration when planning the use of gravity in these studies, that is, accuracy and repeatability of readings, topographic surveying, terrain corrections and readings spacing (Debeglia and Dupond, 2002). In Figure 12, data from a gravity profile carried out over the Ílhavo Landfill (Mantl'ik et al., 2009) are shown.

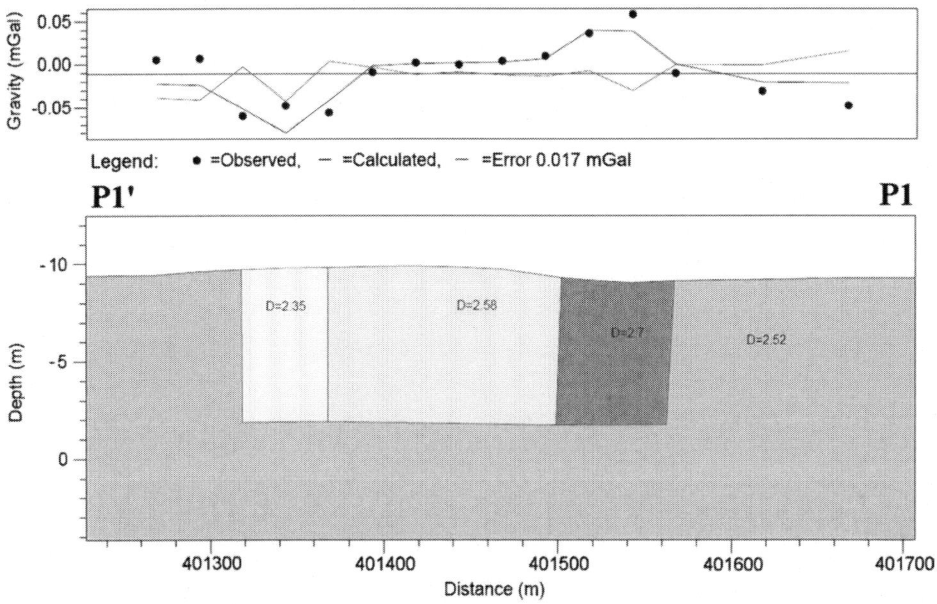

Figure 12. Gravity profile carried out over the Ílhavo Landfill (Mantl'ik et al., 2009): (top) observed and calculated data (mgal) and (below) gravity modeling.

It should be pointed out that the thickness of the landfill is consistent with the interpretation of Figure 5. On this latter figure, the 2D resistivity model (bottom of Figure 5) also allows the zoning of the landfill. The drawing bellow the 2D resistivity model (bottom of Figure 5) shows more resistive and more conductive areas that must correspond to different types of refuse. The top of Figure 5 shows a radargram and in the region corresponding to the landfill some areas of high attenuation, contoured by the dashed line, contrast with regions where GPR signal remains strong, contoured by the dot-dashed line. This contrast is due to the heterogeneities inside the landfill and it is likely that the contoured dashed lines areas could correspond to zones where fluids prevail. On the other hand, the regions corresponding to the green lines show strong GPR signal and events and, therefore, could correspond to a concentration of solid waste.

Although different physical properties are been measured and discussed in Figures 5 and 12, the gravity profile also depicts a lateral zoning and adds more information on the internal structure of the landfill.

Airborne Thermal Studies in the Ílhavo Landfill

Thermal conditions beneath the earth's surface are usually related to natural causes of anomalous thermal activity, such as surface outcrops of deep seated fault zones, volcanic regions, continental rifts or oceanic ridges. Regional heat flow and temperature distribution with depth is both a scientific as well as a geothermal exploration theme and thus surface thermal measurements can be useful in confirming and mapping out structures and lithological contacts (Jessop, 1990). The contrasts between rock temperatures are essentially due to a fundamental thermal property of materials that is also important for the understanding of the thermal behavior of a site: the thermal conductivity (K, W/m°C), a physical property of materials that measures how easily heat can flow through them (Sharma, 1997). K for rocks is generally higher than those for soils and has a relatively wide range of values. Some of the most common K values are easily found in the literature (c.f. Cermak and Rybach, 1982; Sharma, 1997) and can reach wide ranges, from very low (cultivated soils ranging from 0.2 to 1.2 W/m°C), to high values, in granites (ranging from 2.30 to 3.60 W/m°C). Measurements of temperature can either be made directly on the materials, through the use of thermocouple or optical fibre sensors, or remotely through the use of special imagery detectors (Ribeiro et al., 2013). Thermographical cameras use detectors that are sensitive to various spectral bands. The visible band is 0.4-0.65 μm, followed by Near-Infra-Red (0.65-1 μm), Medium-Infra-Red (1-7 μm) and Long-Infra-Red (7-20 μm) bands. The Long Infra-Red band is also known as the thermal band and is the one that is used for thermographical measurements. The thermographical cameras are also known as FLIR (Forward Looking Infra-Red) cameras.

Temperature surveys aiming at detecting temperature variations within waste dumps, have been used in the past. One such case, described by Kobr and Linhart (1994), dealt with an area of disposed coal refuse. However the temperature measurements were carried out directly, with the aid of digital thermometers, at two different depth levels (0.5 and 1 m) and in a regular sampling grid of 10 m by 10 m. The result was a temperature map of isotherms and it was concluded that the temperature variations were affected by both biological as well as chemical thermal processes within the refuse. Remote sensing FLIR imagery was already successfully applied in coal fires in mining waste piles in Portugal (Ribeiro et al., 2013) although the imagery was taken from ground level.

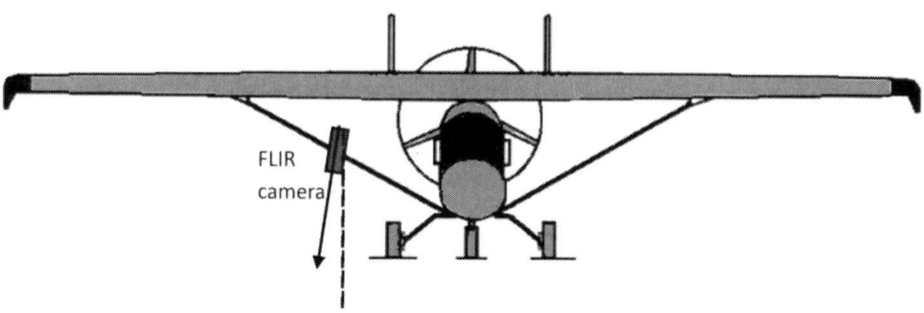

Figure 13. Coupling of the FLIR camera to the wing strut of the aircraft in a slightly slanted configuration and offset from vertical at an angle of 20°.

Slonecker et al. (2010) and Beaumont et al. (2014) consider IR imagery and airborne and spaceborne remote sensing efficient tools for landfill site characterization during the activity phase. In order to apply and test these features, an airborne FLIR survey was planned over the Ílhavo Landfill. The airborne data set was therefore acquired at an altitude of 150 m with a FLIR A655SC, Long Wave IR camera, mounted on the wing of a Streak Shadow aircraft (Figure 13). The flight was carried out on April 22, 2013, between 10h:15m and 11h:00m, local time. The weather conditions during the flight were favorable with a surface temperature of 16 °C, good visibility and mild winds of around 19 km/h. These weather conditions justify, to some extent, adequate conditions for the existence of a contrast between the ground temperature, heated by exothermal processes within the landfill, from other surrounding surface materials and vegetation. The employed camera has a sensor size of 640 x 480 pixels and a thermal resolution of 14 bit, in the range of -40 to +150 °C, thus both the spatial and thermal resolution were quite fine in detail. Georeferencing was conducted based on ground coordinates and thermocouple temperature measurements that were made on the soil to confirm the remotely sensed values.

The resulting image (Figure 2) clearly shows anomalous values of temperature, relative to the lower surrounding temperatures, where the landfill is expected to exist. However there seems to be some sort of heterogeneous distribution in the higher temperature patches that can be related to heterogeneity of the waste and in turn the exothermic landfill processes. In Figure 14, the highest concentration of higher thermal activity is concentrated around the S areas (1 and 2). These are characterized by a more continuous heat agglomeration. A more disperse nature in the distribution of the anomalous thermal patches is verified towards N (3). A colder dirt road (4), at the same topographical level as the landfill, running in the N-S direction, is clearly visible in the thermal images of Figure 14. This fact confirms, to some extent, the correlation of thermal activity with the location of deposited waste in the surrounding cells. To justify some of the other limited colder patches, within the landfill area, one must also bear in mind that the thermal conductivities of sediments are low and since some parts of the landfill might be buried deeper than others, lower temperatures might also be expected to appear over the landfill. Given the surrounding values (averaging around 12.3 °C), as well as a surface air temperature of approximately 16 °C the maximum image temperature value of around 22 °C is indeed anomalous.

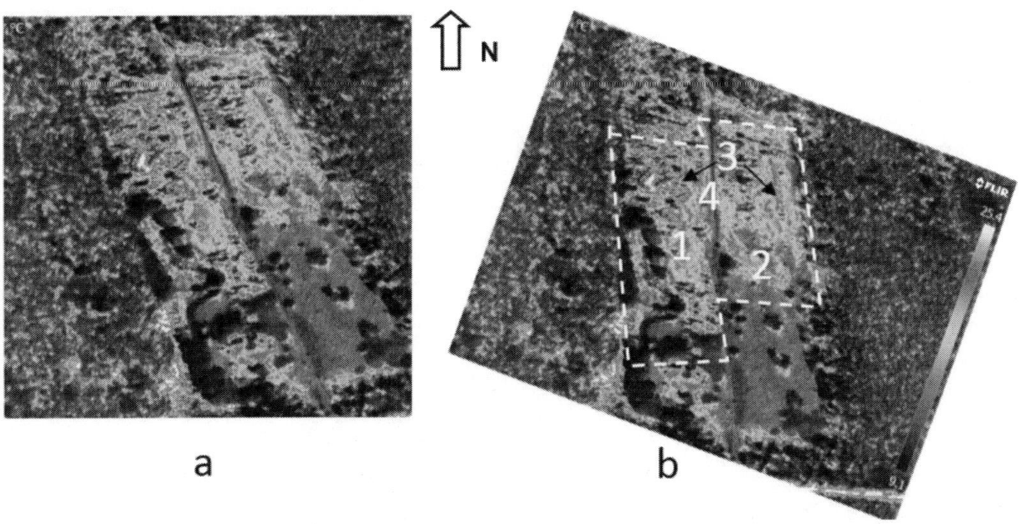

Figure 14. Sidescanned (a) and vertically georeferenced (b) FLIR image of the Ílhavo Landfill. The landfill outline, in dashed white, is depicted in (b). The numbers, 1, 2, 3 and 4 pertain to the various zones that are described in the text.

5. Landfills Monitoring

Geophysical methods can be used to monitor landfills in two ways. First they provide a fast, economic and non-invasive means to investigate time lapse changes in contamination plumes. Secondly, they can be used, together with geological and hydrogeological studies, to determine the optimum locations for the construction of monitoring boreholes. In fact, geophysical methods can be used to project the depth of the boreholes and to locate the best depths for groundwater sampling inside a borehole, with no interference with different aquifers that might exist in the area.

Contamination Plume Evolution in Ílhavo Landfill

GPR studies provide images with both lateral and depth information and are a fast way of obtaining high resolution information. Contaminated areas are known to show high conductive values and will attenuate rapidly GPR signal to a point where no information can be obtained. On Figure 15 two radargrams obtained in the same location in the contamination plume of Ílhavo Landfill, but at different times (1999 and 2005) are depicted.

Contamination is immediately seen by the heavy attenuation of the radargram signal. The attenuation is so important that cancels strong reflections from local mud layers and obscure local any other geological information completely. As shown, contouring of the attenuation zones in both radargrams show the evolution of the contamination plume between 1999 and 2005. The interpreted contamination plume in 1999 is shown by the black solid line in both radargrams whilst the dashed line depicts the extension of the plume in 2005.

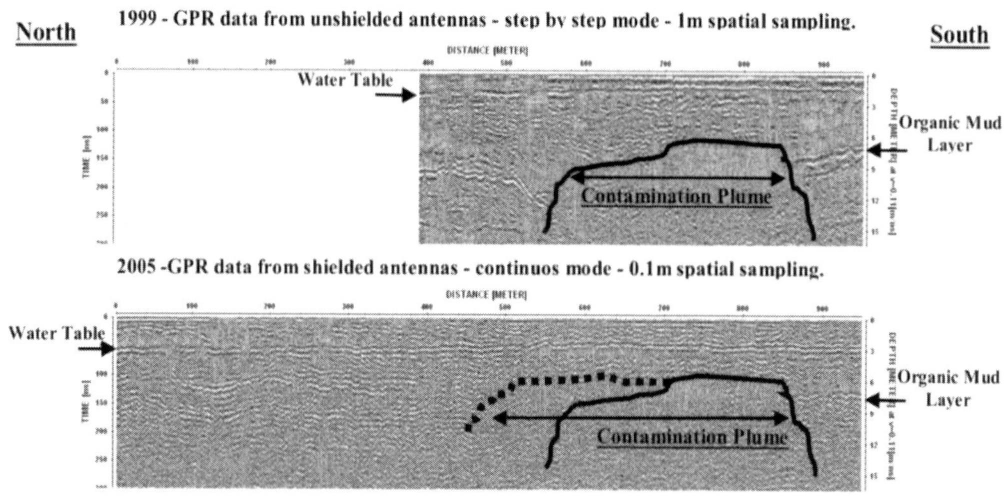

Figure 15. Radargrams obtained in the same location in the contamination plume of Ílhavo Landfill, in 1999 (top) and 2005 (below), where the contamination plume evolution is visible.

Contamination Plume Evolution in the Estarreja Chemical Complex

Figure 16 shows a more detailed FDEM Geonics EM34 (horizontal coils 20 metres apart) carried out in contamination plume area of the Mud Pond (Estarreja Chemical Complex). This survey was carried out firstly in 1993, yellow contour lines, and then repeated by Taunt (2001) (shown in Figure 16 with black contour lines). The hydrochemical data collected in local of wells and boreholes allows to estimate that the contour line 30 mmho/m corresponds roughly to the limit of the contaminated area (Branco, 2007). In Figure 16, the blue crosses indicate the magnitude of displacement of the 40 mmho/m contours and the red crosses indicate the displacement for the 50 mmho/m curve. At it is depicted it is observed a contaminant plume propagation to the West and an overall displacement of about 400 m which corresponds to an average annual displacement of 50 metres during the period of the survey.

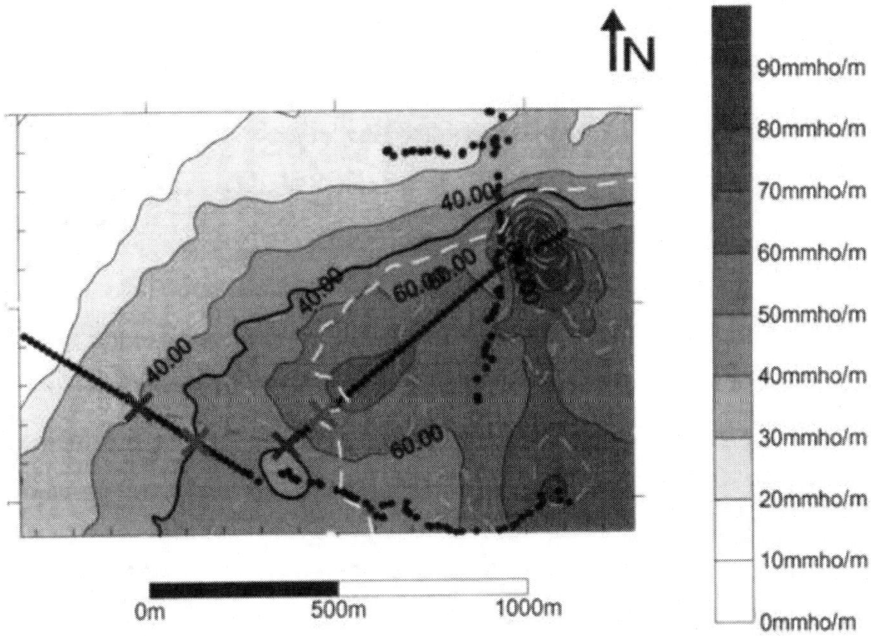

Figure 16. FDEM Geonics EM34 survey in the Estarreja Chemical Complex, using horizontal coils 20 m apart (pseudo-depth of about 15 m). Blue crosses indicate the magnitude of displacement of the 40 mmho/m contours. red crosses indicate the displacement for the 50 mmho/m curve.

Monitoring the Ovar Landfill Contamination

Open landfills leave a heavy burden in the soils and groundwater of the areas. Sealing does not solve the contamination problems and might turn environmental evaluation and mitigation procedures more difficult to implement. Nevertheless, the application of geophysical methods to assess contamination problems and their time evolution has given very positive results that can also be used to complement other data. The Ovar Landfill is located on sand dunes containing a shallow semi-confined aquifer. The landfill is located 4 km S from the main Ovar groundwater supply sources (Matias et al., 1994). Before the beginning of the sealing process, an initial study comprising electrical (VES), electromagnetic (FDEM) and hydrogeological studies (involving the drilling and study of 11 piezometers based on geophysical information) was carried out. The interpreted results confirmed the contamination of the semi-confined aquifer, which could lead to disastrous consequences if refuse deposition were to continue simultaneously with groundwater extraction. A few years after the Ovar Landfill sealing the geophysical survey was repeated to assess the impact of the engineered solution in groundwater (Senos Matias et al., 1999).

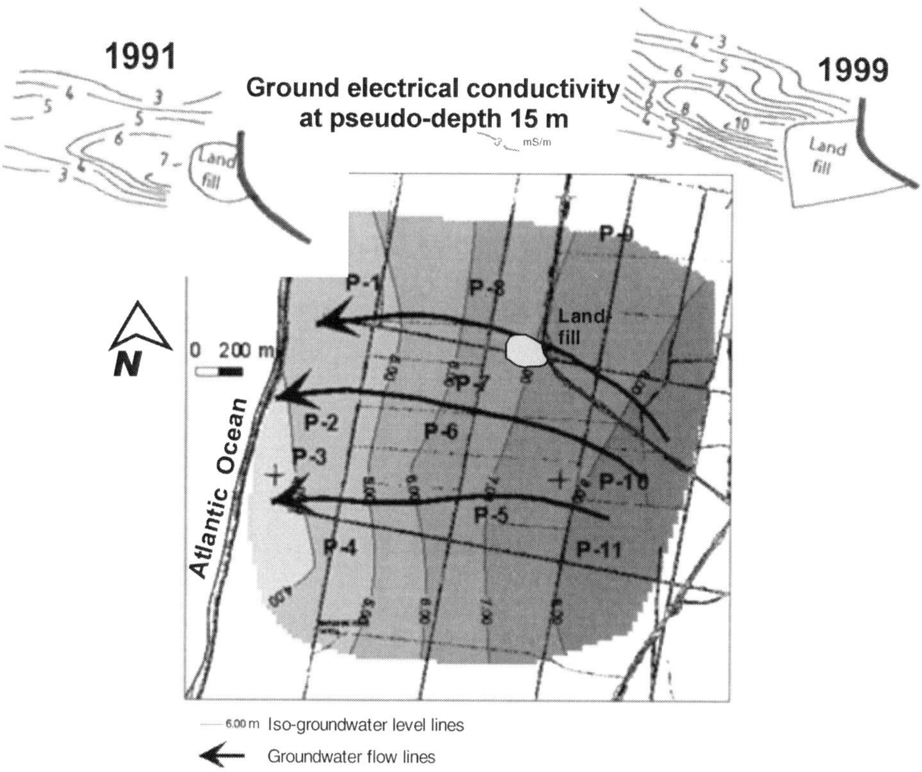

Figure 17. Contamination plume (mS/m) of Ovar Landfill in 1991 and 1999 using FDEM methods (Geonics EM34) at an apparent pseudo-depth of 15 m. Groundwater flow net after the isopiezometry map (m) based on the 11 piezometers (P-#) installed around the landfill in 1991 (Matias et al., 1994; Senos Matias et al., 1999).

Figure 17 shows the electrical conductivity maps and groundwater flow direction based on iso-groundwater levels, depicting the contamination plume of Ovar Landfill and its time evolution from 1991 to 1999. These maps (1991/1999), corresponding to a pseudo-depth of 15 m, were obtained with a FDEM Geonics EM34 device (horizontal coils 20 metres apart). From Figure 17, it can be interpreted that (1) contamination plume is in the E-W direction (that is, towards the Atlantic Ocean), in accordance with groundwater flow, (2) groundwater flow direction remains the same from 1991 to 1999, (3) there is a conductive contrast of less than 3 to more than 10 mS/m between contaminated and non-contaminated areas, (4) monitoring boreholes that were drilled and installed to the W of the landfill confirmed the contamination plume characteristics, (5) the borehole that was drilled E of the landfill was used to check the properties of non-contaminated groundwater, and (4) the contamination plume has increased in area and intensity in 1999 when compared with 1991 data, as it is immediately interpreted from Figure 17. Finally, the contamination plume has suffered a rotational shift of 22° westwards its displacement was about 200 m towards W, meaning that the plume has propagated in that direction at an average velocity of 25 m/y, i.e. 2 m/month (Martinho, 2003; Martinho and Almeida, 2006).

Table 3 shows the comparison between some physical and chemical parameters from 1991 (Matias et al., 1994) and 1999 (Senos Matias et al., 1999) using the 11 piezometers around the Ovar Landfill, located in Figure 17. Piezometer n. 8 is the most contaminated in 1991 and in 1999 surveys and is located immediately W of the landfill.

Table 3. Comparison of some physical and chemical parameters between 1991 and 1991, from groundwater samples analysis in the 11 piezometers around the Ovar Landfill. The most contaminated piezometer in 1991 and in 1999 (P-8) is located immediately W of the landfill and highlighted

	1991				**1999**			
Piezometer	pH	σ (μS/cm)	Temperature (°C)	Alcalinity (HCO_3^- mg/l)	pH	σ (μS/cm)	Temperature (°C)	Alcalinity (HCO_3^- mg/l)
P-3	6.7	340	15.8	170.8	5.4	395	17.6	175.1
P-4	6.1	162	15.0	50.0	5.8	266	16.4	34.8
P-5	5.3	135	15.8	22.1	5.4	228	16.6	28.7
P-6	6.0	165	15.3	45.5	6.0	272	16.5	47.6
P-7	6.4	120	15.1	61.0	6.6	160	16.1	61.6
P-8	**5.8**	**1000**	**17.4**	**142.7**	**5.9**	**773**	**21.4**	**154.9**
P-9	5.5	95	15.6	10.5	5.4	157	16.6	20.7
P-10	5.4	115	15.0	12.2	5.3	175	16.6	20.1
P-11	5.3	130	15.8	25.6	5.9	199	16.3	34.0

This study, covering almost a decade, showed that the ecosystems recovery is slow and thus justifies the application of 4D studies in areas where these problems occur. The work conducted in 1991 allowed the assessment of the reference initial situation, before sealing operations and enabled the construction, location and design of appropriate boreholes for later use in monitoring studies.

Monitoring a 3D Contamination Plume in the Matosinhos Landfill

The case study the Matosinhos Landfill, located on a weathered granitic bedrock, refers to a 3D data integration study and of correlation of resistivity values with the contamination plume (Moura et al., 2012; Gonçalves et al., 2014). Their approach towards determining the anomalous resistivity value, as well as its 3D shape, was based on statistical methods originally employed for determining geochemical backgrounds and anomalies (Lepeltier, 1969; Nakic et al., 2007). Adapting this methodology to the resistivity data obtained from a series of spatially distributed 2D profiles, a statistically defined resistivity limit value was determined, for what should be the anomalous contaminant plume, as opposed to conventionally just defining an arbitrary resistivity value. The reason for devising this approach was based on the fact that low resistivity values can also be a ssociated with normal uncontaminated geological settings and not just the anomalous concentration of contaminants. A measured low resistivity value is also affected by the mineralogy, porosity, the amount and type of natural water content. A certain set of the natural factors can thus be combined to also match the low resistivity values of contaminated soil and rock formations. Therefore, the practical methodology was to carry out numerous 2D resistivity profiles in the surrounding area of a sealed waste dump, both near and as far as possible from the contaminated area. It was soon verified that the total population of all the inverted resistivity values, in all the 2D sections, were distributed in a lognormal manner as is the case of most geochemical data. Thus, in short, if enough profiles are integrated at varying distances from the contaminant source, there is a better chance of gathering a representative population of all possible contamination scenarios. The 2D profiles can then be integrated into a single interpolated 3D volume and the end result is an improved definition of the extent of the anomalous contaminated volume (black portions in Figure 18). In Figure 18, an area can be observed NE of the landfill (1) with higher resistivity values that also extends in depth. This zone was suspected, from borehole water analysis, to be uncontaminated and therefore was surveyed with the goal of contributing to the assessment of the natural average background resistivity values. The same high resistivity behavior can be seen in the SW part (2) although here there were no borehole data by the time of geophysical survey. A dark area (3), generally surrounding the landfill, especially towards the W, was determined and corresponds to the volume of resistivity values that are below the limit considered as the local anomaly, and thus constitute the interpreted contaminated geological formations. This assumption was confirmed by borehole water analysis. Regarding the previously mentioned higher resistivity portion in the SW (2), it was later found to be uncontaminated, justified by a few borehole groundwater analyses. A major contribution for this conditioning and protection from the Southwesterly migrating contaminant plume is the fact that the river runs next to the landfill, separating this portion of the area from the N and W contaminated parts.

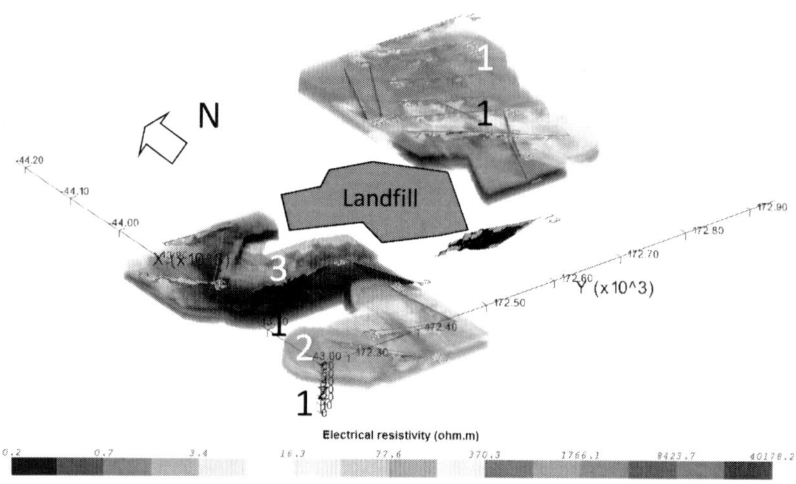

Figure 18. Spatial visualization of the interpolated 3D resistivity data with the anomalous domain in black bellow the anomaly threshold (software adds some level of transparency to the outer parts of the volume in order to visualize the anomaly). The numbers 1, 2 and 3, pertain to the various zones that are described in the text. The landfill stands at the approximate center of the geophysical survey.

Conclusion

The previous discussion demonstrated the usefulness of Surface Geophysical Methods in landfill studies. As shown, Geophysical Methods can be very useful to investigate different aspects, such as geological studies prior to the installation of landfills or to extend the knowledge to local geology and estimate the extension of contamination in existing landfills. These matters can play a major role in the engineering project for the installation of a landfill and for the investigation of effective remediation procedures.

It was also assessed the study and evolution of contamination plumes. In this case, Geophysics provides non-invasive methods, that is, physical invasion or mixing of different aquifers is avoided. Thus, aquifers properties are preserved and it is possible to provide "a priori" information for the installation of monitoring wells and boreholes, as well as to project the best depths for groundwater sampling.

Geophysical methods were also shown to provide good results in the study of the internal properties of landfills, such as refuse zoning and in the identification of leaks and infiltration that can compromise the engineering solutions adopted to mitigate the consequences of landfills on the environment. It is clear that the right combination of Geophysical methods should be the best way to tackle these problems, as information can be checked and interpretation ambiguity reduced.

There is no doubt that improvements are necessary and needed. For example, time lapse techniques and 4D software need development. Remote devices must be engineered to carry out studies in hazardous situations or in areas of difficult access. It is also possible that small aircraft and, possibly, unmanned airborne vehicles can be constructed, tested and used in these studies. Finally, the exploitation of the energy released by landfills will demand the development of on-site and real time techniques to monitor physical and chemical phenomena inside and around landfills.

Acknowledgments

The authors wish to thank the several Town Councils and private companies that gave permission to carry out these studies and to publish the results, the students and postgraduate students that participated in the field work and in data processing and modeling, the University of Aveiro, Environment and Development Institute (IDAD) and National Laboratory for Energy and Geology of Portugal (LNEG).

References

Archie, G. E., 1942. The electrical resistivity log as an aid in determining some reservoir characteristics. *Transactions of the American Institute of Mining and Metallurgical Engineers/Petroleum Division*, 146, 54-62.

Atekwana, E. A., Sauck, W. A., Werkema Jr., D. D., 2000. Investigation of geoelectrical signatures at a hydrocarbon contaminated site. *Journal of Applied Geophysics,* 44, 167–180.

Beaumont, B., Radoux, J., Defourny, P., 2014. Assessment of airborne and spaceborn thermal infrared remote sensing for detecting and characterizing landfills. 7th International Conference on Waste Management and the Environment (Ancona, Italy, du 12/05/2014 au14/05/2014). In: C.A. Brebbia, G. Passerini & H. Itoh, Waste Management and The Environment VII, WIT Press: Ashurt Mpdge, Ashurst, Southampton, S040 7 AA, UK.

Benson, A. K., Frederickson, C., Mustoe, N. B., 1991. Ground penetrating radar, electrical resistivity, soil and water quality studies integrated to determine the source(s) and geometry of hydrocarbon contamination at a site in north-central Arizona. *Proceedings 27th Symposium on Engineering Geology and Geotechnical Engineering,* 38.1–38.13.

Benson, R., Glaccum, R., Noel, M., 1983. Geophysical techniques for sensing buried waste and waste migration. Environ. Monitor. *Syst. Lab., Off. Res.* Develop., US

Bernstone, C., Dahlin, T., Ohlsson, T. and Hogland, W., 2000. DC-resistivity mapping of the internal landfills structure: two pre-excavation surveys. *Environmental Geology* 39, 360–371.

Biandini, G., Santarato, G. and Vaccaro, C., 2004. Geochemical characterization and geophysical mapping of landfills: the Marozzo canal study (NE Italy). *Environmental Geology* 45, 439–447 (45).

Branco, C., 2007. *Estudo da Contaminação do aquífero superior da região de Estarreja,* MSc thesis, University of Coimbra, Portugal, 191p.

Brewster, M.L., Annan, A.P., Greenhouse, J.P., Kueper, B.H., Olhoeft, G.R., Redman, J.D. and Sander, K. A., 1995. Observed migration of a controlled DNAPL release by geophysical methods. *Ground Water,* 33, 6, 977–987.

Carpenter, P. J., Ding, A., Cheng, L., Liu, P. e Chu, F. (2009) – Apparent formation factor for leachate-saturated waste and sediments: examples from the USA and China. *Journal of Earth Science*, 20, (3), 606-617.

Cermak, V. and Rybach, L., 1982. Thermal conductivity and specific heat of minerals and rock. In: G. Angenheister (Editor), *Landolt-Bornstein Numerical Data and Functional Relationships in Science and Technology.* New Series,Group V, Springer-Verlag, Berlin.

Costa, C., Dill, A., Costa, A., Fernandes, J., Sampaio, J., Ramalho, E., Rocha, R., Castro, R., Lopes, M., Avillez, G., Curto, P. and Riscado, J. (2002) – Estudo geológico e hidrogeológico para o controlo dos lixiviados da Lixeira de Gaeiras. *Relatório para a Resioeste efectuado por IGM, CIGA/UNL, AMBIOPOLIS.* 82p. + anexos.

Debeglia, N. and Dupont, F., 2002. Some critical factors for engineering and environmental microgravity investigations *J. Appl. Geophys.* 50, 435–54.

Environ. Protect. Ag., Las Vegas,NV,Rep. 68-03-3050, p. 114.

Galland, P. and Stierman, D., 2005. Gravity investigation of the King Road landfill, Lucas County, Ohio North-Central Section -39th, Annual Meeting, Geological Society of America (Minneapolis, MN) paper 46-4.

Gonçalves, V., Dias, P., Fontoura, M. J., Moura, R. and Sousa Santos, B., 2014. Investigating Landfill Contamination by Visualizing Geophysical Data. *IEEE Computer Graphics and Applications* 34(1): 16-21.

Greenhouse, J., Monier-Williams, M., Ellert, N. and Slaine, D., 1989. Geophysical methods in water contamination studies. *Proc. Exploration'87. Ont. Geol. Surv., Spec,* 3, 666–677.

GSSI Handbook For RADAR Inspection of Concrete, 2006, Published by Geophysical Survey Systems, Inc., 38, www.geophysical.com.

Jessop, A. M., 1990. *Thermal geophysics: Elsevier, Amsterdam, Developments in Solid Earth Geophysics*, 17, 306p.

Kobr, M. and L. Linhart. 1994. Geophysical survey as a basis for regeneration of waste dump Halde 10, Zwickau, Saxony. *Journal of Applied Geophysics* 31:107-116.

Lanz, E., Maurer,H., Ansorge, J., Green, A., 1997.High-resolutiontraveltimetomography used to delineate a shallow waste disposal site. *Proceedings of the 3rd Meeting Environmental and Engineering Geophysics.* Aarhus, Denmark, pp. 33–36.

Lepeltier, C., 1969. A simplified treatment of geochemical data by graphical representation. Economic Geology, 64, 538-550.

Mantl'ik,F., Matias,M., Lourenço,J., Grangeia, C.,Tareco,H., 2009. The use of gravity methods in the internal characterization of landfills — a case study, J. Geophys. Eng. 6 357–364.

Martinho, E. and Almeida, F., 2006. 3D behavior of contamination in landfill sites using 2D resistivity/IP imaging: case studies in Portugal. *Environmental Geology*, 49, 1071–1078.

Martinho, E., 2003. Integrated study of Environmental Problems by Geophysical Methods. Tese de Doutoramento, Universidade Técnica de Lisboa, 269p.

Matias, M. and Hermozilha, H., 2006. A 3D constant offset GPR survey on Ílhavo Landfill. Proceedings of the 12th European Meeting of Environmental and Engineering Geophysics, Helsinky, Finland, vol. A031, p. 5. 90-73781-62-O.

Matias, M., Silva, M., Ferreira, P., Ramalho, E., 1994. A geophysical and hydrogeological study of aquifers contamination by a landfill. *Journal of Applied Geophysics* 32, 155–162.

Monteiro Santos, F. A., 2004. 1-D laterally constrained inversion of EM-34 profiling data. *Journal of Applied Geophysics,* 56, 123-134.

Moura, R., Fontoura, M. J., Goncalves, V., Dias, P., Sousa Santos, B., Guerner Dias, A. and Espinha Marques, 2012. Results from geospatial analysis of resistivity to delineate contamination anomalies. A case study of a Controlled Dump - N Portugal, J., *Symposium on the Application of Geophysics to Engineering and Environmental Problems* (SAGEEP), Tucson, Arizona.

Nakic, Z., Posavec, K. and Bacani, A., 2007. A Visual Basic Spreadsheet Macro for Geochemical Background Analysis. *Ground Water,* 45, (5), 642-647.

Pereira, P. and Lima, O. A. L., 2007. Estrutura elétrica da contaminação hídrica provocada por fluidos provenientes dos depósitos de lixo urbano e de um curtume no município de Alagoinhas, Bahia. *Rev. Bras. Geof.,* 25, 1, 5-19.

Ramalho, E. C., Carvalho Dill, A. and Rocha, R., 2012. Assessment of the leachate movement in a sealed landfill using geophysical methods. *Environmental and Earth Science.* N.68, pp.343-354. DOI 10.1007/s12665-012-1742-8.

Ramalho, E. C., Marques da Silva, M. A., Senos Matias, M. J. and Valente, M. V., 1998. Diagnóstico e recuperação ambiental da zona envolvente da Lixeira de Maceda (Ovar). Actas do V Congresso Nacional de Geologia, (2), E-78-E-81.

Reynolds, J. M., 1997. *An introduction to Applied and Environmental Geophysics.* John Wiley & Sons, 796.

Ribeiro, J., Moura, R., Flores, D., Lopes, D. B., Gouveia, C., Mendonça, S., O. Frazão, 2013. Chapter 16. The Douro Coalfield Fires of Portugal, in Coal and Peat Fires: A Global Perspective, Volume 2 *Photographs and multimedia tours,* Edited by Glenn B. Stracher, Anupma Prakash and Ellina V. Sokol, Elsevier, 313–337.

Roberts, R., Hinze, W., Leap, D., 1990. Application of the gravity method to the investigation of a land fillinglaciated mid-continent USA: a case study. In: Ward, S. H. (Ed.), *Geotechnical and Environmental Geophysics,* 2, Society of Exploration Geophysicists, 253–259.

Sauck, W. A., Atekwana, E. A., Nash, M. S., 1998. High conductivities associated with an LNAPL plume imaged by integrated geophysical techniques. *Journal of Engineering and Environmental Geophysics* (2), 203–212.

Sauck, W. A., 2000. A model for the resistivity structure of LNAPL plumes and their environs in sandy aquifers. *Journal of Applied Geophysics* 44, 151–165.

Senos Matias, M., 2001. Geofísica aplicada ao ambiente: objectivos, âmbito e alguns exemplos. *Comunicações do Instituto Geológico Mineiro*, 88, 223-236.

Senos Matias, M., Marques da Silva, M., Ramalho, E. C., Matos, J. and Teixeira, F., 1999. De volta à Lixeira de Ovar. Actas do 6º Congresso Nacional da Qualidade do Ambiente, Costa da Caparica, FCT-UNL, 20-22 de Outubro de 1999, (3), 143-152.

Sharma, P. V., 1997. *Environmental and Engineering Geophysics*, Cambridge University Press, 475.

Silva, J, Teixeira, W. and Barbosa, V. 2008. *Gravity data as a tool for landfill study Environ.* Geol. doi: 10.1007/s00254-008-1353-6.

Slonecker, T., Fisher, G.B., Aiello, D. P., Haack, B., 2010. Visible and Infrared Remote Imaging of Hazardous Waste: *A Review. Remote Sens.* 2010, 2, 2474-2508.

Soupios, P., Papadopoulos, I., Kouli, M., Georgaky, I., Vallianatos, F., Kokkinou, E., 2007. Investigation of waste disposal areas using electrical methods: a case study from Chania, Crete, Greece. *Environmental Geology* 51, 1249–1261.

Tareco, H., Grangeia, C., Senos Matias, M., 2010. An integrated 3D constant offset GPR and resistivity survey on a sealed landfill – Ílhavo, NW Portugal, Journal of Applied Geophysics, 70, 58-71, doi:10.1016/j.jappgeo.2009.11.004.

Taunt, N., 2001. A time Lapse electromagnetic survey of a contaminated aquifer at Estarreja, Northern Portugal. *MSC thesis,* University of Leeks, UK.

Telford, W. M., Geldart, L. P. and Sheriff, R. E., 1995. *Applied Geophysics.* Cambridge University Press. New York. 770p.

White, C. C., Barker, R. D., 1997. Electrical leak detection system for landfill liners: a case history, *Ground Water Monitor Remediation,* 17, 3, 153-159.

In: Landfills and Recycling Centers
Editor: Carla H. Jackson

ISBN: 978-1-63463-792-3

Chapter 3

MUNICIPAL LANDFILL LEACHATES: WATER CONTAMINANTS OR SOURCE OF NUTRIENTS IN AGRICULTURAL SOILS?

Marco Antonio Nogueira[1,*], Cristiane Alcantara dos Santos[1], Sandra Márcia Cesário Pereira da Silva[2], Daniel Bini[1], Luciana Cadioli Panchoni[1], Vanessa Lázaro Mangieri[2] and Grazielle Favero Valério[2]

[1]Department of Microbiology;
[2]Department of Civil Construction.
Universidade Estadual de Londrina. Londrina Paraná Brazil

ABSTRACT

Municipal landfill leachates are an environmental concern worldwide due to its polluting potential, especially for surface and ground water. Despite available technology for installation of municipal landfills and adequate treatment of leachate, many landfills are improperly established and the formed leachate has no adequate treatment and destination, especially in developing countries. However, even after treated, the resulting residues or the effluent contains nutrients like nitrogen (N), phosphorus (P) and potassium (K), which may cause eutrophication of stream waters. These nutrients are essential for plant growth and thus leachates free from contaminants like heavy metals, organic chemicals or pathogens above safe levels, could be used as source of nutrients in agricultural soils. If a decision is taken for use of landfill leachates in agricultural soils, monitoring of soil chemical, physical and biological properties are essential for a safer destination. In this chapter we discuss on aspects related to leachate formation, its environmental risks, potential use as nutrient source in agricultural soils, and procedures for monitoring the effects of landfill leachates on soil functions to prevent degradation.

* Present address: Embrapa Soybean – Lab. of Soil Biotechnology, C.P. 231 – 86001-970 – Londrina, PR – Brazil. E-mail address: marco.nogueira@embrapa.br

1. INTRODUCTION

One of the main consequences of the human development is generation of large amounts of either solid or liquid residues, which generally results in environmental impacts by polluting soil, air or water. Human activities generate solid residues that have large diversity and complexity, differing in physical, chemical or biological properties according to its origin. In addition, the way such residues are managed may change their characteristics so that the risks for human health and environment can be magnified.

The international tendency, regarded as modern, is to avoid as much as possible the destination of residues to landfills, in order to increase the landfill's lifetime, by using technologies like incineration, composting, recycling, anaerobic digestion, etc. In Brazil, the disposal of municipal solid wastes is below the ideal. According to a study carried out by The National System of Information on Sanitation (Brazil, 2008a) the municipal wastes in 267 Brazilian municipalities in 2007 were disposed as follows: 9.5% in open dump deposits, 26.6% in controlled landfills, and 63.9% in sanitary landfills. However, these results may be biased because the study was carried out in the bigger municipalities all over the country, where the waste destination is usually more adequate than in smaller municipalities. Despite recent technological advances that allowed the development of some efficient techniques for solid wastes treatment, the disposal in landfills and dump deposits are still the most employed processes in Brazil and other developing countries due to the simplicity and lower cost. Nowadays, due to legal exigencies and development of new products like geomembranes, the landfills have advanced in terms of environmental safety. Nevertheless, the problem of leachate formation as a result of natural moisture and biological decomposition of solid wastes, in addition to the rain that infiltrates through the landfill pile, still remains, independently of the system employed for municipal solid waste disposal.

As soon as the residues are disposed in landfills or dump deposits, the formation of leachate takes place, and the water movement through the layers of residues in decomposition carries organic matter and ions in solution, in addition to solids in suspension. Physical, chemical, and microbial processes acting on decomposition of residues will transfer pollutants from the solid phase to the water in movement (Kjedsen et al., 2002) that will finally flow to the soil through the unsaturated zone and reach the saturated zone, causing soil and water pollution. Depending on the amount of water that infiltrates through the landfill, the chemical and physical characteristics of the leachate will change. Variations in organic carbon levels, nitrogen and several other ions, electrical conductivity, pH, and alkalinity will be observed in different seasons and rainfall intensities.

Unfortunately, only a few landfills have an efficient system for collection and treatment of leachate, like in one that was the biggest sanitary landfill in Latin America, at the boundary of São Paulo city, Brazil (Figure 1). In most of cases, particularly in developing countries, landfills are not prepared for retaining and treating leachates, endangering the quality of surface and groundwater.

In several landfills in Brazil, especially in smaller cities, the leachates produced in sanitary landfills are generally recirculated on the pile residues in decomposition or are transported to a sewage treatment station where the microbial degradation of organic components takes place. Afterwards, the effluent is discharged in surface waters and the sludge is led to appropriate disposal. Nevertheless, the effectiveness of such treatment may be unsatisfactory because not all potentially hazardous substances are effectively eliminated (Celere et al., 2007). Moreover, nutrients like nitrogen (N), phosphorus (P), and potassium (K) are not effectively removed and will contribute for eutrophication of surface water that causes excessive growth of aquatic vegetation. In addition, drinkable water contamination by nitrate is a risk for human health causing methemoglobinemia (also called "blue baby syndrome"), disease occurring primarily in children that ingested water with high levels of nitrate.

(a) (b)

Figure 1. General view of the Bandeirantes sanitary landfill in late 1990's, which was the biggest municipal waste landfill in Latin America, located at the boundary of São Paulo city, Brazil (a). This landfill has been inactivated since 2007, but the leachate production still remains. In the lower part of the landfilllandfill, there is a leachate collection system (b), from where it is transported to the sewage treatment station (b). Photos by M.A. Nogueira.

One of the biggest challenges for operation of landfills is the production and adequate destination of leachates due to their complex and variable composition that may include several pollutants of soil, surface and groundwater, and even air, by formation of gases like methane (CH_4) and ammonia (NH_3). It is worth to keep in mind that even when solid wastes are adequately disposed and the landfill reaches its useful lifetime, the leachate production still persists. In general, the organic carbon in the leachate is more recalcitrant in older landfills, i.e., more resistant to biological degradation, what implicate the need for more complex and more expensive treatment systems (Kulikowska & Klimiuk, 2008).

2. Characteristics of Leachates from Municipal Waste Landfills

2.1. Leachate Production

There are at least three features that directly affect the formation of leachate: 1) The intrinsic solid waste properties, like degradability, density, and initial moisture; 2) The climate, like rainfall intensity and frequency, run-off, infiltration, evapotranspiration, and temperature; 3) Characteristics of the disposal site, like permeability, landfill age, depth, degree of compaction of residues, topography, and geology (Qasim & Chiang, 1994).

The formation of leachate is also one of the consequences of microbial decomposition of organic matter in the landfill. The process, initially aerobic is followed by two anaerobic phases, i.e., acetogenic and methanogenic. The limiting amounts of oxygen and the high biological oxygen demand (BOD) results in fast depletion of oxygen, and thus the aerobic decomposition is hampered in landfills. During aerobic decomposition of organic matter, CO_2, water, heat, microbial biomass, and products of organic matter decomposition are formed. The concentrations of CO_2 may reach 90% and as consequence the formation of carbonic acid causes acidification (Qasim & Chiang, 1994). At this stage, the formation of leachate is quite small, but the liquid contains particulate materials, soluble salts, and soluble organic matter. During the anaerobic decomposition, polysaccharides like cellulose, hemicelluloses, and lignin are broken by several microbial groups in succession. At this stage, besides CO_2 and water, methane is also formed by action of methanogenic Archaebacteria that convert organic acids to methane gas and carbon dioxide (Gomes et al., 2006). An increase in pH is observed, in addition to a decrease in BOD due to lower amounts of biodegradable carbon, and accumulation of more recalcitrant substances like fulvic and humic acids. The leachate will start to drain due to gravitational forces when the mass of waste in the landfill accumulate more moisture than its capacity of retention.

2.2. Leachate Composition

The landfill age and the level of biological stabilization of the solid wastes disposed will deeply affect the leachate composition. Additionally, the above-mentioned factors affecting the formation of leachate will also affect some of its physical, chemical, and biological properties. The leachate characterization is essential for knowing its potential for environmental contamination or use as source of nutrients in agricultural soils, for example. Such characterization should include several physical, chemical, and biological parameters, like alkalinity, pH, color, electrical conductivity, BOD, and chemical oxygen demand (COD), total Kjeldahl nitrogen, ammoniacal-N, nitrate-N, chloride, solids in suspension, total solids, pathogens, heavy metals, etc.

According to Kjeldsen et al. (2002), for landfills that receive exclusively municipal wastes, the compounds typically found in the formed leachate are classified in four categories: 1) Dissolved organic carbon, quantified as BOD or COD, that consists of several organic products originated from degradation of organic residues, ranging since short-chain volatile acids up to recalcitrant compounds like humic and fulvic acids; 2) Xenobiotic organic compounds of domestic origin, generally in low concentrations, like aromatic hydrocarbon, phenols, aliphatic chlorinated, and pesticides; 3) Ions: calcium (Ca^{2+}), magnesium (Mg^{2+}), sodium (Na^{+}), potassium (K^{+}), ammoniacal-N (NH_4^{+}/NH_3), iron (Fe^{2+}), manganese (Mn^{2+}), chloride (Cl^{-}), sulfate (SO_4^{2-}), bicarbonate (HCO_3^{-}); 3) Heavy metals: cadmium (Cd^{2+}), chromium (Cr^{3+}), copper (Cu^{2+}), lead (Pb^{2+}), Zinc (Zn^{2+}). Thus, the leachate characterization is essential for understanding its potential as soil and water contaminant and the risks that can be brought to environmental and human health.

Some physical-chemical characteristics of leachates from different sanitary landfills are shown (Table 1). There is a large range of variation among the assessed parameters, which depend on several factors that determines the formation of leachate, as stated above.

Table 1. Physical-chemical characteristics of leachates from different municipal waste landfills

Parameter	Variation range in different landfills			
	[1]	[2]	[3]	[4]
pH	7.9- 8.2	4.5-9.0	7.3-8.8	8.2-8.8
Alkalinity (mg $CaCO_3 \cdot L^{-1}$)	9547 - 13607	-	923-8070	5749-6177
Conductivity ($mS \cdot cm^{-1}$)	21.6-27.5	2.5-35.0	6.2-34.0	13.4-13.7
Total solids ($mg \cdot L^{-1}$)	15083-37439	2000-60000	3100-18000	8260-11724
Volatile solids ($mg \cdot L^{-1}$)	11211-12312	-	10-5900	5800-8746
Total N (mg $N\text{-}NH_3 \cdot L^{-1}$)	-	64-4700	369-1800	920-1220
Ammoniacal-N (mg $N\text{-}NH_3 \cdot L^{-1}$)	-	50-3.700	39-1750	850-1120
COD (mg $O_2 \cdot L^{-1}$)	14598-16800	140-152000	-	2208-2325
BOD (mg $O_2 \cdot L^{-1}$)	1529-2245	20-57000	-	84-150

[1] Jucá et al. (1999) Muribeca's landfill at Recife-PE Brazil, produced in young cells.
[2] Kjeldsen et al. (2002). Different landfills in European countries.
[3] Tatsi & Zouboulis (2002). Municipal landfill at Thessaloniki, Greece.
[4] Valério (2008). Municipal landfill at Londrina, Brazil.

As observed, landfill leachates contain considerable amounts of organic carbon and mineral nitrogen as ammonium, which can reach up to 2200 mg L^{-1} or even more. In addition to potential pollutant, especially for surface and groundwater if not properly treated, so high levels of ammoniacal-N may be harmful for biological treatment systems of liquid wastes. Leachates from younger landfills generally have greater potential as pollutant than leachates from elderly ones due to higher amounts of non-biologically stable organic matter (Jucá et al., 1999).

BOD and COD concentrations in leachates usually tend to decrease along the time, but BOD decreases more rapidly than COD because recalcitrant organic matter still remains (Gomes et al., 2006). The leachate BOD/COD ratio permits to deduce the degradation stage in which the landfill is in, as follows: BOD/COD greater than 0.5 indicates young, non-biologically stable landfills; BOD/COD between 0.1 to 0.5 indicates fairly stable landfills; and finally, BOD/COD lower than 0.1 indicates biologically stable landfills. Leachates from young landfills are more risky to have prohibitive levels of heavy metals due to recent acidogenic stages during the anaerobic phase that causes metal solubilization due to pH variation in an acidic range. On the other hand, in leachates from elderly landfills, generally with more than 4 years, pH ranges from 7 to 8, causing metal precipitation. In young landfills with predominance of acidic phase of anaerobic degradation, more than 95% of the dissolved organic carbon is formed by volatile acids, from which only about 1% are molecules with high molecular weight. In elderly landfills, on the other hand, with predominance of methanogenic phase, volatile acids are almost absent, while 32% of the dissolved organic carbon consists of high molecular weight compounds and more than 60% are humic compounds (Klimiuk & Kulikowska, 2006). Thus, low BOD/COD ratios indicate lower concentrations of volatile organic acids and greater proportion of humic and fulvic acids

(Kjeldsen et al., 2002). Such highly complex humic substances are not easily removed from leachates by traditional biological treatment systems and contribute to decrease the effectiveness of the process (Carrera et al., 2003). Nevertheless, in spite of lower risks of excessive heavy metal availability, ammoniacal-N concentrations in leachates from elderly landfills are still higher than in leachates from younger landfills.

3. Landfill Leachate Motion in Soil

Although sanitary landfills are considered environmentally safer than dumps, a general concern is that, besides domestic wastes, some landfills may also have received commercial and industrial wastes. In addition, many landfills are located in areas where contact with groundwater occurs; thus, without proper underdrains or engineered impermeable layers, the generated leachate may carry contaminants down into the groundwater.

3.1. Interaction between Soil and Contaminant

Pollutants transport in soil depend on the flow of the fluid in which the pollutant is dissolved and the interactions with several physical, chemical or biological processes in soil that may lead to increase or decrease of the pollutant concentration or availability. Understanding the mechanisms of soil-pollutant interactions may help us to understand and forecast processes controlling the pollutant accumulation, transport, and fate.

Some aspects like the amount of contaminants in movement in soil, attenuation of concentration by physical, chemical or biological processes, and the extension of the contaminating plume, must be considered in programs aiming at the attenuation of soil and groundwater contamination. In addition, the soil type, mineralogy, texture, structure, cation exchangeable capacity, pH, and organic matter concentrations are some soil variables that directly affect the pollutant movement.

3.2. Factors Affecting Pollutants Dispersion in Soil

As stated above, physical, chemical, and biological processes are involved in the interaction between soil/pollutant and reflect in the pollutant movement in soil and its consequences. There are three main physical processes affecting the pollutant movement in soil: advection, dispersion, and diffusion. Advection is the mechanism responsible for pollutant migration in the fluid in which it is dissolved. In this case, the speed of pollutant dispersion is the same as the fluid percolation (Freeze & Cherry, 1979). The advection results from a unidirectional flow and do not change the pollutant during transport (Chapra, 1997). Dispersion results from different flow speeds in the space caused by the resistance the soil particles impose on the flow speed. Such effect decreases with the distance from the particle surface, what means that the dispersion speed is greater in soils with larger pores than in soil with smaller pores, as happens in sandy and clayey soils, respectively. Finally, diffusion results from the movement towards a concentration gradient. This is an irreversible process that will cease only when a steady-state condition has reached. The diffusion of pollutants (ions, molecules or other compounds) in soil is slower than in pure solutions due to tortuosity and branching of soil pores. In clayey soils with low hydraulic conductivity, the diffusion may be the main mechanism in the pollutant movement (Freeze & Cherry, 1979).

Chemical processes affect the pollutant movement in soil by reactions like adsorption-dessorption, acid-base, dissolution-precipitation, oxi-reduction, and formation of complexes. Adsorption-dessorption and dissolution-precipitation are involved in pollutants transference from solid to liquid phase. The other processes affect the availability and the chemical form of the pollutant, increasing or decreasing the polluting effect. The adsorption is one of the main chemical processes in the interaction soil-pollutant and depends on the electrostatic interactions between the charges from soil particles and the pollutant charges. The sum of all negative charges in a soil is denominated cation exchangeable capacity (CEC), while the sum of all positive charges are called anion exchangeable capacity (AEC). Soil charges arise mainly on clay surfaces, metallic oxides, and organic matter. In tropical soils, the main sources of charges are organic matter, caulinite (a mineral clay), and iron and aluminum oxides. In most of soils there is predominance of negative charges in the topsoil, while in deeper soil layers, especially in tropical regions, a predominance of positive charges may happen. As a consequence, cations are predominantly retained in the topsoil, while anions are repulsed to the soil solution and may flow to deeper soil layers where they might be retained by positive charges. As a practical example, ammoniacal nitrogen (NH_4^+) is adsorbed in the negative charges in the soil surface, but after conversion to nitrate (NO_3^-) it is repulsed to the soil solution where can be lost to deeper soil layers by leaching and reach groundwater.

Chemical precipitation has as characteristic the transference of solutes from aqueous to a solid phase and depends on the pH and concentration of solutes. For example, higher pH values near or above the neutrality lead to precipitation of metals, transferring them to from the liquid to the solid phase. Increase of pH can be a strategy for avoiding metal leaching from soil surface to groundwater in case of soil pollution.

Oxi-reductions comprise electron flow between two substances and can be mediated entirely by chemical reactions or by biological processes. This is a very important way of transformations of many substances in soil, affecting their potential as pollutants. For example, some metals in their reduced forms like Fe^{2+} and Mn^{2+} are soluble, but are insoluble in their oxidized forms Fe^{3+} and Mn^{4+}.

Acid-base reactions are also chemical processes affecting pollution dynamics in soil. pH variation may strongly affect the mobility of some contaminants in soil like heavy metals, besides effects on CEC, especially in tropical soils. pH values near to neutrality tends to release negative charges on soil charge complex, increasing the CEC. On the other hand, low pH decreases the amount of negative charges, decreasing the CEC. Such changes affect the interaction of anions and cations with the solid phase and consequently the flow of ionic pollutants in soil.

Complexation reactions occur when a metallic cation reacts with an anion by covalent linkage and can form chelates when the central atom (metal) is linked to two or more atoms of one or more molecules or ions. Complexation changes the solubility and may increase or reduce the transport of metal along the soil profile. For example, fulvic, tartaric, citric, and salicylic acids are highly soluble and may displace metals along soil profile up to the groundwater, while the humic acids have low solubility, retaining the metal chelated in it.

Besides physical and chemical processes, biological processes may also affect the pollutant dynamics in soil. Considering that the soil organic matter in the A horizon of tropical soils may range from 0.5 to 5%, part of this carbon (around 2-5%) is composed by living microbial cells which play several oxi-reduction reactions that affect the solubility and dynamic of several substances. Among several other essential roles in soil, microbial community is involved in the formation and mineralization of soil organic matter, nitrogen and sulfur transformations, etc. Along with carbon, nitrogen has its dynamics in soil highly dependent on microbial processes, like ammonification, nitrification, denitrification, and biological nitrogen fixation. Nitrogen is an essential element for all living forms, but at the same time is a potential pollutant when excessively added to the soil. The several oxi-reduction levels that nitrogen can assume affect its dynamics in soil. Organic forms of nitrogen in soil are mineralized to ammonium by heterotrophic microorganisms. In the sequence, ammonium can be oxidized to nitrite and nitrate by autotrophic microorganisms and as consequence, N dynamics in soil is completely changed, given that nitrate can be easily leached to groundwater or oxidized to gaseous forms and lost to the atmosphere as N_2, NO, or N_2O in a process named denitrification.

As stated above, several factors can act concurrently on the dynamics of a pollutant in soil, what make the characterization of the phenomena involved quite complex.

4. Landfill Leachates As Nutrient Source in Agriculture

The treatment of landfill leachates based on physical-chemical processes in sewage treatment stations can be effective for depuration and is based on adsorption, precipitation, and flocculation. Nevertheless, polluting substances are not degraded, and may result in highly contaminated sludges (Bae et al., 1999). Treatments based on biological processes need for long retention periods (up to weeks), are low-effective in eliminating more complex, recalcitrant compounds like humic substances and pigments (Freire et al., 2000), and also bring about production of greenhouse gases like nitrous oxide and methane (Cardoso et al. 2011).

Before the above-mentioned concerns, the leachate disposal on agricultural soils, within certain limits, may make possible the stabilization of its organic content and serve as source of nutrient for plants, especially N and K. This practice, besides being cheaper than conventional treatments, reduces the risk of water contamination because the soil, within certain limits, is able to retain and break down contaminants (Santos et al., 2013). Nevertheless, there is still no enough information on which effects the disposal of landfill leachates on agricultural soils may cause to soil and groundwater properties. Certainly, effects will depend on the soil, leachate and environmental characteristics, and the resulting interaction among them.

The presence of fresh organic material in liquid effluents aggravates the risks of environmental problems in case the liquid effluent reaches a water stream. However, when disposed on agricultural soils the organic matter in the effluent serves as carbon sources to the soil microbial community (Bowen et al., 2009). On the other hand, if the carbon in the effluent is already biologically stabilized as humic substances as in landfill leachates, it may contribute to improve the chemical, physical and biological attributes of soil, in addition to plant nutrition (Piccolo, 2001; Nayak et al., 2007; Cardoso et al., 2011).

Residues disposal in soil is environmentally safer than disposal in hydrosphere or atmosphere, but generally causes many controversies. In soil, pollutants can be better oxidized, or precipitated, and removed from food web more safely than in water or air. Moreover, if hazardous substances or pathogens are not limiting, some residues may serve as nutrient source in agriculture or for reclamation of degraded sites in partial or total substitution of mineral industrialized fertilizers (Tedesco et al., 2008; Santos et al., 2013).

Land farming of wastes is a way of disposal, in which the material is applied onto the soil, and the degradation of the organic fraction is carried out by microbial processes, whereas the mineral fraction is physic-chemically retained. The same characteristics that make the landfill leachate potentially pollutant make it attractive for agricultural use, like its contents of ammoniacal-N, K and biologically stable organic carbon. The use of residues containing high concentrations of N as alternative source of nutrients in agriculture has been studied (Gonzalez et al., 2009; Cardoso et al., 2009). In order to avoid or minimize any environmental risk derived from this practice, and assure the crop yield levels, the concentration of potential pollutants like N, K, Na, organic chemicals, pathogens, heavy metals, etc., must be considered and constantly monitored (Martines et al., 2006; Santos et al., 2013).

Nitrogen is essential for plant growth and is required in higher amounts than other nutrients. Highly N-demanding plants, especially some gramineous species, promptly respond to N addition, resulting in increased yield and protein content in grains. The use of increasing doses of landfill leachate in wheat, under controlled conditions, indicated that the leachate has potential as nitrogen source for plants (Figure 2).

The use landfill leachates as N source for plants must be cautiously planned in order not to cause environmental risks, especially due to N in excess, which is the most concerning substance in leachates from elderly landfills. Thus, the rates of applications must be calculated based on the plant demands, because if N is supplied in amounts above the plant uptake capacity, it may lead to negative environmental impacts like surface and groundwater contamination, especially by nitrate leaching (Ojeda et al., 2006), given that nitrification is a fast process in soil (Jackson et al., 2008; Santos et al., 2013). In addition, there is the risk of nitrous oxide formation (N_2O) by denitrification, contributing for atmosphere pollution and increasing the greenhouse effect (Albritton et al., 1996; Cardoso et al., 2011).

Residues containing predominantly ammoniacal-N, like landfill leachates, or organic N that is further mineralized to ammoniacal-N, are subjected to N losses by ammonia volatilization that may reach 50% of the ammoniacal-N applied (Sommer et al., 2003; Santos et al., 2013) or even 90% (Dohler et al., 1990).

Sodium and, K and Cl are also found in relatively high concentrations in landfill leachates and their accumulation and effects in soil also must be monitored, especially regarding physical properties. While K is a macronutrient and Cl is a micronutrient, Na has no essential role. High salt concentrations increase the electrical conductivity in soil and may hamper the water-plant relationships due very low osmotic potential. Moreover, Na and K excessively accumulated in soil may lead to clay dispersion, causing disruption of soil aggregates and reducing porosity and water infiltration rates (Leal et al., 2009), in addition to plant nutritional imbalance. However, Santos et al. (2013) observed increases in soil electrical conductivity and content of Na and K with successive applications of landfill leachates which resulted in more than 400 kg ha^{-1} of each in three months, but there was no effect on water-dispersible clay. This no effect can be result of the local climate that allow for leaching of Na to deeper soil layers, avoiding accumulation in the topsoil.

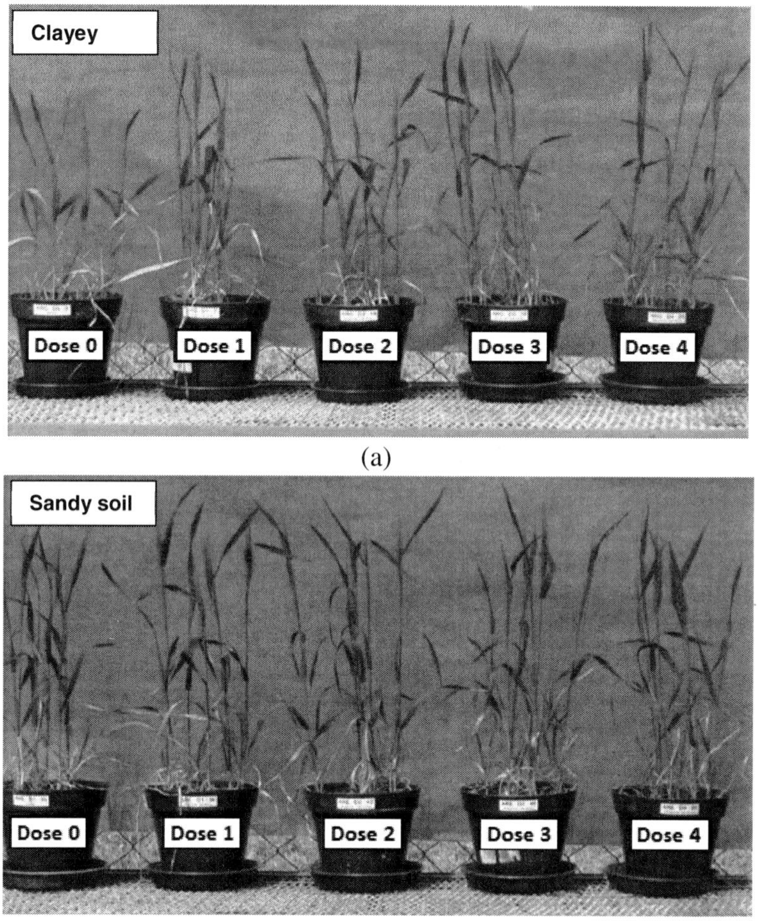

(a)

(b)

Figure 2. A pot experiment showing the effects of increasing doses of municipal waste leachate on the growth of wheat in a clayey (a) and sandy (b) soil. Plants received doses equivalent to 0, 15, 30, 45, and 60 mg pot^{-1} of total N every two weeks for 10 weeks (Kishino et al., under preparation). Photos by N. Kishino.

More researches are necessary to clarify which limits are more appropriate for each soil type, climate, crops, methods for application, and thus reduce the environmental risks associated to disposal of landfill leachates in soil, like nitrate leaching through the soil profile, salinization or clay dispersion. The method of application, for example, may affect the dynamics of N in the system and thus increase environmental risks of ammonia volatilization of nitrate leaching. For example, nitrate is more prone to be lost when the leachate is applied on non-vegetated soils, whereas when applied on cropped soils, the mineral N and other nutrients can be immobilized in the plant biomass, preventing losses by leaching (Figure 3). Such information will be useful for elaboration of technical standards for disposal of landfill leachates in agricultural soils.

(a) (b)

(c) (d)

Figure 3. Experimental application of landfill leachate on agricultural soil. Applications directly on soil surface may cause damages in soil properties like clay dispersion and N losses (a, b). Applications on mulching and crop-covered soils are more feasible because protects soil and immobilizes N in plant biomass, preventing losses to deep soil and groundwater (c, d). Photos by C.A. Santos.

5. Monitoring the Risks of Landfill Leachates Use in Agricultural Soils

The agricultural use of liquid effluents may contribute to improve the soil fertility and plant nutrition and may represent a safer way for disposal of residues in environment. Nevertheless, it must be emphasized the need of technical criteria for safe disposal. Thus, whenever the disposal of liquid effluents in soil is a more feasible (environmentally and economically) way for destination, a monitoring strategy must be previously designed in order to evaluate the possible environmental impacts and risks.

As seen above, one of the main concerns related to the application of landfill leachates in soil is the possibility of groundwater contamination by excess of N, especially as nitrate. Thus, the monitoring of sites that received landfill leachates is necessary for assuring safer applications and look for eventual nitrate movement to deeper soil layers and possibility of losses to groundwater. The use of porous capsules buried at certain soil depth for collecting soil solution can be useful for monitoring the nitrate (and other ions) movement at depths beyond the plant root system reach. When ions in solution overtake the root system depth, there are great chances of losses by leaching, especially in regions (or seasons) where the rain intensity is greater than the evapotranspiration. This is a cheaper alternative for monitoring wells, but is only useful as a preventive method for monitoring the potential for contamination of groundwater.

(a) (b)

Figure 4. General view of an experimental field on landfill leachate disposal on soil (a). Detailed view of tensiometers (right and left) and porous capsules (center) buried up to 1.2 m depth for sampling soil solution (b). Photos by C.A. Santos (a) and A.M. Martines (b).

Results from monitoring water contaminants must be compared with reference standards and the acceptable contamination levels will depend on the water use purposes.

In Brazil, the maximum allowed levels for nitrite, nitrate, and chloride, according to the water use, are established by the Environment National Environment Council, according to resolution CONAMA 396/08 (Brazil, 2008b) (Table 2).

Table 2. Maximum allowed limits for nitrite, nitrate, and chloride, according to resolution CONAMA 396/08 (Brazil, 2008b)

Parameters	Water use			
	Human	Animal	Irrigation	Recreation
	Maximum allowed limits (mg L^{-1})			
Nitrite	1	10	1	1
Nitrate	10	90	-	10
Chloride	250	-	100-700	400

As observed, the most concerning issue regarding landfill leachates in soil are relative to its organic contents and dissolved minerals like ammoniacal-N, Cl, Na, and K, in addition to the possibility of presence of heavy metals and pathogens. As a result, the monitoring the landfill leachate to be used as well as the soil that has received the effluent must be constantly carried out in order to minimize the risks of surface and groundwater contamination and soil degradation. Moreover, more studies must be carried out in order to better establish the appropriate limits for environmentally safer disposal in each situation of soil, climate, and crop, considering the soil-plant system.

ACKNOWLEDGMENTS

The authors would like to thank to National Council for Science and Technology (CNPq (Process # 485156/2007-5), for financial support. M.A. Nogueira is CNPq scholar.

REFERENCES

Albritton, D.; Derwent, R.; Isanksen, I.; Lal, M. Wuebbles, D. Radioactive forcing of climate change. In: Houghton, J.T.; Ding, Y.; Griggs, D.J.; Noguer, M.; Van der Linden, P.J.; Xiaosu, D. (eds). *Climate change 1995. The science of climate change.* New York: Cambridge University Press, p.118-131, 1996.

Bae, B., Jung, E., Kim, Y., Shim, H. Treatment of lLandfill lLeachate uUsing sActivated sSludge pProcess and eElectro-bBean rRadiation. *Water Research.* v. 33, n. 11, p. 2669 – 2673, 1999.

Bowen, S.R.; Gregorich, E.G.; Hopkins, D.W. Biochemical properties and biodegradation of dissolved organic matter from soils. *Biology Fertility of Soils.* v. 45, p. 733-745, 2009.

Brazil. Ministry of Cities (Ministério das Cidades - Secretaria Nacional de Saneamento Ambiental). Sistema Nacional de Informações Sobre Saneamento – SNIS. *Release sobre o Diagnóstico do Manejo de Resíduos Sólidos Urbanos 2007.* 2008a.

Brazil. Environment National Environment Council (Conselho Nacional do Meio Ambiente – CONAMA). Resolução nº 396 de 3 de abril de 2008. *Classificação e diretrizes ambientais para o enquadramento das águas subterrâneas.* Diário Oficial da União nº 66 de 07.04.2008. Brazil. 2008b.

Cardoso, E.J.B.N.; Nakatani, A.S.; Martines, A.M.; Nogueira, M.A.; Paula, A.M.; Santos, C.A. *Recycling Industrial And Urban Wastes In Tropical Agricultural Soils.* In: Nielsen, C.J. Recycling: Processes, Costs and Benefits. Nova Science Publishers, Inc. 2011.

Carrera, J.; Baeza J.A.; Vicent, T.; Lafuente, J. Biological nitrogen removal of high-strength ammonium industrial wastewater with two-sludge systems. *Water Research.* v. 37, p. 4211 – 4221, 2003.

Celere, M.S.; Oliveira, A.S.; Trevilato, T.M.B., Segura-Munoz, S.I. Metais presentes no chorume coletado no aterro sanitário de Ribeirão Preto, São Paulo, Brasil, e sua relevância para saúde pública. *Cadernos de Saúde Pública*, v. 23, p. 939-947, 2007.

Chapra, S.C. *Surface water - quality modeling.* United States of America: McGraw-Hill, 844 p. 1997.

Dohler, H.; Nielsen, V. C.; Vooburg, J. H. Laboratory and field experiments for estimating ammonia losses from pig and cattle slurry. Odour and ammonia emissions from livestock farming. In: *Odour and ammonia seminar farming. Proceedings of a Seminar: Silsoe.* United Kingdom, p. 132-140, 1990.

Freeze, R.A.; Cherry, J.A. *Groundwater.* Englewoods Cliffs, New Jersey: Prentice-Hall, 604 p. 1979.

Freire, R.S. Pelegrini, R., Kubota, L. T., Durán, N., Zamota, P.P. Novas tendências para o tratamento de resíduos industriais contendo espécies organocloradas. *Química Nova*, v. 23, p. 504-511, 2000.

Gomes, L.P.; Lange, L.C.; Jucá, J.F.T; Fernandes, F.; Silva, S.C.C.P. Processos e Monitoramento da Digestão de Resíduos Sólidos Urbanos. In: Castilhos Junior, H. (coordenador). *Gerenciamento de resíduos sólidos urbanos com ênfase na proteção de corpos d'água: prevenção, geração e tratamento de lixiviados de aterros sanitários*. Rio de Janeiro: ABES, 2006.

Gonzalez, C.E.A.; Gil, E.; Fernández-Fálcon, M.; Hernández, M.M. Water leachates of nitrate nitrogen and cations from poultry manure added to an Alfisol Udalf Soil. *Water Soil Pollut*. v. 202, p. 273-288, 2009.

Jackson, L.E.; Burger, M.; Cavagnaro, T.R. Roots, nitrogen transformations, and ecosystem services. *Annual Review of Plant Biology*. v. 59, p. 341-363, 2008.

Jucá, J.F.T., Monteiro, V.E.D., Oliveira, F.J.S.; Maciel F. J. Monitoramento ambiental do aterro de resíduos sólidos da Muribeca. In: *III Seminário Nacional sobre Resíduos Sólidos Urbanos*, Toledo, PR, 1999.

Kulikowska, D.; Klimiuk, E. The effect of landfill age on municipal leachate composition. *Bioresource Technology*. v. 99, p. 5981-5985, 2008.

Kjeldsen, P.; Barlaz, M.A.; Rooker, A.P.; Baun, A.; Ledin, A.; Christensen, T.H. Present and long-term composition of MSW landfill leachate: A review. *Critical Reviews in Environmental Science and Technology*. v. 32, p. 297-336, 2002.

Klimiuk, E.; Kulikowska, D. Organics removal from landfill leachate and activated sludge production in SBR reactors. *Waste Management*. v. 26, p. 1140-1147, 2006.

Leal, R.M.P.; Herpin, U.; Fonseca, A.F.; Firme, L.P.; Montes, C.R.; Melfi, A.J. Sodicity and salinity in a Brazil Oxisol cultivated with sugarcane irrigated with wastewater. *Agricultural Water Management*. v. 96, p. 307-316, 2009.

Martines, A.M.; Andrade, C.A.; Cardoso, E.J.B.N. Mineralização do carbono orgânico em solos tratados com lodo de curtume. *Pesquisa Agropecuária Brasileira*. v. 41, n. 7, p. 1149-1155, 2006.

Nayak, D.R.; Babu, Y.J.; Adhya, T.K. Long-term application of compost influences microbial biomass and enzyme activities in a tropical Aeric Endoaquept planted to rice under flooded condition. *Soil Biology and Biochemistry*, v. 39, p. 1897-1906, 2007.

Ojeda, G.; Terrasón, D.; Ortiz, O.; Alcaniz, J.M. Nitrogen losses in runoff Waters from a loamy soil treated with sewage sludge. *Agriculture, Ecosystems and Environment.* v. 117, p. 49-56, 2006.

Piccolo, A. The supramolecular structure of humic substances. *Soil Science*. v. 225, p. 810-832, 2001.

Qasim, S.R.; Chiang, W. *Sanitary landfill leachate: generation, control and treatment*. Lancaster: Technomic publishin, 339 p. 1994.

Santos, C.A.; Panchoni, L.C.; Bini, D.; Kuwano, B.H.; Carmo, K.B.; Silva, S.M.C.P.; Martines, A.M.; Andrade, G.; Andrade, D.S.; Cardoso. E.J.B.N.; Zangaro, W.; Nogueira, M.A. Land application of municipal landfill leachate: Fate of ions and ammonia volatilization. *Journal of Environmental Quality*. v. 42, p. 523-531, 2013.

Sommer, S.G.; Génermont, S.; Cellier, P.; Hutchings, N.J.; Olesen, J.E.; Morvan, T. Processes controlling ammonia emission from livestock slurry in the field. *European Journal of Agronomy*. v. 19, p. 465-486, 2003.

Tatsi, A.; Zouboulis, A. A field investigation of the quantity and quality of leachate from a municipal solid waste landfill in a Mediterranean climate (Thessaloniki, Greece). *Advances in Environmental Research*. v. 6, p. 207–219. 2002.

Tedesco, M. J.; Selbach, P.A.; Gianello, C.; Camargo, F.A.O. Resíduos orgânicos no solo e os impactos no ambiente. In: G.A. Santos; L.S. Silva; L.P. Canellas; F.A.O. Camargo. (Org.). *Fundamentos da matéria orgânica do solo: ecossistemas tropicais e subtropicais*. 2 ed. Porto Alegre: Metrópole, 2008, v. 1, p. 113-135.

Valério, G.F. *Disposição de chorume em solo visando a benefícios agrícolas: deslocamento pelo perfil do solo e características químicas após precipitações atmosféricas*. Monografia (Graduação). Universidade Estadual de Londrina, Londrina. 59 p. 2008.

In: Landfills and Recycling Centers
Editor: Carla H. Jackson

ISBN: 978-1-63463-792-3

Chapter 4

INTEGRATED TECHNOLOGIES BASED ON THE USE OF ACTIVATED CARBON AND RADIATION TO REMOVE CONTAMINANTS PRESENT IN LANDFILL LEACHATES

R. Ocampo-Perez[1,*], J. Rivera-Utrilla[2], M. M. Abdel daiem[2,3] and M. Sánchez-Polo[2]

[1]Centro de Investigación y Estudios de Posgrado, Facultad de Ciencias Químicas, Universidad Autónoma de San Luis Potosí, San Luis Potosi, SLP, Mexico
[2]Departamento de Química Inorgánica, Facultad de Ciencias, Universidad de Granada, Granada, España
[3]Environmental Engineering Department, Faculty of Engineering, Zagazig University, Zagazig, Egypt

ABSTRACT

Landfills remain a prime concern of waste management systems throughout the world and have been identified as one of the main threats to groundwater resources. The two most serious environmental effects of landfills are related to the leachate formed during decomposition from the percolation of rainwater through the waste layer and to the toxic byproducts generated by chemical and biological processes during waste leaching. Once landfill leachate is released into the environment, it can pollute soils, surface water, groundwater, and wildlife, due to its toxicity. Various organic pollutants and chemical species have been detected in the composition of landfill leachates, including: i) plasticizers and organic esters added to polymers in order to improve the properties of the final product, ii) antioxidants or stabilizers used to improve the durability of materials, and iii) herbicides/pesticides or other widely-used anti-pest compounds. This chapter is concerned with conventional and novel technologies for the elimination of pollutants present in landfill leachates stand out: (a) Photocatalysis in the presence of activated carbon; (b) Photolysis with the participation of $HO^{\bullet}$ y $SO_4^{\bullet}$-

[*] Corresponding author: R. Ocampo-Perez. Centro de Investigación y Estudios de Posgrado, Facultad de Ciencias Químicas, Universidad Autónoma de San Luis Potosí, Av. Dr. M. Nava No.6, San Luis Potosi, SLP 78210, Mexico. Tel: 524448132157, e-mail: raul_iqi@yahoo.com.mx.

radicals; (c) Application of gamma radiation to generate oxidative and reductive species; and (d) adsorption on activated carbon. Thus, the objective of this chapter is to summarize the main technologies applied to the elimination of contaminants present in landfill leachates. The chapter will be divided in four main topics.

- Adsorption/bioadsorption of phthalic acid, an organic micropollutant present in landfill leachates, on activated carbons.
- Comparative study of the photo-degradation of bisphenol A by $HO^{\bullet}$, $SO_4^{\bullet}$- and $CO_3^{\bullet}$-/$HCO_3^{\bullet}$ radicals in aqueous phase
- Role of activated carbon in the photocatalytic degradation of 2, 4- dichloro-phenoxyacetic acid by the UV/TiO_2/activated carbon system.
- Treatment of water contaminated with diphenolic acid by gamma radiation in the presence of different compounds.

1. Introduction

Landfills remain a prime concern of waste management systems throughout the world, including Europe [1], and have been identified as one of the main threats to groundwater resources [2]. The two most serious environmental effects of landfills are related to the leachate formed during decomposition from the percolation of rainwater through the waste layer and to the toxic byproducts generated by chemical and biological processes during waste leaching [3]. Once landfill leachate is released into the environment, it can pollute soils [4, 5], surface water [6, 7], groundwater [8, 9], and wildlife, due to its toxicity [10-12].

Various organic pollutants and chemical species have been detected in the composition of landfill leachates, including: i) plasticizers and organic esters added to polymers in order to improve the properties of the final product [13], ii) antioxidants or stabilizers used to improve the durability of materials [14], and iii) herbicides/pesticides or other widely-used anti-pest compounds [15].

The extensive use of plasticizers worldwide has resulted in the presence of phthalates in multiple environments, with evidence of phthalic acid esters (PAEs) in soils, natural water and wastewater. The most common monomeric plasticizers are Bisphenol A (BPA) and Diphenolic Acid (DPA), which have been detected in landfill leachates and in different aqueous systems [16-18]. Due to their very large production and wide distribution, they have become ubiquitous environmental pollutants; moreover, they are considered to be endocrine disruptors [19, 20].

Novel effects of herbicides at environmentally relevant concentrations were recently reported [21]. Dichlorophenoxy-Acetic Acid (2,4-D) is a widely used herbicide [22] that is commonly preferred worldwide because of its low cost and good selectivity [23]. 4-Chloro-2-Methylphenoxyacetic Acid (MCPA) is frequently utilized to control a wide variety of broadleaf weeds in crops, including corn, grapes, flax, sugarcane, and pulses, and in grasses and orchards, as well as in non-cultivated areas. Both compounds have frequently been detected in drinking water [24].

The following sections provide a brief description of the above pollutants and outline the treatments commonly used for their removal from aqueous solutions.

1.1. Phthalic Acid Esters

Due to their very large production and wide distribution, PAEs have become ubiquitous environmental pollutants [25,26], and some of them are suspected to be mutagens [27] or carcinogens [28]. They are also considered to be endocrine disruptors, a group of contaminants causing special concern [29].

Large amounts of PAEs are leached from plastics dumped at municipal landfills [30]. PAE concentrations in wastewater from chemical plants and nearby rivers ranged from 10 to 300 μg/L, reaching 30 mg/L in wastewater near plasticiser-producing factories [31].

PAEs have been including in the list of priority pollutants compiled by U.S. Environmental Protection Agency (USEPA) and appear in the list of priority pollutants in Chinese waters [32, 33]. In 1998, the Scientific Committee on Toxicity, Ecotoxicity and the Environment (CSTEE) of the European Union suggested different guideline values [34] for extractable phthalates in toys. However, in the absence of a reproducible test able to determine migration levels for regulatory purposes, phthalates were eventually banned, in the European Community in 1999 [35], in those applications intended for children under the age of 3 years, and their use has been restricted in the European Union for use in children toys since then [35]. In US, some phthalates have been restricted in the state of California (for children toys) starting in 2009 [36]. Phthalic acid (PA) is considered as a model PAE compound.

1.2. Bisphenol A (4, 4'-Propane-2.2-Diyldiphenol)

BPA is a representative compound among endocrine disruptors. It causes reproductive damage to aquatic organisms and is widely used as a raw material in the production of polycarbonate plastics and epoxy resins.

Due to the widespread utilization of BPA, there is increasing interest in effective remediation technologies for its removal from contaminated water [18].

BPA has been detected in all types of environmental waters at varying concentrations: up to 17.2 mg/L in hazardous waste landfill leachate [37], 12 μg/L in stream water [38], and 0.1 μg/L in drinking water [39].

The production and application of BPA is rising with the increasing use of plastics, and the global production of BPA was 3.9 million tons in 2006 and increased to 4.7 million tons in 2008 [40, 41]. EU countries are responsible for nearly 30 % of world production.

The US National Institute for Occupational Safety and Health's Registry of Toxic Effects of Chemical substances describes BPA as a primary irritant, a mutagen, and an agent that affects the reproductive system [42].

1.3. Diphenolic Acid (4,4'-Bis(4-Hydroxyphenyl) Valeric Acid)

In the recent decade, it has been observed much attention from researchers to synthesize new polyesters and polycarbonates using DPA instead of bisphenol A (BPA) [43-46]. The research interest resulted from that DPA is commercially available and much cheaper than BPA and it can introduce functional carboxyl groups into the polymer structure [47]. DPA has the same environmental risks of BPA because it is one of BPA derivatives. They cause not only strong estrogenic endocrine disrupting effect but also various diseases including carcinogenesis, liver damage and obesity-promoting effects.

1.4. 2,4- Dichlorophenoxy-Acetic Acid

Among the herbicides, 2,4- dichlorophenoxy-acetic acid (2,4-D) is the most widely used herbicide in the [22], due to its low cost and good selectivity [23]. It is used in cultivated agriculture, in pasture and rangeland applications, forest management, home, garden, and to control aquatic vegetation [48], and it has been detected in waters in various regions [49]. It is considered moderately toxic, with a maximum allowed concentration in drinking water of 100 μg/L.

1.5. 4-Chloro-2-Methylphenoxyacetic Acid

Phenoxy acid pesticides include 4-chloro-2-methylphenoxyacetic acid (MCPA); it is one of the most frequently pesticides found in drinking water [50]. MCPA was developed in the 1940s, and it is widely used to effectively control a wide variety of broadleaved weeds in cornfields, grasses, orchards, grapes, flax, sugarcane, pulses, and non-crop areas [24].

2. Novel Technologies to Remove Contaminants Present in Landfill Leachates

Numerous studies have addressed the removal of PAEs, BPA, DPA, 2,4-D, and MCPA from aqueous solution by means of novel and conventional technologies. Reference [103] describes the most recent methods developed to remove PAEs from water, wastewater, sludge, and soil, whereas, Tables 1-4 summarize some of the most representative studies on the removal of these contaminants from water.

Adsorption on activated carbon has proven the most effective and reliable physicochemical non-destructive technique for landfill leachate treatment, achieving a higher reduction in dissolved organic carbon in comparison to chemical methods [51-53]. In fact, the US Environmental Protection Agency has acknowledged that adsorption on activated carbon, one of the oldest water treatments, is one of the best methods available to remove organic and inorganic pollutants from water intended for human consumption [54].

Numerous studies have addressed the removal of PAEs from water and sediments, and treatments can be classified into three general groups: i) physical/chemical treatments, ii) biological treatments, and iii) advanced oxidation processes [103].

Activated carbons have a high BPA and 2,4-D adsorption capacity, which depends on the characteristics of the activated carbon, the solution pH, and the presence of electrolytes in the water [55].

Biological treatment is effective to eliminate BPA from aqueous solution but requires a long treatment time, and its application is limited to very low concentrations of BPA [56],

However, BPA biodegradation is enhanced by the presence of natural organic matter (NOM), which acts as a substrate for the microorganisms (e.g., green alga) in the system, increasing their growth [56]. The presence of Fe^{3+} in an anaerobic system was also found to improve the bioremediation of BPA, attributed to the action of Fe^{3+} as an alternative electron acceptor [57].

Technologies based on advanced oxidation processes (AOPs) have proven highly effective to degrade organic compounds in aqueous solution. AOPs generate species with high oxidizing power (e.g., $HO^{\bullet}$ radicals) that interact with the pollutant and degrade it into byproducts with lower molecular weight, and they can even achieve its complete mineralization.

Hydroxyl radicals have been used to eliminate BPA from aqueous phase in various studies. Rosenfeldt et al. [58] used ultraviolet radiation (UV) combined with H_2O_2 and reported that BPA degradation was more effective with the UV/H_2O_2 system than with direct photolysis, obtaining a reaction rate constant of hydroxyl radical ($HO^{\bullet}$) with BPA of $1.02 \pm 0.23 \times 10^{10}$ $M^{-1}s^{-1}$. Rivas et al. [59] studied both the elimination and mineralization of BPA with multiple technologies, i.e., UV, O_3, UV/O_3, UV/O_3/carbon, UV/O_3/TiO_2, and UV/O_3/TiO_2/carbon. They found that ozonation or photooxidation effectively remove BPA from water but that the mineralization levels are rather low, whereas the combination of O_3/UV/carbon completely mineralized the organic matter. Katsumata et al. [60] studied the photodegradation of BPA in the presence of the Fenton reagent and found the degradation rate to be strongly influenced by the solution pH and the initial H_2O_2 and Fe (II) concentrations. Total BPA degradation was achieved after 9 min of treatment under optimal conditions, and the formation of CO_2 due to BPA mineralization was observed during the photo-Fenton process, with >90% conversion of BPA to CO_2 after 36 h of treatment.

Unlike the hydroxyl radical, the sulfate and carbonate radicals are selective transient species against organic compounds with a high oxidization potential (E^0 = 2.6 V and E^0 = 1.78 V, respectively). The most frequent methods to generate sulfate radicals, $SO_4^{\bullet-}$, are by the photochemical, thermal, or chemical decomposition of $S_2O_8^{2-}$. The $CO_3^{\bullet-}$/$HCO_3^{\bullet}$ radical is usually formed *via* a reaction between bi/carbonate ions and hydroxyl radicals or by quenching the aromatic ketone triplet excited state [61, 62].

Various studies have found the UV/$K_2S_2O_8$ system to be more effective than the UV/H_2O_2 system to treat organic compounds in aqueous solution, offering a higher percentage degradation within a shorter treatment time and reducing the medium toxicity to a greater extent [63]. For its part, the carbonate radical is one of the most extensively studied inorganic radicals, due to its ubiquitous nature in the environment and relatively long lifetime. However, there has been little research on the use of this radical species to degrade organic compounds [64].

BPA photodegradation of up to 100% has been reported using electrocatalytic, electrochemical, ultrasonic, gamma radiation, photo Fenton, UV, O_3, UV/O_3, UV/TiO_2, O_3/TiO_2, UV/O_3/carbon, UV/O_3/TiO_2, and UV/O_3/TiO_2/carbon processes [59, 60, 65-69]. However, the mineralization levels are rather low, except in the case of UV/O_3/activated carbon, UV/O_3/TiO_2/activated carbon, and UV/O_3/TiO_2, which completely mineralize the organic matter after 2 h of treatment [59].

Few studies have been published on the elimination of DPA from aqueous solution by adsorption or photocatalytic processes [70-72]. These have shown a low effectiveness in DPA degradation in aqueous solution, and there is a need for research into other processes that could mineralize DPA and completely remove it from waters. Table 2 summarizes the most representative studies on the removal of DPA from aqueous phase by different processes.

An alternative to the generation of free radicals is photocatalysis on the surface of a semiconductor. Heterogeneous photocatalysis consists of the direct or indirect absorption of visible or ultraviolet (UV) radiant energy by a solid that, in its excited form, acts as a catalyst for degradation reactions with compounds at the solid-liquid interface. Various semiconductor materials possess suitable characteristics for their utilization as photocatalysts, i.e., stability, economically affordability and no requirement for excessively energetic radiation, and those that are activated by the radiation of the visible spectrum are of particular interest. The literature includes studies of photocatalysts such as TiO_2, ZnO, SnO_2, WO_3, ZrO_2, CeO_2, CdS, and Fe oxides, which have all evidenced a good performance in the removal of numerous pollutants [73-78].

It was recently reported that carbon materials can markedly improve the photocatalytic process, largely through one of the three following mechanisms: i) minimization of the recombination of photogenerated electron-hole pairs; ii) modification of the band gap of the photocatalyst to higher wavelengths; and iii) presence of adsorption centers that accelerate contact between pollutant and catalyst [79-86].

A possible mechanism by which activated carbon increases the photocatalytic activity of TiO_2 was proposed by Silva et al. [81], who reported that the presence of mild concentrations of oxygen-containing surface groups, mainly carboxylic acids and phenols, may improve the photocatalytic activity of multi-walled carbon nanotubes–TiO_2 composite catalysts in the photocatalytic oxidation of benzene derivatives in aqueous suspensions.

In a previous study, Ocampo-Perez et al. [79] investigated the photocatalytic degradation of cytarabine in aqueous phase (UV/TiO_2) in the presence of activated carbon with different chemical and textural properties. Results obtained suggested that the reaction of e_{aq}^-, from the UV-TiO_2 interaction, with the activated carbon carboxylic acids are responsible for the increased photocatalytic activity of TiO_2. We therefore hypothesized that the interaction between e_{aq}^- and surface carboxylic acids on oxidized carbons reduces carboxylic acids to aldehydes, generating H_2O_2 that can be decomposed into $HO^\bullet$ radicals, and that these aldehydes can be transformed into a surface alcohol group that generates additional $HO^\bullet$ radicals. The proposed mechanism is given by the following reactions.

$$\text{AC-COOH} + H_2O \xrightarrow{e_{aq}^-} \text{AC-COH} + H_2O_2 \tag{1}$$

$$H_2O_2 + h\nu(UV) \rightarrow 2HO^\bullet \tag{2}$$

$$\text{AC-COH} + 2H_2O \xrightarrow{2e_{aq}^-} \text{AC-CH}_2\text{OH} + 2HO^\bullet \tag{3}$$

Photocatalysis has been proposed as a good option for the degradation of 2,4-D, and UV/TiO_2 and UV/TiO_2/activated carbon processes achieved removal rates of 75% and 100%, respectively [87, 88]. The improved 2,4-D photodegradation obtained with the presence of activated carbon in the photocatalytic process can be attributed to its synergistic effects with TiO_2. Table 3 summarizes some of the most representative articles on 2,4-D removal from aqueous solution. The MCPA degradation rate was only 1% with UV radiation and 12% with adsorption on TiO_2 particles, but it was 90% when UV radiation and TiO_2 were combined [89]. Ionizing radiation is a promising water treatment option because it simultaneously generates oxidizing and reductive radical species ($HO^{\bullet}$, $H^{\bullet}$ and e_{aq}^{-}) from water radiolysis (Equation (4)) [90], allowing the degradation of a wide variety of compounds. Guo et al. [66] found that bisphenol A (BPA) degradation with gamma radiation was a feasible and effective to remove BPA from aqueous solution, which was largely attributed to $HO^{\bullet}$ radical oxidation. Other authors have used gamma radiation to degrade EDCs [66, 71, 72], pesticides [91, 92], nitroaromatic compounds [93], herbicides [94], and pharmaceutical compounds [95-97].

$$H_2O \rightsquigarrow [2.8]HO^{\bullet} + [2.7]e_{aq}^{-} + [0.6]H^{\bullet} + [0.72]H_2O_2 + [2.7]H_3O^{+} + [0.45]H_2 \quad (4)$$

The values in square brackets in Equation (4) are the number of molecules formed with the absorption of 100 eV of energy in an air-free medium at pH 7.0 [90].

Gamma radiation obtained an optimal percentage MCPA removal (up to 100%) at a solution pH of 1.5 and dose of 3000 Gy, MCPA can be completely removed by gamma radiation and partially removed by photooxidation and photocatalytic processes [94]. Table 54 summarizes the main studies on the elimination of MCPA from water.

3. Results and Discussion

The results showed in this section were obtained during the development of a wider project involved in the removal of pollutants present in landfill leachates (PA, BPA, DPA, 2,4-D, and MCPA) from aqueous solution by means of different technologies. Some of these results have already been published [102-107].

3.1. Phthalic Acid Adsorption Process

Figure 1 depicts the PA adsorption isotherms for the two activated carbon samples. They both showed the L form in the classification of Giles et al. [108, 109], indicating that the aromatic ring of the PA molecule is adsorbed parallel to the carbon surface and there is no major competition between PA and water molecules for active adsorption centres on the carbon.

The Langmuir isotherm model was applied to the experimental adsorption isotherm data. It is the most widely used model for this type of process and is represented mathematically as:

$$\frac{C_{eq}}{X_{eq}} = \frac{C_{eq}}{X_m} + \frac{1}{BX_m} \quad (5)$$

where C_{eq} is the equilibrium concentration (mg/L), X_{eq} is the amount of adsorbate adsorbed at the equilibrium (mg/g), X_m is the adsorption capacity (mg/g), and B is a constant related to the adsorption energy.

Table 1. Removal of BPA from aqueous solution

Type of Treatment	Concentration	Experimental conditions	Removing ratio	References	Observation
Adsorption onto activated carbon	[50-350 mg/L]	Three activated carbons with different chemical nature were used to adsorb BPA from aqueous solution. Analyzing the influence of solution pH (2-12), ionic strength (NaCl (0.01 M and 0.10 M)), and the presence of mineral matter in carbons on the adsorption of BPA.	129.6-263.1 mg/g	[55]	The adsorption of BPA fundamentally depends on the chemical nature of the carbon surface and the solution pH. Furthermore, the presence of electrolytes in solution favored the adsorption process due to a screen effect; however, the presence of mineral matter in carbons reduced their adsorption capacity because of the hydrophilic nature of this matter
Electrocatalytic degradation (Ti-based PbO_2-ionic liquids (ILs) electrode)	[30-80 mg/L]	The influence of solution current density, pH, concentration of supporting electrolyte, and initial concentration of BPA were studied on the BPA degradation.	100% within 150 min	[67]	The electrode showed a high stability and reusability due to the better lectroconductivity and reusability of ILs. The hydroxyl radicals ($HO^{\bullet}$) generated in the electrocatalytic process played a key role in oxidizing BPA to form CO_2 and H_2O. Moreover, these results illustrated that PbO_2-ILs/Ti electrode was effective for pollutants degradation and had a great potential application. The optimal conditions were (current density (30 mA/cm^2), pH (9), and supporting catalyst (0.07 mol/L)).
Electrochemical degradation (EC)	[100 mg/L]	EC degradation of BPA was tested for four types of anode materials, Pt, Ti/Boron-doped diamond (BDD), Ti/Sb–SnO_2 and Ti/RuO_2.	Pt (100%) Ti/BDD (100%) Ti/Sb-SnO_2 (100%) Ti/RuO_2 (75%)	[65]	BPA was readily destructed at the Ti/Sb–SnO_2 and Ti/BDD anodes, the Pt anode had a moderate ability to remove BPA, and the Ti/RuO_2 anode was incapable of effectively oxidising BPA. The BPA degradation based on the $HO^{\bullet}$ radicals produced by water electrolysis. In comparison, with its high durability and good reactivity for organic oxidation, the Ti/BDD anode appears to be the more promising one for the effective EC treatment of BPA and similar endocrine disrupting chemical (EDCs) pollutants.

Type of Treatment	Concentration	Experimental conditions	Removing ratio	References	Observation
Ultrasonic irradiation	[1 mg/L]	The presence of different additives (H_2O_2, air bubbles and humic acid) were investigated on the ultrasonic degradation of BPA under various operating conditions, i.e., ultrasonic frequency, power intensity and power density.	(70-100%)	[69]	The ultrasonic irridiation is directly related to the $HO^{\bullet}$ production as a result of BPA degradation. The optimal experimental conditions were determined to be the frequency of 800 kHz and the power intensity of 3.0 W/cm^2. The addition of H_2O_2 with a low concentration could facilitate BPA ultrasonic degradation efficiently. However, the further increased dosage of H_2O_2 was shown to decrease the degradation efficiency of BPA. It was also found that the BPA ultrasonic degradation was inhibited by aeration and the presence of humic acids.
Advanced oxidation processes	[0.10 mmol/L]	Studied both the elimination and mineralization of BPA by using several technologies such as UV, O_3, UV/O_3, UV/TiO_2, O_3/TiO_2, UV/O_3/carbon, UV/O_3/TiO_2, UV/O_3/TiO_2/carbon,	BPA: (100%) for all TOC: UV (15%) O_3 (25%) UV/O_3 (75%) UV/TiO_2 (80%) O_3/TiO_2 (60%) UV/O_3/carbón, UV/O_3/carbon /TiO_2, and UV/O_3 /TiO_2 (100%)	[59]	The processes such as ozonation or photooxidation are capable of efficiently removing BPA from water, however, mineralization levels are rather low. On the other hand, the combination of UV/O_3/carbon UV/O_3/carbon/TiO_2, and UV/O_3/TiO_2 completely mineralized the organic matter
Photocatalytic (TiO_2/UV) in a self-designed horizontal circulating bed photocatalytic reactor (HCBPR)	[10-50 mg/L]	Photocatalytic degradation of BPA in the presence of TiO_2 and UV illumination was performed in HCBPR. TiO_2 catalyst was immobilized on the surface of polyurethane foam (PF). The effects of initial BPA concentration, initial pH, TiO_2 dosage and temperature on BPA photodegradation were investigated	BPA (97%) TOC (95%) within 6 h	[68]	The BPA degradation efficiency can be effectively improved by increasing pH from 3.4 to 12.3. The optimum TiO_2 carrier dosage (the ratio of the volume of PF carriers to the effective reaction volume of HCBPR) was about 1%. Besides, the effect of temperature on BPA photodegradation was found to be unremarkable in the range of 21.2-30.5 °C.

Table 2. Removal of DPA from aqueous solution

Type of Treatment	Concentration	Experimental conditions	Removing ratio	References	Observation
Photocatalytic (UV/ β-cyclodextrin (β-CD))	[10-100 μ mol/L]	The influence of β-CD on the photodegradations of DPA radiated under UV lamp was investigated, moreover, to investigate the potential of β-CD photocatalysis for treatment and removal of EDCs from aqueous solution.	UV (20%) UV/β-CD (33.5%)	[71]	The photocatalytic degradation was enhanced in the presence of (β-CD) mainly results from the moderate inclusion depth of DPA molecule in the (β-CD) cavity, this kind of inclusion structure allows the DPA molecule sufficient proximity to secondary hydroxyl groups of the (β-CD) cavity, and these hydroxyl groups could be activated and converted to hydroxyl radicals under UV irradiation, which can enhance the photooxidation of DPA.
Photochemical (β-CD/TiO_2)	[10-80 μ mol/L]	Studied the influence of β-CD on the photochemical behaviour of DPA in the presence of TiO_2 suspensions under a 250 W metal halide lamp (MHL).	MHL (2%) MHL/TiO_2 (58%) MHL/β-CD (68%) TiO_2 (2.93 μmol/g) TiO_2/β-CD (30.00 μmol/g)	[72]	The enhancement of DPA photochemical degradation in the presence of TiO_2 and (β-CD) mainly due to the adsorption of DPA on TiO_2, which makes DPA degrade more easily in the presence of hydroxyl radicals photoproduced by TiO_2
Adsorption and photodecomposition	[10-80 μ mol/L]	Comparison between the influence of β-CD or α-CD on the adsorption and photodecomposition of DPA in the presence of TiO_2 suspensions under a 250 W metal halide lamp (MHL).	MHL/TiO_2/β-CD (78%) MHL/TiO_2/α-CD (40%) TiO_2/α-CD (0.63 μmol/g)	[70]	The inhibition of photodegradation of DPA toward α-CD mainly because α-CD and DPA did not form a stable inclusion compounds and decreased the adsorption amount of DPA on the TiO_2 surface.

Table 3. Removal of 2,4 -D from aqueous solution

Type of Treatment	Concentration	Experimental conditions	Removing ratio	References	Observation
Adsorption (activated carbon)	[50-400 mg/L]	Activated carbon derived from date stones (DSAC) was studied with respect to solution pH (2-11) and initial 2,4-D concentration on the adsorption capacity at 30 °C.	238.10 mg/g	[98]	The equilibrium data were fitted to the Langmuir, Freundlich and Temkin isotherm models, and the equilibrium data were best described by the Langmuir isotherm model. The adsorption kinetics of 2,4-D obeyed pseudo-first-order adsorption kinetics.
Adsorption (activated carbon)	[100-600 mg/L]	Adsorption equilibrium, kinetics and thermodynamics of 2,4-D onto granular activated carbon were studied in aqueous solution in a batch system with respect to pH, temperature and initial 2,4-D concentration.	518.00 mg/g at 45 °C and pH = 2	[23]	Equilibrium data fitted very well to the Freundlich and Koble-Corrigan isotherm models. It was found that both the boundary layer and intraparticle diffusion played important roles in the adsorption mechanisms of 2,4-D. The activation energy of adsorption (*E*A) was determined as 8.46 kJ mol^{-1} using the Arrhenius equation.
Photooxidation (UV)	[5-40 mg/L]	The effects of various parameters such as light intensity, exposure time, and the effects of some other parameters such as O_2, wavelength of light, pH, other commonly occurring ions (e.g. Cl^-, SO_4^{2-}, Ca^{2+}/Mg^{2+}, Fe^{3+}, NO_3^-) were studied.	81%	[99]	The percentage of degradation of 2,4-D can be increased more either by increasing the light intensity or by irradiating the sample for longer time, moreover, the presence of O_2 improved the degradation of 2,4-D and at pH up to 7, however, the presence of different ions decrease the degradation of 2,4-D.
Photocatalytic (UV/(Ca–Ce–W–TiO_2))	[10-50 mg/L]	Preparation of (Ca–Ce–W–TiO_2): solution A: was prepared by dissolving appropriate amount of Ca, Ce, and W in ethanol and water. Solution B: the doped TiO_2 was prepared by the sol-gel method using $[Ti(OBu)_4]$ as the precursor, which dissolved in ethanol under stirring. Solution A was gradually poured into solution B and allowed to stir for sometimes until sols were formed.	100% at reaction time (hr): 10 mg/L (1.0) 20 mg/L (1.5) 30 mg/L (2.5) 40 mg/L (2.5) 50 mg/L (3.5)	[87]	The degradation of 2,4-D was a result of the properties of the composite photocatalyst (large specific surface area, reduced band gap which also inhibited electron–hole recombination, high crystallinity). A comparison of the efficiency of the photocatalyst with a commercially available photocatalyst, TiO_2–Sigma product and other photocatalysts in literature placed the composite photocatalyst at an excellent position above the commercial and others available in literature for the degradation of 2,4-D.

Table 4. Removal of MCPA from aqueous solution

Type of Treatment	Concentration	Experimental conditions	Removing ratio	References	Observation
Photochemical (UV and sunlight)	[0.11- 0.55 μM]	Study the phototransformation of MCPA in aqueous solution irradiated at 254 nm and in the range of solar light ($\lambda > 300$ nm) as direct photolysis and indirect photolysis in the presence of silica and TiO_2.	UV (1 %) TiO_2 (12%) UV/TiO_2 (90%)	[89]	Direct photolysis of MCPA on silica and the phototransformation in the presence of TiO_2 are almost specific. The phototransformation of MCPA is not significantly affected by O_2 nor by the irradiation wave length between 254 and 310 nm, but it is highly depending on the solution pH.
Solar photoelectro-Fenton	[186 mg/L]	To optimize the degradation of MCPA, the current, Fe^{2+} concentration and pH were selected as independent variables, where the TOC, degradation, and energy consumption were taken as responses.	23% to 80% after 120 min	[100]	The optimum variables found were 5.0 A, 1.0mM Fe^{2+} and pH 3.0 after 120 min of electrolysis. Under these conditions, 75% of mineralization with 71% of current efficiency and 87.7 kWh/kg TOC of energy consumption were obtained. Hydroxyl radicals also destroyed 4-chloro-2-methylphenol, methylhydroquinone and methyl-p-benzoquinone detected as aromatic by-products.
Photooxidation (KrCl and XeBr excilamps irradiation)	[0.2- 2.0 M]	The photodegradation of MCPA was carried out in the presence of UV radiation and (KrCl and a XeBr excilamps), and also to estimate biodegradability (COD/BOD) of phototreated MCPA solutions.	Up to 55%	[101]	The application of excilamps significantly improved the efficiency of MCPA phototransformation in aqueous solutions. Irradiation of MCPA with a KrCl excilamp emitting mainly at 222 nm, however, the irradiation of MCPA solution with a XeBr excilamp emitting mainly at 283 nm. Biological processes are not suitable for MCPA removal due to low or total absence of biodegradability of this class of pollutants
Gamma radiation	[100 mg/L]	The radiolytic degradation of MCPA was carried out using gamma radiation and was optimized in terms of irradiation dose, solution pH (1.5, 7.0, and 11.5) and H_2O_2.	100 % at a 3000 Gy dose	[94]	Radiolytic degradation of MCPA should be carried out in solution pH 1.5. The presence of H_2O_2in the irradiation of MCPA does not affect the efficiency of decomposition of MCPA, however, it increases the efficiency of formation of carboxylic acids.

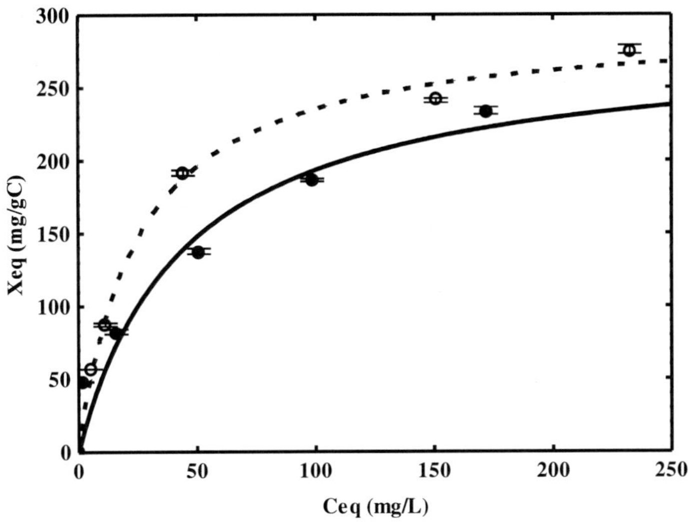

Figure 1. Adsorption isotherms of PA on activated carbons S (•) and C (○). pH 4, T 298 K. PA

Table 5. Parameters obtained by applying the Langmuir equation to PA adsorption isotherms on activated carbon samples

Carbon	Xm[a] (mg/g)	X'm (mg/m^2)	B[b] (L/mg)	BX_m[c] (L/g)
S	242.9 ± 1.6	0.20	0.15	36.4
C	274.5 ± 1.3	0.23	0.08	21.6

[a]X_m: Adsorption capacity.
[b]B: Langmuir constant.
[c]BX_m: Adsorbent-adsorbate relative affinity.

Langmuir model assumes the presence of a finite number of binding sites, homogeneously distributed over the adsorbent surface, presenting the same affinity for sorption of a single layer, and with no interaction between adsorbent species.

Results obtained are shown in Table 5. The adsorption capacity (X_m) of the carbon samples was high: 274.5 mg/g for carbon C and 242.9 mg/g for carbon S. Langmuir constant (B) values, related to the adsorption energy, and relative affinity (BX_m) values were lower than those reported for the adsorption of aromatic compounds on activated carbon [110, 111].

The PA adsorption capacity of the activated carbons was expressed per unit of carbon surface area, X'_m, to permit comparisons (Table 5). PA adsorption capacity was higher for carbon C than for carbon S, despite the former's lower surface area, external surface area, water-accessible pore volume, and hydrophobicity.

Therefore, the higher adsorption capacity of carbon C may be related to its greater content of phenolic groups, which are electronic activators of the aromatic rings of carbon graphene planes, favouring the adsorption of aromatic compounds like PA, which can be adsorbed by dispersion interactions between π electrons of its aromatic ring with π electrons of the carbon graphene planes [112].

Hence, the presence of two electron-withdrawing carboxylic groups in the aromatic ring of PA reduces its electronic density and, therefore, the adsorbent-adsorbate interactions, which may explain the low B values obtained (Table 5).

Because the solution pH during the PA adsorption process was around 4, the charge density of both activated carbon samples was positive (pH_{PZC} of both activated carbons >4, see Table 6), while the PA molecules were deprotonated and therefore had a negative charge. Under these experimental conditions, attractive electrostatic interactions between adsorbent and adsorbate would be established.

Table 6. Chemical characteristics of the activated carbon

Activated Carbon	Carboxyl Groups[a] (μeq/g)	Lactone Groups[b] (μeq/g)	Phenol Groups[c] (μeq/g)	Carbonyl Groups[d] (μeq/g)	Total Acidic Groups[e] (μeq/g)	Total Basic Groups[f] (μeq/g)	pH_{pzc}[g]	Oxygen (% wt)	Ashes (%)
S	0.00	56.00	244.0	146.7	446.7	1080.0	9.0	9.80	6.07
C	44.60	156.10	557.6	149.4	907.7	96.0	6.0	10.00	8.30

[a]Concentration of carboxyl groups determined by titration with $NaHCO_3$ (0.02 N).
[b]Concentration of lactone groups determined by titration with Na_2CO_3 (0.02 N) - $NaHCO_3$ (0.02 N).
[c]Concentration of phenol groups determined by titration with NaOH (0.02 N) - Na_2CO_3 (0.02 N).
[d]Concentration of carbonyl groups determined by titration with NaOH (0.1 N) - NaOH (0.02 N).
[e]Concentration of acidic groups determined by titration with NaOH (0.1 N).
[f]Concentration of basic groups determined by titration with HCl (0.1 N).
[g]pH of the point of zero charge.

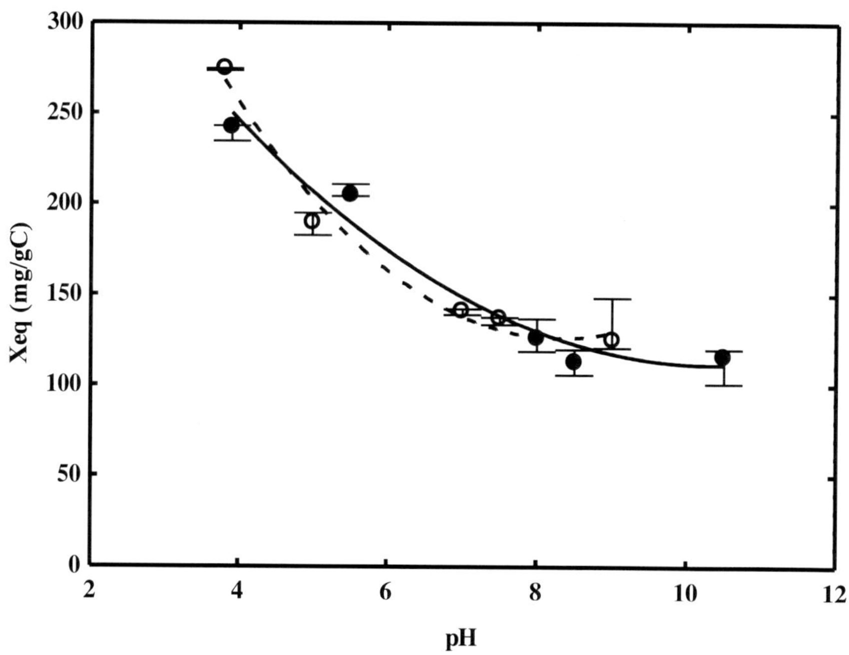

Figure 2. Adsorption capacity of activated carbons S (•) and C (○) as a function of the final solution pH, T 298 K.

According to the above observations, we conclude that PA adsorption on these activated carbons was largely determined by dispersion and by electrostatic adsorbent-adsorbate interactions.

Figure 2 shows the influence of solution pH on PA adsorption on both activated carbon samples. The adsorption process was highly dependent on the solution pH, finding that the adsorption capacity was elevated at acidic pH, markedly decreased with higher pH up to pH 6, and then remained virtually constant at higher pH values.

Adsorption on activated carbons in aqueous phase is governed by electrostatic and non-electrostatic interactions [110].

The electrostatic interactions mainly derive from the surface charges generated on carbon after water immersion and from the ions in solution [104]. Non-electrostatic interactions are largely of the Van der Waals type [110, 113]. Results in Figure 2 demonstrate that adsorbent-adsorbate electrostatic interactions play a major role in this adsorption process and are affected by the solution pH, which is attributable to two factors: i) the pH_{pzc} of the carbon samples (Table 6), and ii) the species distribution of PA as a function of solution pH.

Thus, both activated carbons have a positive charge density at pH values lower than their pH_{pzc} and a negative charge density at higher pH values. PA is deprotonated at pH >2.90; therefore, PA molecules show a negative charge density at pH >3.

According to the above data, the behavior shown in Figure 2 is due to electrostatic interactions between carbon surface and PA molecules with increased solution pH. Oxygenated acidic groups on the carbon surface are progressively ionized with higher solution pH, increasing the negative surface charge of the carbon, enhancing repulsive electrostatic carbon-PA interactions, and consequently reducing PA adsorption.

The presence of microorganisms has a major influence on the effectiveness of activated carbon in water treatments because they can be adsorbed on the carbon and form bacteria colonies on its surface (biological activated carbon) [114, 115]. Their presence on the carbon surface can have beneficial effects, including: 1) prolongation of carbon bed life by transforming organic biodegradable matter into biomass, carbon dioxide, and waste products, thereby avoiding carbon saturation [115]; and 2) formation by the adsorbed microorganisms of a biofilm that changes the texture of the carbon and its surface charge, which can improve the adsorption of some cation/metal contaminants [114].

Bacteria range in size from 0.3 to 30 μm [116, 117], therefore their adsorption on the carbon has a direct effect on the largest micropores, although they also exert an indirect effect by blocking the entrance of smaller pores.

Figure 3 depicts the adsorption/bioadsorption isotherms of PA on carbon S in the presence and absence of microorganisms. Application of the Langmuir equation to these isotherms (Table 7) showed that the presence of microorganisms during PA adsorption increased the adsorption capacity of carbon S by 15% and slightly enhanced the adsorbate-adsorbent relative affinity value (BX_m) by around 3%. PA biodegradation kinetics with the bacteria under study were investigated before obtaining the bioadsorption isotherms, observing a slight biodegradation (around 5 mg of PA per mL of bacteria solution) of this compound under the present experimental conditions.

According to previous reports [114, 115], the effects of microorganism adsorption on the chemical and textural properties of activated carbon S include: i) a decreased external surface area, due to pore blocking; and ii) a reduced pH_{PZC} value, increasing the negative charge density on the activated carbon surface.

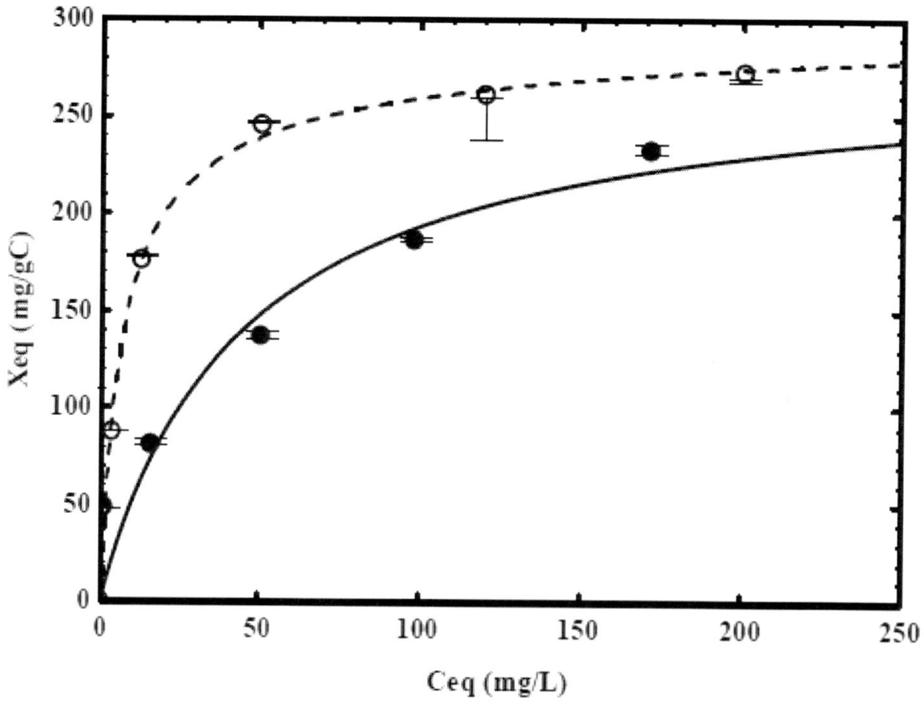

Figure 3. Adsorption isotherms of PA on carbon S in the presence (○) and absence (•) of microorganisms. pH 4, T 298 K.

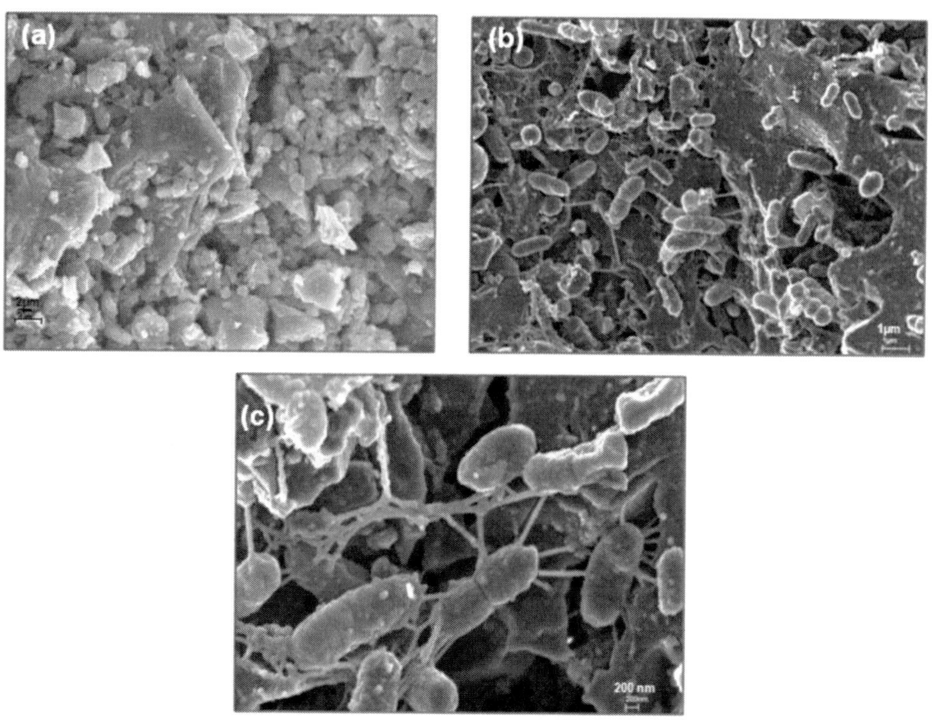

Figure 4. SEM images of carbon S before and after PA bioadsorption. a) Carbon S; b) and c) Carbon S + bacteria + PA.

Table 7. Results obtained by applying the Langmuir equation to PA adsorption isotherms on carbon S in the presence and absence of microorganisms

Activated carbon	Without Bacteria		With bacteria	
	X_m (mg/g)	BX_m (L/g)	X_m (mg/g)	BX_m (L/g)
S	242.9±1.6	36.4	279.5±2.5	37.5

Moreover, because the external walls of bacteria are formed by phospholipids [118, 119], their adsorption on activated carbon (Figure 4) increases the hydrophobicity of the carbon surface. The results depicted in Figure 3 and Table 7 may be explained by this increased hydrophobicity of the activated carbon surface, known to considerably favour the adsorption of pollutants in aqueous phase. Moreover, bacteria on activated carbon produced extracellular polymeric substances (Figures 4, b and c) that could be involved in PA adsorption.

3.2. Photo-Degradation of Bisphenol A by $HO^•$, $SO_4^{•-}$ and $CO_3^{•-}/HCO_3^•$ Radicals in Aqueous Phase

Results for the direct photodegradation of BPA showed that only 15.11 % of the initial BPA was degraded after 180 min of treatment (Exp. No. 1, Table 8). The BPA absorption spectrum shows a maximum absorption peak at a wavelength of 275 nm (ε = 329 m^2/mol), and only 1.48 % of the total energy emitted by the lamp corresponded to this wavelength, implying that the amount of radiation absorbed at 275 nm was inadequate to induce electron state transitions capable of transforming BPA, explaining the low degradation rate. Previous studies using low pressure lamps for BPA photolysis reported similar results [58, 59].

The results in Table 8 (Exp. Nos. 1-7) demonstrate that BPA degradation was enhanced by the presence of H_2O_2. For example, the addition of 1000 μM of H_2O_2 produced around a 15-fold increase in the reaction rate constant, k_t, from $0.91 \pm 0.23 \times 10^{-3}$ to $13.85 \pm 0.03 \times 10^{-3}$ min^{-1}, producing an increase in BPA degradation from 15.11% to 91.73%. Moreover, there was a linear increase in percentage degradation (from 26.63 to 91.73 %) with a rise in the initial H_2O_2 concentration from 100 to 1000 μM. This is because a higher dose of H_2O_2 increases the $HO^•$ generation rate, enhancing BPA oxidation.

The effect of BPA concentration on the BPA degradation rate was investigated by varying the initial BPA concentration from 2 to 50 mg/L while keeping the H_2O_2 concentration constant (1000 μM). Table 8 exhibits the k_t values and percentage degradation values obtained (Exp. Nos. 7-11), showing that 100 % degradation was achieved at low BPA concentrations. The main reason for the rise in reaction rate constant at lower initial BPA initial concentrations is the increase in the $[HO^•]/[C_{BPA}]_0$ ratio.

The influence of solution pH on BPA degradation was studied in the pH range from 2 to 9 (Exp. Nos. 7, 12-14, Table 8). The highest k_t value was observed at pH 7. The BPA degradation rate was markedly reduced at acidic and basic pH values, because the PO_4^{3-}, HPO_4^{2-}, and $H_2PO_4^-$ ions in the system under these conditions act as scavengers of $HO^•$ radicals, with rate constants of $k = 1.0\times10^7$ M^{-1} s^{-1}, $k = 1.5\times10^5$ M^{-1} s^{-1}, and $k = 2.0\times10^4$ M^{-1} s^{-1}, respectively [120]. $HO^•$ radicals can also be recombined at acidic pH to form H_2O_2, reducing their effectiveness in BPA oxidation. Furthermore, most of the oxidation reactions of organic compounds with $HO^{\bullet}$ in aqueous phase are reduced at acidic pH due to the presence of H^+, because H^+ ions act as $HO^{\bullet}$ radical scavengers with a rate constant of k = 7×10^9 $M^{-1}s^{-1}$[120]. A solution at basic pH favors the decomposition of H_2O_2 to form water and oxygen instead of $HO^{\bullet}$ radicals, and, moreover, the conjugate base of H_2O_2 (HO_2^-) acts as an $HO^{\bullet}$ radical scavenger by the following reactions [120]:

$$H_2O_2 + HO^- \leftrightarrow HO_2^- + H_2O \tag{6}$$

$$HO_2^- + HO^{\bullet} \rightarrow H_2O + O_2^{\bullet -} \tag{7}$$

$$O_2^{\bullet -} + HO^{\bullet} \rightarrow O_2 + OH^- \tag{8}$$

$$OH^{\bullet} + OH^- \rightarrow O^{\bullet -} + H_2O \tag{9}$$

$$O^{\bullet -} + O_2^{\bullet -} + H_2O \rightarrow 2OH^- + O_2 \tag{10}$$

The effect of reaction temperature on BPA photodegradation with the UV/H_2O_2 system was investigated in experiments conducted at 25, 35, and 45 °C. The results in Table 8 (Exp. Nos. 7, 15, 16) show that k_t values were very similar at the three temperatures, indicating that the temperature had no substantial effect on the BPA degradation rate.

The effect of the initial $K_2S_2O_8$ concentration was assessed by carrying out experiments at an initial BPA concentration of 10 mg/L (44 μM) and $K_2S_2O_8$ concentrations of 100, 200, 300 400, 500, and 1000 μM (Table 8, Exp. Nos. 17-22). In most cases, the addition of $K_2S_2O_8$ to the system produced a slight increase in the BPA degradation rate in comparison to the rate obtained for the same concentration of H_2O_2 in the UV/H_2O_2 system. Thus, under the same conditions, the percentage BPA degradation was 1.3-fold higher with $UV/K_2S_2O_8$ than with UV/H_2O_2 (Table 8, Exp. 6, 21), attributed to the generation of both $HO^{\bullet}$ and $SO_4^{\bullet -}$ radicals in the $UV/K_2S_2O_8$ system. The results in Table 8 also show that the addition of elevated concentrations of $K_2S_2O_8$ (e.g., 1000 μM) did not substantially increase the BPA degradation rate due to an excessive generation of $SO_4^{\bullet -}$ radicals, which can recombine according to the following equations [120].

$$SO_4^{\bullet -} + OH^- \rightarrow SO_4^{2-} + HO^{\bullet} \quad k \simeq 7.3\times10^7\ M^{-1}\ s^{-1} \tag{13}$$

$$SO_4^{\bullet -} + OH^{\bullet} \rightarrow HSO_4^- + \frac{1}{2}O_2 \quad k = 1.0\times10^{10}\ M^{-1}\ s^{-1} \tag{14}$$

The working temperature is an important factor that influences the decomposition of persulfate to generate $SO_4^{\bullet -}$ radicals [121]; therefore, the influence of temperature on BPA degradation was examined. The results (Table 8, Exp. Nos. 22, 30, 31) show an increase in BPA degradation rates with a rise in temperature from 25 to 45 °C.

Figure 5 depicts the BPA photodegradation kinetics in the presence of different doses of Na_2CO_3. It shows that the addition of Na_2CO_3 produced an increase in the BPA degradation rate, which rose from 24.89% to 68.23% with an increase in initial Na_2CO_3 dose from 100 to 1000 μM. The initial solution pH of the above experiments ranged from 9.4 to 10.3. Under these experimental conditions, carbonate anions are in equilibrium with hydrogen carbonate anions according to the following reaction [122]:

$$HCO_3^- \leftrightarrow CO_3^{2-} + H^+ \quad pK_a = 10.25 \tag{15}$$

Hence, the solution contains 85 % of bicarbonate at a solution pH of 9.4, which reduces to 45 % at a pH of 10.3.

Experiments were conducted under the same experimental conditions as above but in the absence of UV radiation to investigate the possibility of $CO_3^{\bullet -}/HCO_3^{\bullet}$ radicals generation in the UV/Na_2CO_3 system due to the formation of complexes between BPA and Na^+, CO_3^{2-} and/or HCO_3^- anions. Results obtained showed that the absorbance of BPA solutions remained constant over time, indicating that these complexes are not formed. Therefore, the explanation for the results in Figure 5 is the formation of carbonate radicals $CO_3^{\bullet -}$ and bicarbonate radicals $HCO_3^{\bullet}$ *via* the following reactions:

$$OH^- + h\nu \xrightarrow{+36.0\ Kcal/mol} HO^{\bullet} \tag{16}$$

$$HO^{\bullet} + HCO_3^- \xrightarrow{+50.9\ Kcal/mol} HCO_3^{\bullet} + OH^- \tag{17}$$

$$HCO_3^{\bullet} \underset{H_2O}{\longleftrightarrow} H_3O^+ + CO_3^{\bullet -} \tag{18}$$

This mechanism was corroborated by means of the program Gaussian09 (CCSD/aug-cc-pVTZ) obtaining the corresponding reaction enthalpies. Thus, the energy of our UV system was enough to carry out the above reactions.

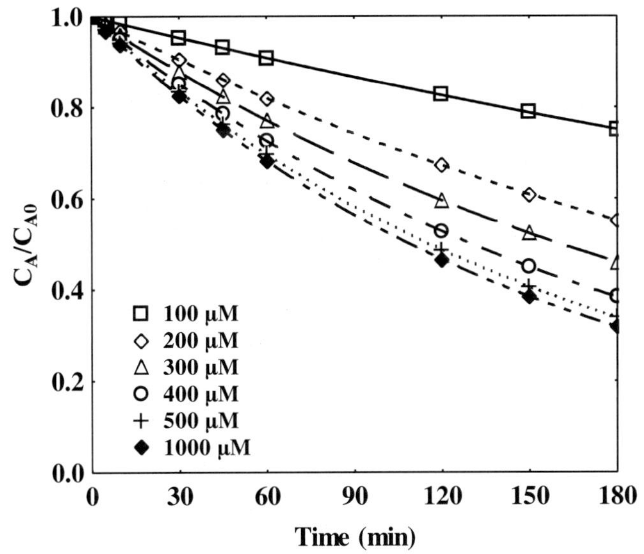

Figure 5. Effect of initial concentration of Na_2CO_3 on BPA degradation with the UV/Na_2CO_3 system. $[BPA]_0 = 10$ mg/L, T = 25°C.

k_t values obtained at different initial concentrations of Na_2CO_3 are given in Table 8. These values were lower than those obtained for the UV/H_2O_2 and $UV/K_2S_2O_8$ systems, which may be due to the lesser reactivity of $CO_3^{\bullet-}/HCO_3^{\bullet}$ radicals in comparison to $HO^{\bullet}$ and $SO_4^{\bullet-}$ radicals. To test this hypothesis, the value of the radical reaction rate constant ($k_{CO_3^{\bullet-}/HCO_3^{\bullet}BPA}$) was determined yielded a value of $k_{CO_3^{\bullet-}/HCO_3^{\bullet}BPA} = 3.89 \pm 0.09 \times 10^6\ M^{-1}\ s^{-1}$. This value is within the range of values reported for the degradation of aromatic compounds with $CO_3^{\bullet-}$ radicals [123]. The reaction rate constant was lowest between BPA and $CO_3^{\bullet-}/HCO_3^{\bullet}$ radicals, followed by $SO_4^{\bullet-}$ radicals and $HO^{\bullet}$ radicals; i.e., the capacity to attack BPA was lower for $CO_3^{\bullet-}/HCO_3^{\bullet}$ radicals than for $SO_4^{\bullet-}$ and $HO^{\bullet}$ radicals.

The influence of the initial BPA concentration on its photodegradation with the UV/Na_2CO_3 system was investigated (Figure 6 and Table 8, Exp. Nos. 37-41). The BPA degradation rate constant was lower at higher initial concentrations of BPA. Thus, it was $22.32 \pm 2.03 \times 10^{-3}\ min^{-1}$ for a concentration of 2 mg/L and $2.92 \pm 0.02 \times 10^{-3}\ min^{-1}$ for one of 50 mg/L, respectively. Once again, BPA degradation rates were lower with this system than with the UV/H_2O_2 or $UV/K_2S_2O_8$ systems.

The effect of temperature on BPA photodegradation with the UV/Na_2CO_3 system was investigated. As shown in Table 8 (Exp. Nos. 37, 45, 46), the BPA degradation rate decreased with higher temperatures, indicating that, like the UV/H_2O_2 system, this process is exothermic.

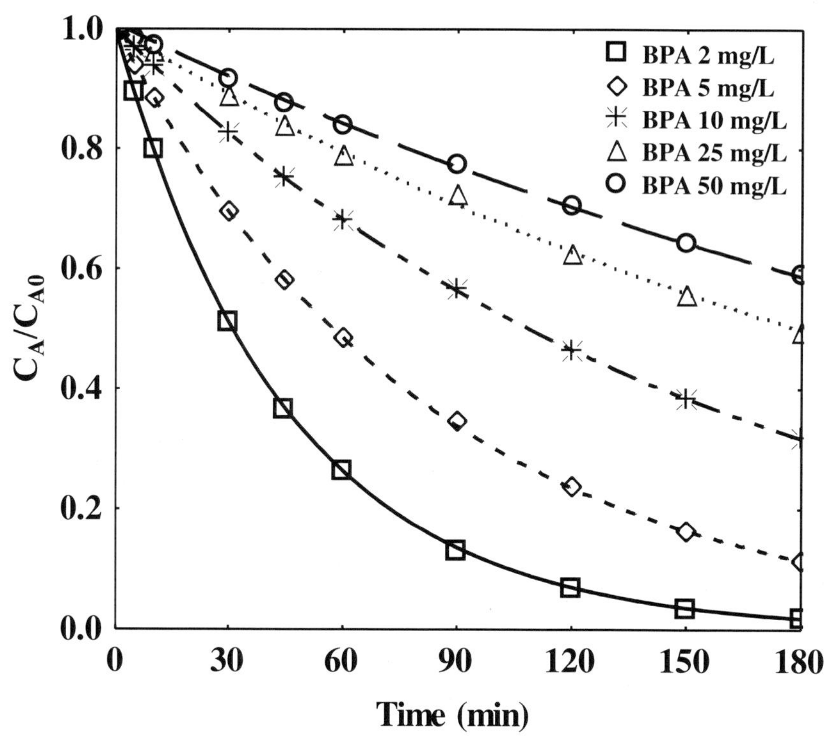

Figure 6. Effect of initial concentration of BPA on its degradation with the UV/Na_2CO_3 system in ultrapure water. $[Na_2CO_3]_0 = 1000\ \mu M$, T = 25°C, pH = 10.3.

Table 8. Photodegradation of bisphenol A under different experimental conditions for a treatment time of 180 min

Exp. No.	[BPA] (mg/L)	[H_2O_2] (μM)	[$K_2S_2O_8$] (μM)	[Na_2CO_3] (μM)	T (°C)	pH	$1-\left(\frac{C_e}{C_0}\right)\times 100$ (%)	$k_t\times10^3$ (min^{-1})	%D
1	10	0	--	--	25	7.0	15.11	0.91±0.23	1.79
2	10	100	--	--	25	7.0	26.63	1.72±0.00	0.01
3	10	200	--	--	25	7.0	56.15	4.58±0.03	0.01
4	10	300	--	--	25	7.0	61.48	5.30±0.71	0.01
5	10	400	--	--	25	7.0	66.16	6.02±0.02	0.01
6	10	500	--	--	25	7.0	67.53	6.25±0.00	0.00
7	10	1000	--	--	25	7.0	91.73	13.85±0.03	0.09
8	2	1000	--	--	25	7.0	100.00	61.16±1.54	1.40
9	5	1000	--	--	25	7.0	99.98	48.08±2.08	6.45
10	25	1000	--	--	25	7.0	71.84	7.04±0.15	1.75
11	50	1000	--	--	25	7.0	39.91	2.83±0.06	0.72
12	10	1000	--	--	25	2.0	33.18	2.24±0.32	6.32
13	10	1000	--	--	25	4.0	29.98	1.98±0.11	1.93
14	10	1000	--	--	25	9.0	28.32	1.85±0.50	4.09
15	10	1000	--	--	35	7.0	90.99	13.37±0.17	8.28
16	10	1000	--	--	45	7.0	90.28	12.95±0.25	2.22
17	10	--	100	--	25	7.0	30.98	2.06±0.17	0.85
18	10	--	200	--	25	7.0	49.18	3.76±0.27	1.48
19	10	--	300	--	25	7.0	69.02	6.51±0.34	1.78
20	10	--	400	--	25	7.0	81.11	9.26±0.49	2.71
21	10	--	500	--	25	7.0	89.80	12.68±0.88	3.86
22	10	--	1000	--	25	7.0	91.46	13.67±0.64	2.77
23	2	--	1000	--	25	7.0	100.00	199.59±11.60	9.62
24	5	--	1000	--	25	7.0	100.00	125.12±35.72	4.21
25	25	--	1000	--	25	7.0	78.73	8.60±0.21	3.02
26	50	--	1000	--	25	7.0	59.92	5.08±0.19	2.55
27	10	--	1000	--	25	2.0	99.04	25.80±2.22	15.76
28	10	--	1000	--	25	4.0	97.25	19.97±0.85	6.22
29	10	--	1000	--	25	9.0	65.92	5.98±0.68	10.64
30	10	--	1000	--	35	7.0	100.00	61.80±0.00	7.46
31	10	--	1000	--	45	7.0	100.00	66.89±0.00	14.21
32	10	--	--	100	25	9.4	24.89	1.59±0.15	0.48
33	10	--	--	200	25	9.5	44.89	3.31±0.12	0.46
34	10	--	--	300	25	9.7	54.05	4.32±0.11	0.65
35	10	--	--	400	25	10.0	61.69	5.33±0.48	2.34
36	10	--	--	500	25	10.1	66.16	6.16±0.22	1.23
37	10	--	--	1000	25	10.3	68.23	6.37±0.12	0.68
38	2	--	--	1000	25	10.3	98.20	22.32±2.03	18.34
39	5	--	--	1000	25	10.3	88.47	12.00±1.39	11.90
40	25	--	--	1000	25	10.3	50.53	3.91±0.16	2.35
41	50	--	--	1000	25	10.3	40.88	2.92±0.02	0.92
42	10	--	--	1000	25	2.0	25.43	1.63±0.21	3.65
43	10	--	--	1000	25	4.0	26.23	1.69±0.01	1.53
44	10	--	--	1000	25	7.0	67.65	6.27±0.25	0.37
45	10	--	--	1000	35	10.3	50.44	4.14±0.00	1.02
46	10	--	--	1000	45	10.3	51.50	3.67±0.00	2.54

3.3. Role of Activated Carbon in the Photocatalytic Degradation of 2, 4-Dichlorophenoxyacetic Acid by the UV/TiO_2/Activated Carbon System

Figure 7a depicts the variation in percentage 2,4-D degradation as a function of treatment time with the UV/TiO_2 system in the presence of carbons S, M, and W, while Figure 7b depicts this variation in the presence of carbons W_{O3-30} and W_{O3-120}.

Figure 7a shows that the highest percentage 2,4-D removal after 60 min of treatment was obtained with carbon S (80%), followed by carbons M (72%) and W (59%).

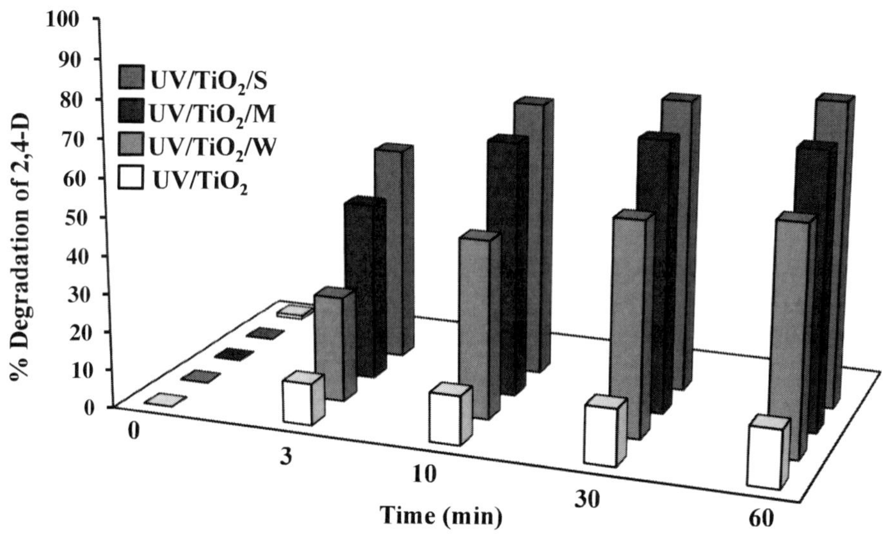

a

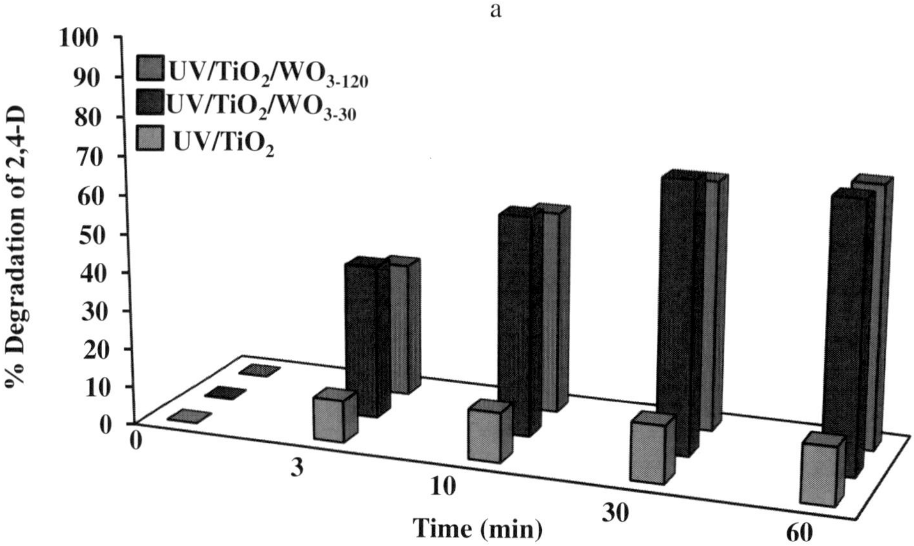

b

Figure 7. 2,4-D photodegradation by means of the UV/TiO_2 system in the presence of different activated carbons. T = 25° C, pH = 7, [2,4-D]= 50 mg/L, mass of TiO_2 and activated carbon = 5 mg.

Table 9. Adsorptive and photocatalytic contributions to the removal of 2,4-D with the UV/TiO_2/activated carbon system

Activated Carbon	Total (%)	Adsorption (%)	UV/TiO_2 (%)	Synergistic contribution (%)
S	80	72	14	0
M	72	56	14	2
W	59	23	14	22
$W_{O3\text{-}30}$	70	3	14	53
$W_{O3\text{-}120}$	70	4	14	52

Comparing with the results with those obtained for adsorption process, the removal of 2,4-D was 1.1, 1.38, and 2.57-fold higher, respectively, with the combined UV/TiO_2/activated carbon process than with the adsorption process.

The role of activated carbon in the UV/TiO_2/activated carbon system was analyzed by determining the contributions of adsorption and photocatalysis (UV/TiO_2) to the removal of 2,4-D at 60 min of the combined treatment. The results in Table 9 show that: i) the greatest adsorptive contribution to the global removal process is made with carbons S and M; ii) with these two carbons (S and M), the contribution of the photocatalytic process (UV/TiO_2) is very low in comparison to the contribution of the adsorptive process, and iii) with carbon W, the total percentage removal obtained by the UV/TiO_2/activated carbon system is markedly higher than the result of summing the adsorptive and catalytic contributions, indicating that its presence has a synergistic effect on the 2,4-D removal process.

The results in Table 9 were compared with the chemical and textural properties of the activated carbons. No clear relationship was observed between the surface area and the synergistic contribution to the global removal process, given that the three activated carbons have a similar surface area (≅1200 m^2/g).

However, the content of carboxyl and phenol groups is higher in carbon W than in carbons S or M. Hence, the synergistic activity of carbon W in the removal of 2,4-D by the UV/TiO_2/activated carbon system may be related to the elevated carboxyl group content on its surface (0.32 meq/g vs. 0.00 meq/g on carbon S and 0.04 meq/g on carbon M), as proposed in a previous study [79].

The role of carboxyl groups in the photocatalytic process was examined by increasing the carboxyl group content of carbon W through oxidation with ozone for 30 min ($W_{O3\text{-}30}$) and 120 min ($W_{O3\text{-}120}$) and then texturally and chemically characterizing these samples (Table 10). The ozone treatment reduced the surface area value by 50%, increased the content of carboxyl and lactone groups, and reduced the content of phenol groups. The mechanism by which ozone interacts with the activated carbon surface has been reported in various publications [124, 125].

Figure 7b depicts the photocatalytic degradation of 2,4-D in the presence of these ozonated samples ($W_{O3\text{-}30}$ and $W_{O3\text{-}120}$), demonstrating that the degradation was markedly enhanced by their presence, achieving 70% removal of 2,4-D after 60 min in both cases. This is much higher than the percentage removal obtained with the untreated carbon (W). As shown in Table 9, i) these ozonated carbon samples ($W_{O3\text{-}30}$ and $W_{O3\text{-}120}$) make a very low adsorptive contribution (<4%) to the global 2,4-D removal process, and ii) they make the highest synergistic contribution to this global process (>50 %).

Table 10. Chemical characteristics of the W_{O3-30} sample after treatment with UV or UV/TiO_2

Activated carbon treatment	pH_{PZC}	Basic groups (meq/g)	Carboxyl groups (meq/g)	Phenol groups (meq/g)	Lactone groups (meq/g)	Acid groups (meq/g)
W_{O3-30}	3.3	0.05	1.42	0.15	0.99	2.56
UV/W_{O3-30}	3.2	0.02	1.40	0.11	0.97	2.48
UV/ TiO_2/W_{O3-30}	3.8	0.05	1.12	0.23	1.21	2.61

In order to identify the mechanism by which ozonated carbon samples enhance 2,4-D photocatalysis, we chemically characterized sample W_{O3-30} by exposing it to UV and UV/ TiO_2 radiation in the absence of 2,4-D. The results in Table 10 show: i) similar values of surface acid groups between UV-treated and untreated samples, and ii) a lower concentration of carboxyl groups and markedly higher content of phenol and lactone groups in the UV/ TiO_2-treated *versus* untreated samples.

These results verify the proposed mechanism (Reactions 1-3), indicating that the main pathway by which activated carbons enhance 2,4-D removal by the UV/TiO_2/carbon system is the reduction of superficial carboxyl groups to aldehyde groups and finally to alcohol groups. Alcohol groups formed on the carbon surface may in turn react with adjacent carboxyl groups in the graphene planes of the activated carbon, generating lactone groups. This would explain the increase in lactone groups in sample W_{O3-30} with the UV/TiO_2/W_{O3-30} system. The increase in phenol groups in sample W_{O3-30} would result from the decarboxylation of carboxyl groups to CO_2 by $HO^•$ radicals, followed by oxidation of the nascent surface carbons to phenol groups. Key aspects of the effectiveness of treatments of organic compound-polluted water are (i) the transformation of dissolved organic carbon into CO_2, and (ii) the toxicity of the degradation byproducts. We therefore studied these parameters during 2,4-D photodegradation.

Figure 8a depicts the variation in TOC concentration as a function of treatment time. According to these results: i) the UV/W_{O3-30} system does not markedly reduce the TOC, confirming the low adsorption capacity of the ozone-treated carbon; and ii) the greatest percentage removal was achieved with the UV/TiO_2/W_{O3-30} system, due to the higher concentration in the medium of radical species, mainly $HO^•$ radicals.

Figure 8b depicts the percentage inhibition of *Vibrio Fischeri* bacteria during 2,4-D photodegradation with UV/W_{O3-30}, UV/TiO_2, and UV/TiO_2/W_{O3-30} treatments. This percentage was 97 % for the 2,4-D aqueous solution, demonstrating the high toxicity of 2,4-D, but it was drastically diminished with the UV/TiO_2, and UV/TiO_2/W_{O3-30} systems for 5 min o treatment, indicating that the degradation byproducts obtained have a lower toxicity in comparison to 2,4-D; however for longer treatment times the byproducts toxicity increased.

3.4. Remove of Contaminants Present in Landfill Leachates by Gamma Radiation

Highly reactive radical species, mainly $HO^•$, $H^•$ and e_{aq}^- are produced in water radiolysis (Equation (4)).

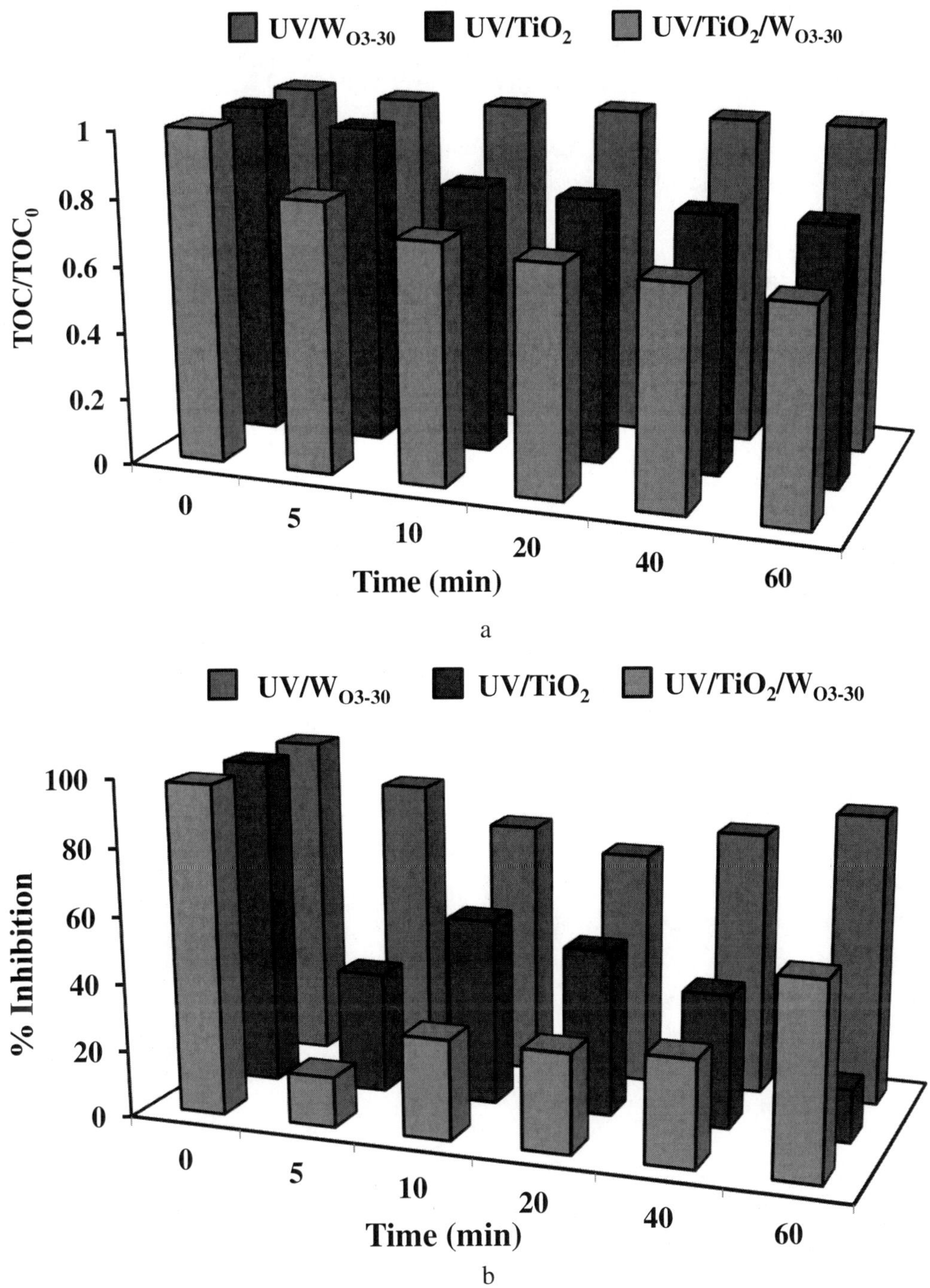

Figure 8. Variation in (a) TOC and (b) toxicity during 2,4-D photodegradation with UV/TiO_2, UV/W$_{O3-30}$, and UV/TiO_2/W$_{O3-30}$ systems. T = 25° C, pH = 7, [2,4-D]=50 mg/L, mass of TiO_2 and activated carbon = 5 mg.

The extent of their production depends on various factors, including solution pH, temperature, energy emitted per unit of distance travelled by the incident radiation in aqueous medium, absorbed dose, dose rate, and presence of dissolved gases.

Figure 9 shows the degradation kinetics of the five pollutants by gamma irradiation as a function of the dose absorbed, at initial concentrations of 20 mg/L and at a dose rate of 1.66 Gy/min. As shown in this figure, the relative concentration (C_A/C_{Ao}) decreased with the increase in the absorbed dose for all compounds, and it is noticed that the degradation kinetic of these compounds exponentially decreases with higher absorbed dose which can be well represented by a first order kinetic model. Thus, the dose constant, k, for each experiment was obtained from the slope of the plot of $Ln(C_A/C_{A0})$ *versus* adsorbed dose. The k values obtained are given in Table 11, where it is observed that PA presented the highest values of k and MCPA the lowest ones. Furthermore, k values markedly decreased with higher initial pollutant concentration. An explanation of this behavior is that a rise in the initial solute concentration increases the byproducts produced, which compete to react with the radicals species formed in radiolysis and, therefore, the number of radical species available to react with the pollutant is reduced.

The dose constant, k, was used to calculate the dose required to produce 50% and 90% pollutant degradation (D_{50} and D_{90} values) by means of Equations (3) and (4). D_{50} and D_{90} values (Table 11) were different for all pollutants studied, which confirms the high influence of the chemical structure of compounds on their degradation by gamma radiation.

The efficiency of gamma radiation in the degradation of the pollutants was studied evaluating both degradation percentage and G-values. As an example, the results obtained for PA (Table 11, Exp. No. 2) are plotted in Figure 10. The G-values for the radiolytic decomposition of PA were in the range from 0.57 to 0.05 for absorbed doses lower than 1000 Gy. Interestingly, the G-value diminished with augmenting accumulated radiation dose but the removal percentage increased.

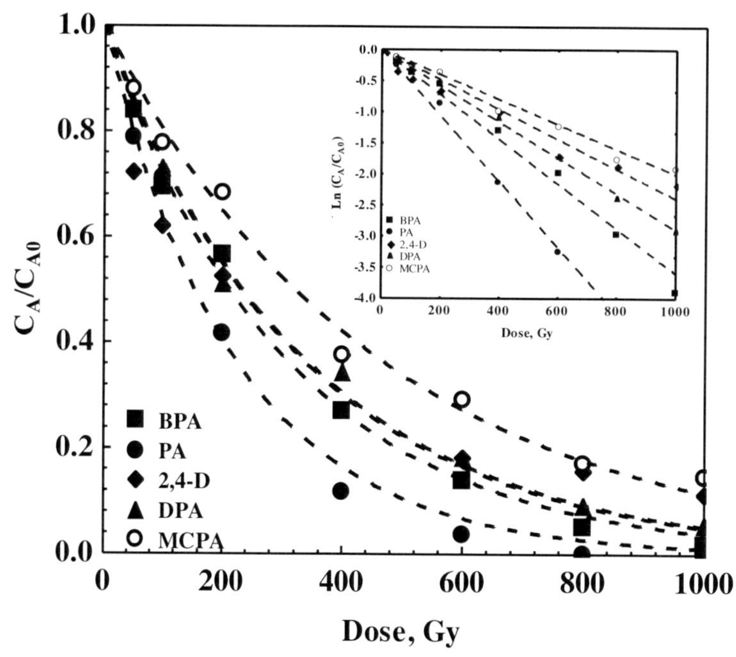

Figure 9. Degradation of all contaminants as a function of radiation dose. [C_{A0}] = 20 mg/L. Dose rate = 1.66 Gy/min, T = 25°C, and pH = 4.0.

Table 11. Experimental conditions and kinetic parameters obtained by gamma radiation at a dose rate of 1.66 Gy/min

Exp. No	$[PA]_0$ (mg/L)	$[BPA]_0$ (mg/L)	$[DPA]_0$ (mg/L)	$[2,4\text{-}D]_0$ (mg/L)	$[MCPA]_0$ (mg/L)	$G_0 \times 10$ (µmol/J)	$k \times 10^3$ (Gy^{-1})	D_{50} (Gy)	D_{90} (Gy)	% Degradation
1	10	-	-	-	-	5.7	12.1±0.6	57.1	189.7	100.0
2	20	-	-	-	-	5.7	4.5±0.3	154.0	511.7	100.0
3	50	-	-	-	-	6.7	2.9±0.2	239.8	796.7	98.0
4	-	10	-	-	-	1.7	4.3±0.3	162.3	539.3	100.0
5	-	20	-	-	-	2.8	3.3±0.2	212.6	706.3	98.1
6	-	50	-	-	-	2.5	1.4±0.1	481.4	1599.0	76.5
7	-	-	10	-	-	1.9	5.4±0. 2	129.1	428.8	100.0
8	-	-	20	-	-	2.2	3.0±0.1	233.4	775.3	94.6
9	-	-	50	-	-	4.9	2.3±0.2	304.0	1009.9	86.7
10	-	-	-	10	-	5.4	6.7±1.3	103.8	344.7	99.8
11	-	-	-	20	-	5.0	3.0±0.4	233.4	775.3	88.7
12	-	-	-	50	-	5.3	1.2±0.1	572.9	1903.0	64.0
13	-	-	-	-	10	1.5	3.2±0.2	220.1	731.0	99.5
14	-	-	-	-	20	2.4	2.2±0.1	319.4	1061.1	85.6
15	-	-	-	-	50	2.4	0.9±0.0	761.7	2530.3	55.5

According to these results, the efficiency of the gamma irradiation process decreases with longer exposure time. This trend can be explained by (i) the competition for solute molecules between the reactive radicals produced by water radiolysis, (ii) the competition for reactive radicals between parent compound and reaction by-products and (iii) radical–radical recombination reactions, including $HO^{\bullet}$, e_{aq}^- and $H^{\bullet}$ [95, 96].

Because the radical species increase at higher radiation dose rates, the absolute rates for radical-radical recombination reactions also increase, reducing the effective radical concentrations for reaction with PA. Similar results were obtained for the rest of pollutants, whose corresponding plots showed a cross point of both curves for dose values in between 100-150 Gy; therefore, gamma radiation is an efficient process for degradation of these pollutants when using doses of around 100 Gy, because, for these doses high degradation percentages and radiation-chemical yield are obtained.

Figure 11 shows, as an example, a typical graph of a competitive kinetics where $Ln([PA]_t/[PA]_0)$ is plotted versus $Ln([3\text{-}Ampy]_t/[3\text{-}Ampy]_0)$. As described in the experimental section, from this type of figures, the corresponding reaction rate constants were determined. Table 12 shows the reaction rate constants of the five pollutants with hydroxyl radical $HO^{\bullet}$, hydrated electron e_{aq}^-, and hydrogen atom $H^{\bullet}$. The values of these constants are similar to those reported by other authors [126-131] for organic compounds of analogous characteristics to these studied in the present work. It is noteworthy that the highest rate constant for the degradation of these compounds is $k_{HO^{\bullet}}$. Thus, the order found was $k_{HO^{\bullet}} > k_{e_{aq}^-} > k_{H^{\bullet}}$, which indicates that the elimination process of these pollutants by gamma radiation mainly follows an oxidative pathway through reaction with $HO^{\bullet}$ radical.

The $k_{HO^{\bullet}}$ values follow the order 2,4-D > PA > DPA > MCPA > BPA, which is similar to the order found for $k_{e_{aq}^-}$ and $k_{H^{\bullet}}$.

This trend could be explain assuming that the radical reactions of these pollutants started with the attack of the radical electron to the benzylic positions of the pollutant molecules, in such a way that C atom in the benzylic position stabilizes the radicals initially formed. For 2,4-D this possibility does not exist because there is not C atoms in that position, and, therefore, the reaction progress will be more rapid than in the rest of pollutants. In the case of BPA, the presence of two aromatic rings stabilizes the benzylic radicals formed after the attack to the pollutant molecules, and, therefore, the reaction progress will be slower.

The radiation-chemical yield of species generated in water radiolysis is a function of the medium pH [132, 133]. Therefore, the irradiation was studied in acidic and basic media (Figure 12), irradiating DPA aqueous solutions at pH = 2.0, 7.0, and 12.0. The lowest DPA degradation was observed at pH = 2.0, with a marked reduction in the dose constant, k = 0.48 ± 0.05×10^{-3} Gy^{-1}, which was not observed at the other pH values studied.

Table 12. Rate constants for the reaction of the selected compounds with species from water radiolysis

Pollutant	$k_{HO\bullet}$ (M^{-1} s^{-1})	$k_{e^-_{aq}}$ (M^{-1} s^{-1})	$k_{H\bullet}$ (M^{-1} s^{-1})
PA	$4.8\pm0.12\times10^{10}$	$9.3\pm0.03\times10^{9}$	$5.5\pm0.25\times10^{7}$
BPA	$1.7\pm0.21\times10^{10}$	$7.0\pm0.07\times10^{9}$	$4.7\pm0.18\times10^{7}$
DPA	$4.5\pm0.08\times10^{10}$	$7.6\pm0.12\times10^{9}$	$3.1\pm0.31\times10^{7}$
2,4-D	$6.6\pm0.13\times10^{10}$	$9.8\pm0.35\times10^{9}$	$6.8\pm0.01\times10^{7}$
MCPA	$3.3\pm0.02\times10^{10}$	$1.5\pm0.17\times10^{9}$	$5.3\pm0.09\times10^{7}$

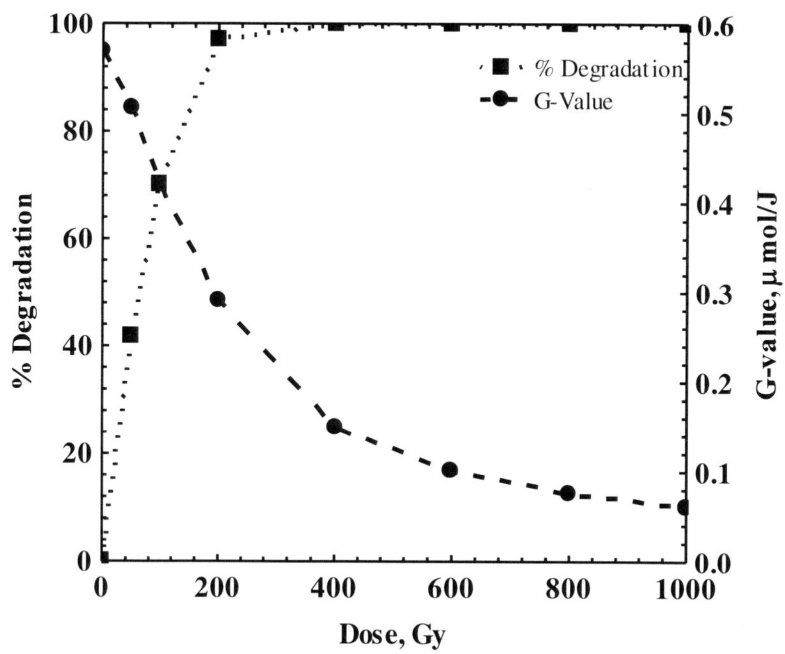

Figure 10. G-values and degradation percentage of PA as a function of radiation dose.

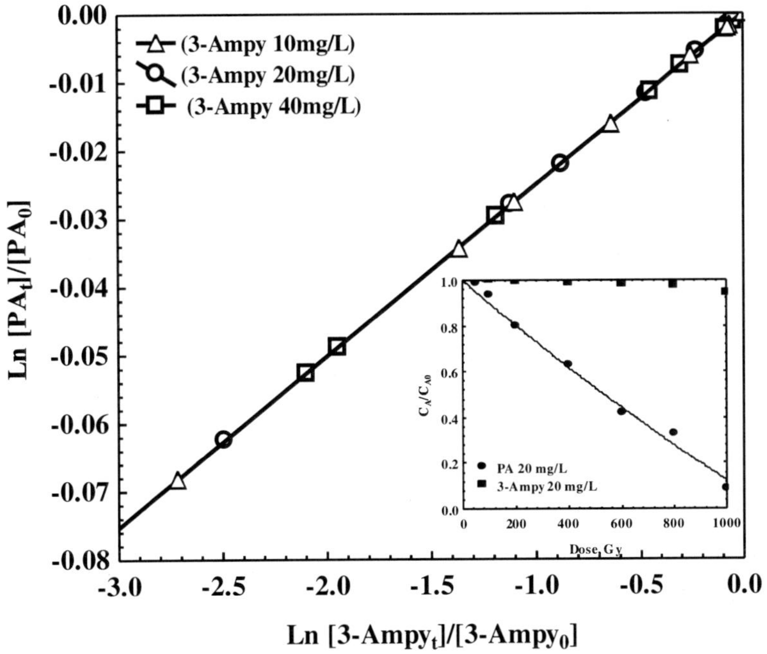

Figure 11. Competitive kinetics data obtained for PA [20 mg/L] and 3-Ampy [10, 20 and 40 mg/L]. Dose rate = 1.66 Gy/min, T = 25°C, and pH=1.0.

One explanation for this drastic decrease in DPA degradation rate is that reaction (19) is favored at pH = 2.0 and the G-value of e_{aq}^- tends to zero, increasing the radiation-chemical yield of $H^\bullet$; this supports the hypothesis that $H^\bullet$ radicals are not involved in DPA degradation.

Moreover, an increase in the $H^\bullet$ concentration favors the recombination of radicals (reaction (20)) and reduces the concentration of $HO^\bullet$ radicals in the medium, which contributes to decreasing the k value.

$$H_3O^+ + e_{aq}^- \rightarrow H^\bullet + H_2O\ k = 2.25\times10^{10}\ M^{-1}\ s^{-1} \quad (19)$$

$$HO^\bullet + H^\bullet \rightarrow H_2O\ k = 7.00\times10^{9}\ M^{-1}\ s^{-1} \quad (20)$$

Joseph et al. [133] established that the concentration of e_{aq}^- increases between pH = 6.0 and pH = 8.0 by reaction (21), explaining why the highest dose constant (k = 2.74±0.03×10-3 Gy-1) was observed at pH = 7.0.

$$H^\bullet + HO^- \rightarrow H_2O + e_{aq}^-\ k = 2.0\times10^{7}\ M^{-1}\ s^{-1} \quad (21)$$

Finally, the equilibrium in reaction (22) gains in importance at a pH value of 12.0, when $HO^\bullet$ radicals transform into the less reactive $O^{-\bullet}$ species [133], explaining the lower dose constant (k = 1.93±0.01×10-3 Gy-1) at this pH.

$$HO^{\bullet} + OH^{-} \rightleftharpoons O^{-\bullet} + H_2O \; pK_a = 11.9 \tag{22}$$

DPA degradation byproducts were determined at pH = 7 using different irradiation doses, with an initial DPA concentration of 50 mg/L. The results obtained showed that the only stable byproduct found was monohydroxylated diphenolic acid, whose concentration gradually increased with the absorbed dose.

Two sets of experiments were performed to examine the influence of O_2 on DPA degradation: one with no initially dissolved O_2 in the solution (samples were bubbled with N_2), and another with a fully aerated solution. Results obtained are shown in Figure 13. The slightly reduction in DPA degradation observed in the presence of O_2 may be due to the reaction of O_2 with e^{-}_{aq} species, which shows a reaction rate constant, $k = 1.9\times10^{10}\ M^{-1}\ s^{-1}$ [120].

Radiolytic treatment of water polluted with organic products changes the structure of the original compounds, generating byproducts with a lower molecular weight. These can be mineralized to CO_2, producing a corresponding decrease in TOC. The usefulness of radiolysis to treat DPA in water depends not only on the effectiveness of the degradation but also on the mineralization of organic matter present and the reduction in toxicity. Therefore, variations in TOC and toxicity were analyzed as a function of the irradiation dose.

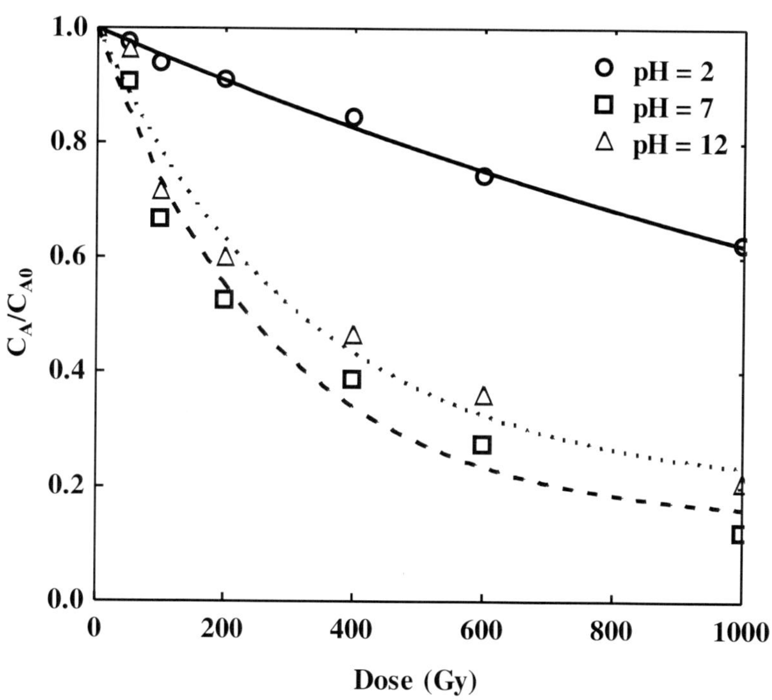

Figure 12. Effect of solution pH on DPA degradation during gamma radiation. Dose rate = 1.66 Gy/min, $[DPA]_0 = 50$ mg/L, and T = 25°C.

Figure 14 a) depicts TOC values in ultrapure water, surface water, and wastewater as a function of the irradiation dose. TOC concentrations decreased at higher doses in all water types.

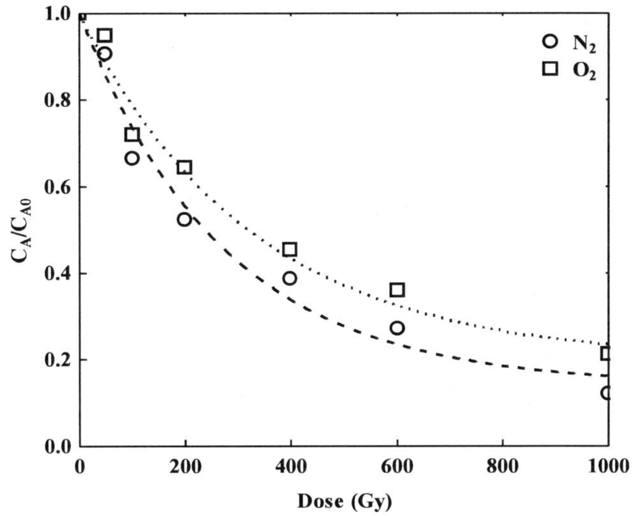

Figure 13. Effect of the presence of O_2 on DPA degradation during gamma radiation. Dose rate = 1.66 Gy/min, $[DPA]_0$ = 50 mg/L, pH = 7.0, and T = 25°C.

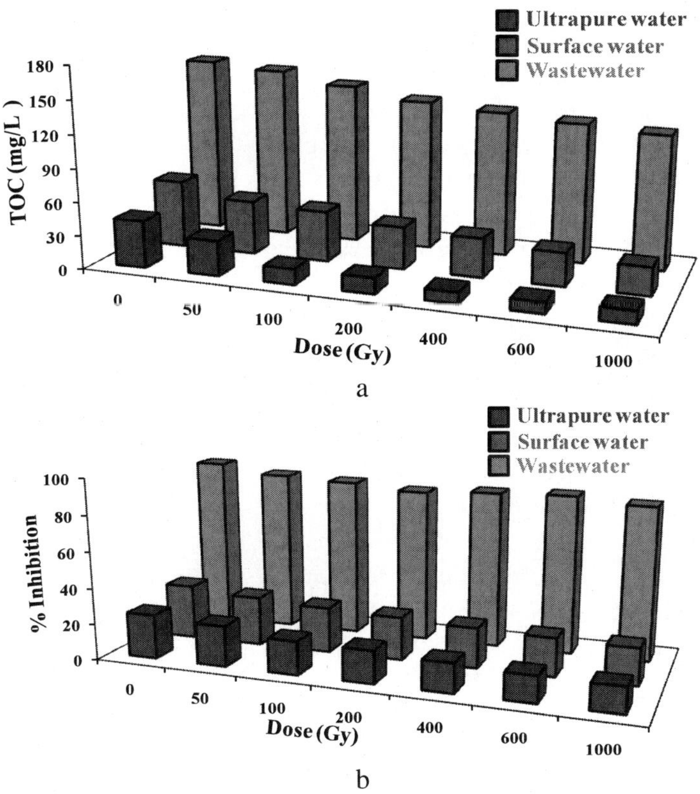

Figure 14. Variation in TOC (a) and inhibition of Vibrio Fischeri (b) in ultrapure water, surface water, and wastewater as a function of the irradiation dose. Dose rate = 1.66 Gy/min, $[DPA]_0$ = 50 mg/L, and T = 25°C.

Thus, at a dose of 1000 Gy, the highest percentage mineralization (73%) was obtained in ultrapure water, followed by surface water (58%), and wastewater (25%), as expected given the latter's high concentration of dissolved organic matter. Figure 14 b) depicts the inhibition percentage of *Vibrio Fischeri* bacteria in ultrapure water, surface water, and wastewater as a function of the irradiation dose. Percentage bacteria inhibition values were lower after than before the treatment; thus, at a dose of 1000 Gy, the toxicity was reduced by around 42%, 32%, and 10% in ultrapure water, surface water, and wastewater, respectively. These results confirm that the radiolytic treatment of DPA is most effective in ultrapure water because of the presence of radical scavengers in natural water samples.

CONCLUSION

The high PA adsorption capacity of the activated carbon samples used in this study is largely attributable to dispersion and electrostatic adsorbent-adsorbate interactions. The PA adsorption process is highly dependent on the solution pH, with an elevated capacity at acidic pH that markedly reduces with increased pH up to pH 6 and then remains practically constant at higher pH values. This behavior is explained by the electrostatic interactions between carbon surface and PA molecules with increased solution pH.

The presence of microorganisms during PA adsorption increases the adsorption capacity of activated carbon by around 15% and slightly enhances the adsorbate-adsorbent relative affinity values (by ~ 3%.) These results can be explained by an increase in carbon surface hydrophobicity produced by adsorption of the bacteria. PA removal varies as a function of the water type, increasing in the order: ground water < surface water ≃ ultrapure water < wastewater. The high removal of PA from wastewater would be due to the formation of organic phthalates and complex species.

UV light has inadequate energy to degrade Bisphenol A, and the quantum yield of this system was close to zero, demonstrating its low effectiveness.

The UV/H_2O_2 system was adequate to degrade BPA, yielding a reaction rate constant of BPA with $HO^{\bullet}$ radicals of $k_{HO^{\bullet}BPA} = 1.70 \pm 0.21 \times 10^{10}$ M^{-1} s^{-1}.

The percentage BPA degradation depended on the solution pH. The photoxidation process of BPA with UV/H_2O_2 was slightly exothermic.

The UV/$K_2S_2O_8$ system was more effective than UV/H_2O_2 to degrade BPA, achieving a higher percentage removal in a shorter time. Moreover, both removal of TOC and reduction of toxicity were the highest when using the UV/$K_2S_2O_8$ system. The reaction rate constant of BPA with $SO_4^{\bullet-}$ radicals was $k_{SO_4^{\bullet-}BPA} = 1.37 \pm 0.15 \times 10^9$ M^{-1} s^{-1}. The solution pH had a major influence on BPA degradation in the UV/$K_2S_2O_8$ system, and the lowest degradation percentage was at pH = 9. The photoxidation of BPA with UV/$K_2S_2O_8$ was endothermic.

The UV/Na_2CO_3 system was less effective to degrade BPA in comparison to the UV/H_2O_2 and UV/$K_2S_2O_8$ systems. The reaction rate constant of BPA with $CO_3^{\bullet-}$/$HCO_3^{\bullet}$ radicals was$k_{CO_3^{\bullet-}/HCO_3^{\bullet}BPA} = 3.89 \pm 0.09 \times 10^6$ M^{-1} s^{-1}. This reaction was exothermic.

BPA oxidation varied as a function of the water type when the UV/H_2O_2, UV/$K_2S_2O_8$, and UV/Na_2CO_3 systems were used, decreasing in order: wastewater > surface water > ultrapure water > ground water. The high degradation of BPA in wastewater may be related to its large content of organic matter, which generates additional hydroxyl radicals from the photolysis of organic matter. BPA degradation was lowest in ground water, due to its high content of metallic species and bicarbonate ions, which act as radical scavengers for both $HO^{\bullet}$ and $SO_4^{\bullet -}$ radicals.

The presence of ozonated activated carbons with a high carboxyl groups content enhances 2,4-D photodegradation by the UV/TiO_2 system. Carboxyl groups in the graphene planes of the activated carbon participate in the additional generation of $HO^{\bullet}$ radicals by interacting with the electrons produced by the UV/TiO_2 system. Consequently, the contribution of $HO^{\bullet}$ radicals to the global 2,4-D degradation process is greater at the beginning of the photocatalytic treatment.

The UV/TiO_2/$W_{O3\text{-}30}$ system mineralized 40 % of the organic matter present in the medium, and the toxicity of the degradation byproducts was considerably lower than that of 2,4-D.

Gamma radiation is an effective treatment for degradation of PA, BPA, DPA, 2,4-D and MCPA in aqueous solutions. The reaction rate constants $k_{HO\bullet}$, k_{e-} and $k_{H\bullet}$ were determined for these compounds by using competition kinetics. The reaction rate constants values obtained are very high and they demonstrated that the degradation process of these micropollutants mainly takes place via oxidation pathway.

The order followed by the reaction rate constants of these pollutants with $HO^{\bullet}$, hydrated electron e_{aq}^-, and hydrogen atom $H^{\bullet}$ can be explained by the stability of the benzylic radicals formed after the attack of these species.

DPA degradation is reduced in the presence of Br^-, Cl^-, CO_3^{2-}, NO_2^-, NO_3^-, or SO_4^{2-}, and this decrease is greater at high concentrations of these anions, largely due to their competition with DPA for the reactive radicals generated, particularly $HO^{\bullet}$

Dose constants are lower in wastewater and surface water than in ultrapure water because they contain organic matter and HCO_3^-, Cl^-, SO_4^{2-}, and NO_3^- ions that react with the radiolytically generated reactive radical species ($HO^{\bullet}$, $H^{\bullet}$ and e_{aq}^-).

REFERENCES

[1] Salem, Z., Hamouri, K., Djemaa, R., Allia, K. Evaluation of landfill leachate pollution and treatment. *Desalination* 2008; 220:108-114.

[2] ground water quality of the greater area. *Environ. Geochem. Hlth.* 1999; 1:175-190.

[3] Primo, O., Rivero, M. J., Ortiz, I. Photo-Fenton process as an efficient alternative to the treatment of landfill leachates. *J. Hazard. Mater.* 2008; 153:834-842.

[4] Islam, J., Singhal, N. A one-dimensional reactive multi-component landfill leachate transport model. *Environ. Modell. Soft* 2002; 17:531-543.

[5] Hernández, A. J., Adarve, M. J., Gil, A., Pastor, J. Soil salivation from landfill leachates: Effects on the macronutrient content and plant growth of four grassland species. *Chemosphere* 1999; 38:1693-1711.

[6] Mangimbulude, J. C., Breukelen, B. Mv., Krave, A. S., Straalen, N. Mv., Röling, W. F. M. Seasonal dynamics in leachate hydrochemistry and natural attenuation in surface run-off water from a tropical landfill. *Waste Manage.* 2009; 29:829-838.

[7] North, J. C., Frew, R. D., Hale, R. V. Can stable isotopes be used to monitor landfill leachate impact on surface waters?. *J. Geochem. Explor*. 2006; 88:49-53.
[8] Mikac, N., Cosovic, B., Ahel, M., Andreis, S., Toncic, Z. Assessment of groundwater contamination in the vicinity of a municipal solid waste landfill (Zagreb, Croatia). *Water Sci. Technol*. 1998; 37:37-44.
[9] Reyes-López, J. A., Ramírez-Hernández, J., Lázaro-Mancilla, O., Carreón-Diazconti, C., Garrido, M. M. Assessment of groundwater contamination by landfill leachate: A case in México. *Waste Manage*. 2008; 28:S33-S39.
[10] Clément, B., Merlin, G. The contribution of ammonia and alkalinity to landfill leachate toxicity to duckweed. *Sci. Total Environ*. 1995; 170:71-79.
[11] Silva, A. C., Dezotti, M., Sant'Anna, G. L. Treatment and detoxification of a sanitary landfill leachate. *Chemosphere* 2004; 55:207-214.
[12] Pivato, A., Raga, R. Tests for the evaluation of ammonium attenuation in MSW landfill leachate by adsorption into bentonite in a landfill liner. *Waste Manage*. 2006; 26:123-132.
[13] Nascimento Filho, Id., von Mühlen, C., Schossler, P., Bastos Caramão, E. Identification of some plasticizers compounds in landfill leachate. *Chemosphere* 2003; 50:657-663.
[14] Marttinen, S. K., Kettunen, R. H., Rintala, J. A. Occurrence and removal of organic pollutants in sewages and landfill leachates. *Sci. Total Environ*. 2003; 301:1-12.
[15] Slack, R. J., Gronow, J. R., Voulvoulis, N. Household hazardous waste in municipal landfills: contaminants in leachate. *Sci. Total Environ*. 2005; 337:119-137.
[16] Öman, C., Hynning, P. Identification of organic compounds in municipal landfill leachates. *Environ. Pollut*. 1993; 80:265-271.
[17] Smith, E. H., Weber Jr., W. J. Comparative assessment of the chemical and adsorptive characteristics of leachates from a municipal and an industrial landfill. *Water Air Soil Pollut*. 1990; 53:279-295.
[18] Laganá, A., Bacaloni, A., De leva, I., Faberi, A., Fago, G., Marino, A. Analytical methodologies for determining the occurrence of endocrine distrusting chemicals in sewage treatment plant and natural waters. *Anal. Chem. Acta* 2004; 501:79-88.
[19] Gültekin, I., Ince, N. H. Synthetic endocrine disruptors in the environment and water remediation by advanced oxidation processes. *J. Environ. Manage*. 2007; 85:816-832.
[20] Tanaka, S., Nakata, Y., Kimura, T., Yustiawati, Kawasaki, M., Kuramitz, H. Electrochemical decomposition of bisphenol A using Pt/Ti and SnO_2/Ti anodes. *J. Appl. Electrochem*. 2001; 32:197-201.
[21] Orton, F., Lutz, I., Kloas, W., Routledge, E. J. Endocrine Disrupting Effects of Herbicides and Pentachlorophenol: In Vitro and in Vivo Evidence. *Environ. Sci. Tech*. 2009; 43:2144-2150.
[22] Han, D., Jia, W., Liang, H. Selective removal of 2,4-dichlorophenoxyacetic acid from water by molecularly-imprinted amino-functionalized silica gel sorbent. *J. Environ. Sci*. 2010; 22:237-241.
[23] Aksu, Z., Kabasakal, E. Batch adsorption of 2,4-dichlorophenoxy-acetic acid (2,4-D) from aqueous solution by granular activated carbon. *Sep. Purif. Technol*. 2004; 35:223-240.
[24] Peixoto, F. P., Lopes, M. L., Madeira, V. M. C., Vicente, J. A. F. Toxicity of MCPA on non-green potato tuber calli. *Acta physiol. Plant* 2009; 31:103-109.

[25] Wu, D., Hu, B., Zheng, P., Qaisar, M. Anoxic biodegradation of dimethyl phthalate (DMP) by activated sludge cultures under nitrate-reducing conditions. *J. Environ. Sci.* 2007; 19:1252-1256.

[26] Jianlong, W., Lujun, C., Hanchang, S., Yi, Q. Microbial degradation of phthalic acid esters under anaerobic digestion of sludge. *Chemosphere* 2000; 41:1245-1248.

[27] Kozumbo, W. J., Kroll, R., Rubin, R. J. Assessment of the mutagenicity of phthalate esters. *Environ. Health Perspect.* 1982; Vol. 45:103-109.

[28] Huff, J. E., Kluwe, W. M. Phthalate esters carcinogenicity in F344/N rats and B6C3F1 mice. *Prog. Clin. Biol. Res.* 1984; 141:137-154.

[29] Zheng, Z., Zhang, H., He, P.-, Shao, L.-, Chen, Y., Pang, L. Co-removal of phthalic acid esters with dissolved organic matter from landfill leachate by coagulation and flocculation process Ref 42. *Chemosphere* 2009; 75:180-186.

[30] Bauer, M. J., Herrmann, R., Martin, A., Zellmann, H. Chemodynamics, transport behaviour and treatment of phthalic acid esters in municipal landfill leachates. *Water Sci. Technol.* 1998; 38:185-192.

[31] Mailhot, G., Sarakha, M., Lavedrine, B., Cáceres, J., Malato, S. Fe(III)-solar light induced degradation of diethyl phthalate (DEP) in aqueous solutions. *Chemosphere* 2002; 49:525-532.

[32] Zhao, R., Wang, X., Yuan, J., Lin, J. Investigation of feasibility of bamboo charcoal as solid-phase extraction adsorbent for the enrichment and determination of four phthalate esters in environmental water samples. *J. Chromatogr. A* 2008; 1183:15-20.

[33] Yuan, B.-, Li, X.-, Graham, N. Aqueous oxidation of dimethyl phthalate in a Fe(VI)-TiO_2-UV reaction system. *Water Res.* 2008; 42:1413-1420.

[34] CSTEE, *Proceedings of the Third Plenary Meeting. Phthalate migration from soft PVC toys and children care articles*, Brussels, 1998.

[35] *OJEC, Official Journal of the European Communities*. L 315/46, 9 December 1999.

[36] Bette, H.California Bans Phthalates In: *Toys For Children. Chemical and Engineering News*, 2007, P. 12. http://pubs.acs.org/cen/news/85/i43/8543news4.html.

[37] Yamamoto, T., Yasuhara, A., Shiraishi, H., Nakasugi, O. Bisphenol A in hazardous waste landfill leachates. *Chemosphere* 2001; 42:415-418.

[38] Liu, G., Ma, J., Li, X., Qin, Q. Adsorption of bisphenol A from aqueous solution onto activated carbons with different modification treatments. *J. Hazard. Mater.* 2009; 164: 1275-1280.

[39] MHLWJ, Ministry of health, labour and welfare of Japan. *Exposure and behavior researches of endocrine disrupting chemicals in tap water, 2000.*

[40] Nempo, K., Chosa-Kai, T. *Ministry of international trade and industry of Japan,* 1997, Japan.

[41] Jiao, F. R., Sun, X. J., Pang, Z. T. Production and market analysis of Bisphenol A. *Chem. Ind.* 2008; 26:21-33.

[42] NIOHS, National Institute for Occupational safty and health., *Registry of toxic effects of chemical substances, 1998.*

[43] Zhang, R., Moore, J. A. Synthesis, Characterization and Properties of Polycarbonate Containing Carboxyl Side Groups. *Macromol. Sympos.* 2003; 199:375-390.

[44] Fischer, P., Hartraxft, G. R. Diphenolic acid ester polycarbonates, *J. Appl. Polym. Sci.* 1966; 10:245-252.

[45] Zhang, R. F., Moore, J. A. Synthesis, characterization and properties of polycarbonate containing diphenolic acid. *Polym. Prepr.* 2002; 43:1007-1008.

[46] Cerbai, B., Solaro, R., Chiellini, E. Synthesis and characterization of functional polyesters tailored for biomedical applications. *J. Polym. Sci. Part A* 2008; 46:2459-2476.

[47] Ping, Z., Linbo, W., Bo-Geng, L. Thermal stability of aromatic polyesters prepared from diphenolic acid and its esters. *Polym. Degrad. Stab.* 2009; 94:1261-1266.

[48] Benli, A. Ç. K., Sarıkaya, R., Sepici-Dincel, A., Selvi, M., Şahin, D., Erkoç, F. Investigation of acute toxicity of (2,4-dichlorophenoxy)acetic acid (2,4-D) herbicide on crayfish (Astacus leptodactylus Esch. 1823). *Pestic. Biochem. Physiol.* 2007; 88:296-299.

[49] Gold, A. J., Morton, T. G., Sullivan, W. M., Mcclory, J. Leaching of 2,4-D and dicamba from home lawns. *Water Air Soil Pollut.* 1988; 37:121-129.

[50] Sun, Y., Pignatello, J. Evidence for a surface dual hole-radical mechanism in the TiO_2 photocatalytic oxidation of 2,4-Dichlorophenoxyacetic acid. *Environ. Sci. Technol.* 1995; 29:2065-2072.

[51] Imai, A., Onuma, K., Inamori, Y., Sudo, R. Effects of pre-ozonation in refractory leachate treatment by the biological activated carbon fluidized bed process. *Environ. Technol.* 1998; 19:213-221.

[52] Kargi, F., Yunus Pamukoglu, M. Simultaneous adsorption and biological treatment of pre-treated landfill leachate by fed-batch operation. *Process Biochem.* 2003; 38:1413-1420.

[53] Venkata Mohan, S., Shailaja, S., Rama Krishna, M., Sarma, P. N. Adsorptive removal of phthalate ester (Di-ethyl phthalate) from aqueous phase by activated carbon: A kinetic study Ref no 7. *J. Hazard. Mater.* 2007; 146:278-282.

[54] Zhang, C., Wang, Y. Removal of dissolved organic matter and phthalic acid esters from landfill leachate through a complexation–flocculation process. *Waste Manage.* 2009; 29:110-116.

[55] Bautista-Toledo, I., Ferro-García, M. A., Rivera-Utrilla, J., Moreno-Castilla, C., Fernández, F. J. V. Bisphenol A removal from water by activated carbon. Effects of carbon characteristics and solution chemistry. *Environ. Sci. Technol.* 2005; 39:6246-6250.

[56] Gattullo, C. E., Bährs, H., Steinberg, C. E. W., Loffredo, E. Removal of bisphenol A by the freshwater green alga Monoraphidium braunii and the role of natural organic matter. *Sci. Total Environ.* 2012; 416:501-506.

[57] Li, G., Zu, L., Wong, P., Hui, X., Lu, Y., Xiong, J., et al. Biodegradation and detoxification of bisphenol A with one newly-isolated strain Bacillus sp. GZB: Kinetics, mechanism and estrogenic transition. *Bioresour. Technol.* 2012; 114:224-230.

[58] Rosenfeldt, E. J., Linden, K. G. Degradation of endocrine disrupting chemicals bisphenol A, ethinyl estradiol, and estradiol during UV photolysis and advanced oxidation processes. *Environ. Sci. Technol.* 2004; 38:5476-5483.

[59] Rivas, F. J., Encinas, Á., Acedo, B., Beltrán, F. J. Mineralization of bisphenol A by advanced oxidation processes. *J. Chem. Technol. Biotechnol.* 2009; 84:589-594.

[60] Katsumata, H., Kawabe, S., Kaneco, S., Suzuki, T., Ohta, K. Degradation of bisphenol A in water by the photo-Fenton reaction. *J. Photochem. Photobiol. A* 2004; 162:297-305.

[61] Wu, C., Linden, K. G. Phototransformation of selected organophosphorus pesticides: Roles of hydroxyl and carbonate radicals. *Water Res.* 2010; 44:3585-3594.

[62] Canonica, S., Kohn, T., Mac, M., Real, F. J., Wirz, J., Von Gunten, U. Photosensitizer method to determine rate constants for the reaction of carbonate radical with organic compounds. *Environ. Sci. Technol.* 2005; 39:9182-9188.

[63] Ocampo-Pérez, R., Sánchez-Polo, M., Rivera-Utrilla, J., Leyva-Ramos, R. Degradation of antineoplastic cytarabine in aqueous phase by advanced oxidation processes based on ultraviolet radiation. *Chem. Eng. J.* 2010; 165:581-588.

[64] Huang, J., Mabury, S. A. A new method for measuring carbonate radical reactivity toward pesticides. *Environ. Toxicol. Chem.* 2000; 19:1501-1507.

[65] Cui, Y., Li, X., Chen, G. Electrochemical degradation of bisphenol A on different anodes. *Water Res.* 2009; 43:1968-1976.

[66] Guo, Z., Dong, Q., He, D., Zhang, C. Gamma radiation for treatment of bisphenol A solution in presence of different additives. *Chem. Eng. J.* 2012; 183:10-14.

[67] Ju, P., Fan, H., Guo, D., Meng, X., Xu, M., Ai, S. Electrocatalytic degradation of bisphenol A in water on a Ti-based PbO_2–ionic liquids (ILs) electrode. *Chem. Eng. J.* 2012; 179:99-106.

[68] Wang, R., Ren, D., Xia, S., Zhang, Y., Zhao, J. Photocatalytic degradation of Bisphenol A (BPA) using immobilized TiO2 and UV illumination in a horizontal circulating bed photocatalytic reactor (HCBPR). *J. Hazard. Mater.* 2009; 169:926-932.

[69] Zhang, K., Gao, N., Deng, Y., Lin, T. F., Ma, Y., Li, L., et al. Degradation of bisphenol-A using ultrasonic irradiation assisted by low-concentration hydrogen peroxide. *J. Environ. Sci.* 2011; 23:31-36.

[70] Guo, L., Wang, B., Huang, W., Wu, F., Li, Y. Adsorption and photodecomposition of diphenolic acid α- or β- cyclodextrins and TiO_2 suspensions. *S O P O*, 2009a, 1-4.

[71] Guo, L., Liu, Y., Wang, B., Wu, F., Li, Y. Photochemical behavior of diphenolic acid in β-Cyclodextrin and TiO_2 suspensions. *I C B B E*, 2009b, 1-4.

[72] Guo, L., Wang, B., Huang, W., Wu, F., Huang, J. Photocatalytic degradation of diphenolic acid in the Presence of beta-Cyclodextrin under UV Light. *E S I A T*, 2009c, 322-324.

[73] Kim, S. K., Chang, H., Cho, K., Kil, D. S., Cho, S. W., Jang, H. D., et al. Enhanced photocatalytic property of nanoporous TiO2/SiO2 micro-particles prepared by aerosol assisted co-assembly of nanoparticles. *Mater. Lett.* 2011; 65:3330-3332.

[74] Gao, J., Luan, X., Wang, J., Wang, B., Li, K., Li, Y., et al. Preparation of Er3+:YAlO3/Fe-doped TiO2–ZnO and its application in photocatalytic degradation of dyes under solar light irradiation. *Desalination* 2011; 268:68-75.

[75] Ashkarran, A. A., Afshar, S. A. A., Aghigh, S. M., Kavianipour, M. Photocatalytic activity of ZrO2 nanoparticles prepared by electrical arc discharge method in water. *Polyhedron* 2010; 29:1370-1374.

[76] Ohno, T., Tanigawa, F., Fujihara, K., Izumi, S., Matsumura, M. Photocatalytic oxidation of water on TiO2-coated WO3 particles by visible light using Iron(III) ions as electron acceptor. *J. Photochem. Photobiol. A* 1998; 118:41-44.

[77] Faisal, M., Khan, S. B., Rahman, M. M., Jamal, A., Akhtar, K., Abdullah, M. M. Role of ZnO-CeO2 Nanostructures as a Photo-catalyst and Chemi-sensor. *J. Mat. Sci. Technol.* 2011; 27:594-600.

[78] Antoniadou, M., Daskalaki, V. M., Balis, N., Kondarides, D. I., Kordulis, C., Lianos, P. Photocatalysis and photoelectrocatalysis using (CdS-ZnS)/TiO_2 combined photocatalysts. *Appl. Cat. B: Environ.* 2011; 107:188-196.
[79] Ocampo-Pérez, R., Sánchez-Polo, M., Rivera-Utrilla, J., Leyva-Ramos, R. Enhancement of the catalytic activity of TiO2 by using activated carbon in the photocatalytic degradation of cytarabine. *Appl. Cat. B: Environ.* 2011; 104:177-184.
[80] Sellappan, R., Zhu, J., Fredriksson, H., Martins, R. S., Zäch, M., Chakarov, D. Preparation and characterization of TiO2/carbon composite thin films with enhanced photocatalytic activity. *J. Mol. Cat. A: Chemical.* 2011; 335:136-144.
[81] Silva, C. G., Faria, J. L. Photocatalytic oxidation of benzene derivatives in aqueous suspensions: Synergic effect induced by the introduction of carbon nanotubes in a TiO2 matrix. *Appl. Cat. B: Environ.* 2010; 101:81-89.
[82] Maldonado-Hódar, F. J., Moreno-Castilla, C., Rivera-Utrilla, J. Synthesis, pore texture and surface acid–base character of TiO2/carbon composite xerogels and aerogels and their carbonized derivatives. *Appl. Cat. A: General.* 2000; 203:151-159.
[83] Wang, X., Liu, Y., Hu, Z., Chen, Y., Liu, W., Zhao, G. Degradation of methyl orange by composite photocatalysts nano-TiO2 immobilized on activated carbons of different porosities. *J. Hazard. Mater.* 2009; 169:1061-1067.
[84] Xue, G., Liu, H., Chen, Q., Hills, C., Tyrer, M., Innocent, F. Synergy between surface adsorption and photocatalysis during degradation of humic acid on TiO2/activated carbon composites. *J. Hazard. Mater.* 2011; 186:765-772.
[85] Wang, W., Silva, C. G., Faria, J. L. Photocatalytic degradation of Chromotrope 2R using nanocrystalline TiO2/activated-carbon composite catalysts. *Appl. Catal. B Environ.* 2007; 70:470-478.
[86] Li, Y., Li, X., Li, J., Yin, J. Photocatalytic degradation of methyl orange in a sparged tube reactor with TiO2-coated activated carbon composites. *Cat. Comm.* 2005; 6:650-655.
[87] Akpan, U. G., Hameed, B. H. Photocatalytic degradation of 2,4-dichlorophenoxyacetic acid by Ca–Ce–W–TiO_2 composite photocatalyst. *Chem. Eng. J.* 2011; 173:369-375.
[88] Matos, J., Laine, J., Herrmann, J.-. Effect of the Type of Activated Carbons on the Photocatalytic Degradation of Aqueous Organic Pollutants by UV-Irradiated Titania. *J. Cat.* 2001; 200:10-20.
[89] Zertal, A., Sehili, T., Boule, P. Photochemical behaviour of 4-chloro-2-methylphenoxyacetic acid: Influence of pH and irradiation wavelength. *J. Photochem. Photobiol. A* 2001; 146:37-48.
[90] Spinks, R. J. *Wood, Introduction to radiation chemistry*, Willey, New York, 1990.
[91] Zhao, R., Bao, H., Xia, L.?-Irradiation Degradation of Methamidophos. *Chinese J. Chem.* 2009; 27:1749-1754.
[92] Mohamed, K. A., Basfar, A. A., Al-Shahrani, A. A. Gamma-ray induced degradation of diazinon and atrazine in natural groundwaters. *J. Hazard. Mater.* 2009; 166:810-814.
[93] Bao, H., Gao, J., Liu, Y., Su, Y. A study of biodegradation/γ-irradiation on the degradation of p-chloronitrobenzene. *Radiat. Phys. Chem.* 2009; 78:1137-1139.
[94] Bojanowska-Czajka, A., Drzewicz, P., Kozyra, C., Nałęcz-Jawecki, G., Sawicki, J., Szostek, B., et al. Radiolytic degradation of herbicide 4-chloro-2-methyl phenoxyacetic acid (MCPA) by γ-radiation for environmental protection. *Ecotoxicol. Environ. Saf.* 2006; 65:265-277.

[95] Ocampo-Pérez, R., Rivera-Utrilla, J., Sánchez-Polo, M., López-Peñalver, J. J., Leyva-Ramos, R. Degradation of antineoplastic cytarabine in aqueous solution by gamma radiation. *Chem. Eng. J.* 2011; 174:1-8.
[96] Sánchez-Polo, M., López-Peñalver, J., Prados-Joya, G., Ferro-García, M. A., Rivera-Utrilla, J. Gamma irradiation of pharmaceutical compounds, nitroimidazoles, as a new alternative for water treatment. *Water Res.* 2009; 43:4028-4036.
[97] Razavi, B., Song, W., Cooper, W. J., Greaves, J., Jeong, J. Free-radical-induced oxidative and reductive degradation of fibrate pharmaceuticals: Kinetic studies and degradation mechanisms, *J. Phys. Chem. A*, 2009; 113:1287-1294.
[98] Hameed, B. H., Salman, J. M., Ahmad, A. L. Adsorption isotherm and kinetic modeling of 2,4-D pesticide on activated carbon derived from date stones. *J. Hazard. Mater.* 2009; 163:121-126.
[99] Kundu, S., Pal, A., Dikshit, A. K. UV induced degradation of herbicide 2,4-D: kinetics, mechanism and effect of various conditions on the degradation. *Sep. Purif. Technol.* 2005; 44:121-129.
[100] Garcia-Segura, S., Almeida, L. C., Bocchi, N., Brillas, E. Solar photoelectro-Fenton degradation of the herbicide 4-chloro-2-methylphenoxyacetic acid optimized by response surface methodology. *J. Hazard. Mater.* 2011; 194:109-118.
[101] Tchaikovskaya, O. N., Karetnikova, E. A., Sokolova, I. V., Mayer, G. V., Shvornev, D. A. The phototransformation of 4-chloro-2-methylphenoxyacetic acid under KrCl and XeBr excilamps irradiation in water. *J. Photochem. Photobiol. A* 2012; 228:8-14.
[102] Méndez-Díaz, J. D., Abdel daiem, M. M., Rivera-Utrilla, J., Sánchez-Polo, M., Bautista-Toledo, I. Adsorption/bioadsorption of phthalic acid, an organic micropollutant present in landfill leachates, on activated carbons. *J. Colloid Interf. Sci.* 2012; 369: 358-365.
[103] Abdel daiem, M. M., Rivera-Utrilla, J., Ocampo-Pérez, R., Méndez-Díaz, J. D., Sánchez-Polo, M. Environmental impact of phthalate acid esters and their removal from water and sediments by different technologies - A review. *J. Environ. Manag.* 2012; 109: 164-178 (2012).
[104] Ocampo-Pérez, R., Abdel daiem, M. M., Rivera-Utrilla, J., Méndez-Díaz, J. D., Sánchez-Polo, M. Modeling adsorption rate of organic micropollutants present in landfill leachates onto granular activated carbon. *J. Colloid Interf. Sci.* 2012; 385: 174-182.
[105] Rivera-Utrilla, J., Sánchez-Polo, M., Abdel daiem, M. M., Ocampo-Pérez, R. Role of activated carbón in the photocatalytic degradation of 2,4-dichlorophenoxyacetic acid by the UV/TiO_2/activated carbon system. *Appl. Cat. B: Environ.* 2012; 126: 100-107.
[106] Abdel daiem, M. M., Rivera-Utrilla, J., Ocampo-Pérez, R., Sánchez-Polo, M., López-Peñalver, J. J. Treatment of water contaminated with diphenolic acid by gamma radiation in the presence of different compounds. *Chem. Eng. J.* 2013; 219: 371-379.
[107] Sánchez-Polo, M., Abdel daiem, M. M., Ocampo-Pérez, R., Rivera-Utrilla, J., Mota, A. J. Comparative study of the photodegradation of bisphenol A by $HO^{\bullet}$, $SO_4^{\bullet-}$ and $CO_3^{\bullet-}$/$HCO_3^{\bullet}$ radicals in aqueous phase. *Sci. Total Environ.* 2013; 463-464: 423-431 (2013).
[108] Giles, C. H., Smith, D., Huitson, A. A general treatment and classification of the solute adsorption isotherm. I. Theoretical. *J. Colloid Interface Sci.* 1974; 47:755-765.

[109] Giles, C. H., D'Silva, A. P., Easton, I. A. A general treatment and classification of the solute adsorption isotherm part. II. Experimental interpretation. *J. Colloid Interface Sci.* 1974; 47:766-778.

[110] Radovic, L. R., Moreno-Castilla, C., Rivera-Utrilla, J. Carbon materials as adsorbents in aqueous solutions. *Chem. Phys. Carbon* 2001; 27:227-405.

[111] Xu, R., Wang, Y., Tiwari, D., Wang, H. Effect of ionic strength on adsorption of As(III) and As(V) on variable charge soils. *J. Environ. Sci.* 2009; 21:927-932.

[112] Coughlin, R. W., Ezra, F. S. Role of surface acidity in the adsorption of organic pollutants on the surface of carbon. *Environ. Sci. Technol.* 1968; 2:291-297.

[113] Lyklema, J. *Solid-liquid interfaces: Fundamentals of interface and colloid science*, Academic press, New York, 1995.

[114] Rivera-Utrilla, J., Bautista-Toledo, I., Ferro-García, M. A., Moreno-Castilla, C. Bioadsorption of Pb(II), Cd(II), and Cr(VI) on activated carbon from aqueous solutions. *Carbon* 2003; 41:323-330.

[115] Rivera-Utrilla, J., Bautista-Toledo, I., Ferro-Garca, M. A., Moreno-Castilla, C. Activated carbon surface modifications by adsorption of bacteria and their effect on aqueous lead adsorption. *J. Chem. Technol. Biotechnol.* 2001; 76:1209-1215.

[116] Beveridge, T. J., Doyle, R. J. *Metal Ions and Bacteria*, John Wiley and Sons, New York, 1989.

[117] Wilcox, D. P., Chang, E., Dickson, K. L., Johansson, K. R. Microbial growth associated with granular activated carbon in a pilot water treatment facility. *Appl. Environ. Microbiol.* 1983; 46:406-416.

[118] Hughes, M. N., Poole, R. K. Metal speciation and microbial growth - The hard (and soft) facts. *J. Gen. Microbiol.* 1991; 137:725-734.

[119] Mikaido, H., Vaara, H. Cellular and Molecular Biology, American Society for Microbiology, In: F. C. Neidhardt (Ed.), Washington DC, 1987, p. 3.

[120] Buxton, B. R., Greenstock, C. L., Helman, W. P., Ross, A. B. Critical review of rate constants for reactions hydrated electrons, hydrogen atoms and hydroxyl radicals ($^{\bullet}OH/^{\bullet}O$-) in aqueous solution. *J. Phys. Chem. Ref. Data*, 1988; 17:513-886.

[121] Huie, R. E., Clifton, C. L. Temperature dependence of the rate constants for reactions of the sulfate radical, $SO_4^{-\bullet}$, with anions. *J. Phys. Chem.* 1990; 94:8561-8567.

[122] Haygarth, K. S., Marin, T. W., Janik, I., Kanjana, K., Stanisky, C. M., Bartels, D. M. Carbonate radical formation in radiolysis of sodium carbonate and bicarbonate solutions up to 250 °C and the mechanism of its second order decay. *J. Phys. Chem. A* 2010; 114:2142-2150.

[123] Wu, C., Linden, K. G. Phototransformation of selected organophosphorus pesticides: Roles of hydroxyl and carbonate radicals. *Water Res.* 2010; 44:3585-3594.

[124] Sánchez-Polo, M., Rivera-Utrilla, J. Effect of the ozone–carbon reaction on the catalytic activity of activated carbon during the degradation of 1,3,6-naphthalenetrisulphonic acid with ozone. *Carbon* 2003; 41:303-307.

[125] Valdés, H., Sánchez-Polo, M., Rivera-Utrilla, J., Zaror, C. A. Effect of ozone treatment on surface properties of activated carbon. *Langmuir* 2002; 18:2111-2116.

[126] Zona, R., Solar, S., Sehested, K. OH-radical induced degradation of 2,4,5-trichlorophenoxyacetic acid (2,4,5-T) and 4-chloro-2-methylphenoxyacetic acid (MCPA): A pulse radiolysis and gamma-radiolysis study. *Radiat. Phys. Chem.* 2012; 81:152-159.

[127] Wu, M., Liu, N., Xu, G., Ma, J., Tang, L., Wang, L., et al. Kinetics and mechanisms studies on dimethyl phthalate degradation in aqueous solutions by pulse radiolysis and electron beam radiolysis. *Radiat. Phys. Chem.* 2011; 80:420-425.

[128] Haag, W. R., Yao, C. C. D. Rate Constants for Reaction of Hydroxyl Radicals with Several Drinking Water Contaminants, *Environ. Sci. Technol.*, 1992; 26:1005-1013.

[129] Mabury, S. A., Crosby, D. G. Pesticide Reactivity toward Hydroxyl and Its Relationship to Field Persistence, *J. Agric. Food Chem.*, 1996; 44:1920-1924.

[130] Peller, J., Kamat, P. V. Radiolytic transformations of chlorinated phenols and chlorinated phenoxyacetic acids, *J. Phys. Chem. A,* 2005; 109:9528-9535.

[131] Szutka, A., Thomas, J. K., Gordon, S., Hart, E. J. Rate Constants of hydrated electron reactions with some aromatic acid, alkyl halides, heterocyclic compounds, and wener complexes, *J. Phys. Chem.*, 1965; 69:289-292.

[132] Ershov, B. G., Gordeev, A. V. A model for radiolysis of water and aqueous solutions of H_2, H_2O_2 and O_2, *Radiat. Phys. Chem.*, 2008; 77:928-935.

[133] Joseph, J. M., Choi, B. S., Yakabuskie, P., Wren, J. C. A combined experimental and model analysis on the effect of pH and O_2(aq) on γ-radiolytically produced H_2 and H_2O_2, *Radiat. Phys. Chem.*, 2008; 77:1009-1020.

In: Landfills and Recycling Centers
Editor: Carla H. Jackson

ISBN: 978-1-63463-792-3

Chapter 5

PHYTOCAPPING OF MUNICIPAL LANDFILLS

***Kartik Venkatraman*[1*], *Nanjappa Ashwath*[1†] *and Ninghu Su*[2‡]**
[1]East Gippsland Shire Council, Victoria, Australia
School of Medical and Applied Science, CQUniversity, Rockhampton, Australia
[2]School of Earth and Environmental Sciences, James Cook University,
Cairns, Australia

ABSTRACT

Landfill capping is a part of the post closure management protocol to minimise environmental problems such as leachate generation from landfills. The costs of conventional caps (e.g. clay caps, HDPE) and remediation of landfills have escalated over the last decade due to increase in the cost of capping material and with the introduction of stringent environmental laws. Research is therefore required to identify alternative capping systems that will not only minimise the costs, but also demonstrate increased environmental performance and community benefits. An alternative landfill capping technique known as 'Phytocapping' (establishment of plants on a layer of soil placed over the waste) was trialled at Lakes Creek Landfill, Rockhampton, Australia. This particular system checks on the seepage as well as allows preferential amount of water to promote biodegradation.

In this technique, trees were used as 'bio-pumps" and soil cover as "water storage". The trees and soil together help minimise percolation of water into the landfill. Phytocaps also know as Evapotranspiration (ET) covers have been very successful in the US. This study is the first of its kind in Australia to evaluate the performance of 21 different tree species over two soil thicknesses (1400 mm and 700 mm). This paper reviews the critical points of the current knowledge and study conducted on phytocapping till date. The paper also reports the progress done on phytocapping till date; identifies the missing links and recommends further research that needs to be done to make this system more sustainable.

Keywords: Phytocapping, landfill, native species, leachate, methane emission, transpiration, canopy rainfall interception, site water balance, literature review, Australia

---

* kartikv@egipps.vic.gov.au; waste1977@gmail.com.
† n.ashwath@cqu.edu.au.
‡ Ninghu.su@jcu.edu.au.

INTRODUCTION

Modern living inevitably results in waste generation. This waste needs to be managed appropriately, as it causes environmental problems. Local Government "Councils" in Australia strive hard to minimise these problems through various technologies, predominantly landfilling of Municipal Solid Waste (MSW). The term "Landfill" is used to describe a physical facility that contains solid wastes and waste residue on the earth's surface (Tchobanoglous and O'Leary 1994). Councils use landfills, as land disposal of waste has been found to be the most economic means of dealing with MSW (Scott et al. 2005, Izzo et al. 2009, and Tonini et al. 2009).

There are two types of MSW landfills: - (i) open dumps, which are covered with a layer of soil (Themelis and Ulloa 2007); and (ii) engineered or regulated landfills that have liners, gas and leachate collection systems and clay caps at the top (Tchobanoglous and O'Leary 1994, Themelis and Ulloa 2007). Both types of landfills currently exist in Australia.

THE PREDICAMENT

Australia is one amongst the ten highest solid waste generators in the Organisation for Economic Cooperation and Development (OECD) (Scott et al. 2005, Weikhardt 2006), as 50% to 60% of the total MSW generated is placed in landfills (Weikhardt 2006, EPA 2007). In Queensland alone, around 1.7 million tonnes of domestic waste was deposited in landfills in 2006, of which 60,000 tonnes of waste was generated in Central Queensland (CQ) alone (EPA 2006). Due to increasing quantities of wastes being produced, landfill owners and operators have great concerns about environmental problems such as gas emissions and leachate generation from these landfills.

In Australia, more than 70% of the population live in coastal areas (EPA 2006). Thus construction of landfills has predominantly occurred in low lying areas and closer to urban developments. This inadvertent practice of siting landfills in the past combined with no or minimum environmental considerations has introduced numerous environmental problems such as groundwater contamination and methane generation which contributes towards global warming and affects urban populations that dwell around landfills (CSIRO 2001, EPA 2002).

Municipal solid waste landfills predominantly contain putrescible wastes which produce leachate and methane gas (CSIRO 2001) when the wastese comes in contact with water. "Leachate is the medium by which soluble materials contained in the landfill are transported to the environment" (Christensen et al. 1994). Leachate generation occurs due to water entering into the landfill. Rain is the major source of water which moves through the landfill system through condensation, evaporation, precipitation, runoff, infiltration and transpiration (Figure 1).

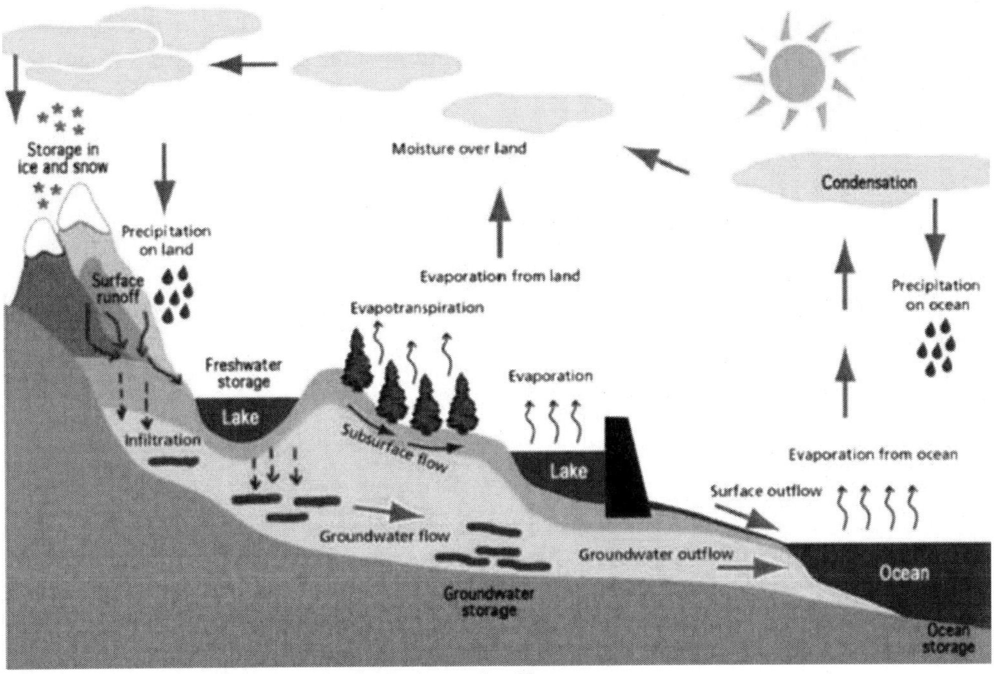

(Source: http://dardel.info/images/hydrologic_cycle.gif)

Figure 1. Hydrological Cycle.

Today several policies and regulations are in place to reduce leachate generation and release of landfill gases into the environment. Waste management practices such as landfill capping, lining, waste segregation, leachate recycling and gas recovery systems have been employed in Australia to reduce adverse environmental impacts of landfills on the surrounding environment (Scott et al. 2005). A report by the Environmental Protection Agency (EPA) reveals that only 2% of the landfills in Queensland have gas collection systems and only 28% have leachate collection systems (EPA 2006). In CQ, only 12% of the landfills have leachate collection systems, but none of the landfills have been installed with a gas collection system (EPA 2006).

STATE OF LANDFILLS IN AUSTRALIA

Landfilling practice in Australia has undergone significant changes over the last few decades (Scott et al. 2005). In the early 19th century, wastes were collected and disposed into open dumps. Concerns over public health and other environmental issues arising due to open dumping led to the introduction of sanitary/engineered landfills by the 20th century in Australia (Scott et al. 2005), and US and Canada (US EPA 2002). This also contributed to significant changes to requirements of siting, management and operation of landfills (Scott et al. 2005).

Many of the MSW landfills that were built in Australia prior to 1993 have liner systems (Friends of the Earth 2000) and have become major sources of local groundwater contamination (Janechek and Prosser 1995, Clister et al. 1997) and gas emission (Bogner et al.

1995). The Resource Conservation and Recovery Act (RCRA) instituted in 1976 in the US made it mandatory to cap the landfills as closure remediation to prevent water from percolating into the waste (Licht et al. 2004).

In the past, landfills in Australia were operated with minimal consideration of long term environmental impacts (Thornton 2002). A survey conducted by Xu et al. (1999) revealed that only one-third of landfills in Australia were lined, just over half had gas collection systems and only two-third had installed leachate treatment systems. Landfill lining is generally accomplished using a low hydraulic conductivity material such as compacted clay, Geosynthetic Clay Liners (GCLs) and/or artificial liners such as High Density Polyethylene (HDPE) (Benson 2000). Capping of landfills has also been undertaken in addition to placing liners at the bottom of the landfill. Leachate recirculation systems have been extensively used in the US and in other parts of the world (Reinhart 1996). Research into potential application of leachate recirculation systems have also been conducted in Australia (Yuen 1999).

There are more than 600 landfills in Australia (Bateman 2005, Johnston 2009) and these include MSW landfill, Co-disposed landfills (MSW and Industrial); and Monofills/Monocells (industrial and hazardous wastes) (Scott et al. 2005).

Environmental Impacts of Landfills

Waste Collection and Separation

Waste is separated at source through kerbside recycling and/or via self-haulage of green waste, recyclables such as cardboards, glass and plastics and regulated waste such as tyres, oil, gas bottles and batteries to increase landfill capacity and reduce the impact of landfills on the environment. Although several efforts have been made to reduce waste going to landfill through awareness programs, waste reduction programs and introduction of State and Federal policies and strategies; a significant amount of waste is still generated and landfilled every month via kerbside collection, special burial, commercial skip bins and self-hauled household and commercial waste to transfer stations. These wastes attract vermin, birds, generate odour, and produce leachate and gases that have an adverse impact both on the environment and on the community.

Vermin, Birds and Odour

Wastes deposited in landfills attract vermin, rodents and birds which in turn create nuisance to landfill operators and the local community. Ibis, an Australian native bird is a great nuisance in and around landfills in Queenslandacross Australia. These birds are a threat to the surrounding flora and fauna. Odour is another issue faced by landfill operators. Several landfill operation technologies such as deodorising units, landfill lids and waste compaction are being adopted by landfill operators to reduce the impact of vermin, birds and odour on the community and environment. Many Councils; predominantly in Queensland and across have adopted the Ibis management program to control these birds within and around landfill sites.

Leachate

Landfills have adverse environmental impacts as theyse produce landfill leachate and generate landfill gases; predominantly methane (Scott et al. 2005). These impacts may give rise to explosions, vegetation damage, odour, groundwater pollution, contamination of water bodies and air pollution (El-Fadel et al. 1997). Landfill leachate not only contaminates groundwater but also deposits organic solvents, salts and heavy metals in the soil and groundwater (Al-Yaquot 2003). Leachate composition and concentration varies between landfills depending on the type of waste deposited (Kalyuzhnyi et al. 2003) and the age of the landfill (Ragle et al. 1995). Leachates have high Chemical Oxygen Demand (COD) (700 mg COD L^{-1}) (Kalyuzhnyi et al. 2003). High levels of COD are also due to slow degradation of organic waste (Fatta et al. 1999). Leachates from Australian landfills could have a Biological Oxygen Demand (BOD_5) of 11,400 mg L^{-1}; conductivities of 240 to 22,000 μS cm^{-1} and lead concentrations up to 0.1 mg L^{-1} (Scott et al. 2005). Scott et al. (2005) reviewed environmental impacts of various landfills in Australia, and a summary of constituents of leachate are presented in Table 1.

Table 1. Leachate composition in Australian landfills

Landfill leachate composition	Values in mg L^{-1} except where indicated	Landfill leachate composition	Values in mg L^{-1} except where indicated
pH	5.3 - 8.9	K	62 - 1200
COD	50 - 14000	Na	71 - 3900
BOD_5	6.8 - 11400	As	0.008 - 0.07
DO	0-4	Ba	< 1 - 22.5
EC (μS/cm)	240 - 22000	B	3 - 2.2
Temp (°C)	15 – 38	Cu	< 1
SS	8.6 - 2600	Cd	< 0.05
TDS	270 - 14000	Cr	0.13
TOC	59 - 6200	Fe	0.56 – 235
TKN	0.14 - 42.7	Pb	< 0.1
NO_3^-	< 0.05 -2.3	Mn	< 0.05 - 0.95
NO_2^-	< 0.05 - 0.21	Hg	< 0.0002
N (organic)	< 2 – 70	Ni	0.07 - 0.08
PO_4 (Inorganic)	35 μg/L	Se	< 0.01
P (total)	80.6 μg/L	Ag	< 0.05
Cl^-	40 - 6400	Zn	0 – 12
SO_3^{2-}	0.2 - 1.03	Phenols	0 – 244
SO_4^{2-}	4 - 295	Toluene	0 – 1
$CaCO_3$	490 - 4500	Xylene	0 - 0.5
Ca	70 – 350	Benzene	0 - 0.04
Mg	0.74 - 540	Ethyl Benzene	0 - 0.12

COD: Chemical Oxygen Demand, BOD_5: Biological Oxygen Demand, DO: Dissolved Oxygen, EC: Electrical Conductivity, SS: Suspended Solids, TDS: Total Dissolved Solids, TOC: Total Organic Content, TKN: Total Kjeldahl Nitrogen.

Source: (Scott et al. 2005)

Many cases of groundwater pollution have been reported globally due to poorly-designed landfill sites and the malfunctioning of landfill capping. For example, Assmuth and Strandber (1993) in Finland, Badv (2000) in Argentina, Neufeld (2000) in Canada, Bocanegra et al. (2001) in Iran, Van Nooten et al. (2009) in Belgium, Modin et al. (2009) in Sweden, Haarstad et al. (2009) in Norway, Aivalioti et al. (2009) in Greece and Jeevanrao and Shantaram (2003) in India reported groundwater contamination from landfills. Similar problems were cited at the South Fremantle landfill in Western Australia (Dunnet 2004). Adverse impacts of landfill leachate on groundwater have prompted numerous studies since 1980, which in turn has led to the evolution of new technologies (Fatta et al. 1999). Flyhammer (1995) examined leachate quality in several Swedish landfills and Sanchez et al. (1993) in Madrid. Similar studies were undertaken on leachate quality by Zheng et al. (1991) in New Jersey, Gailey and Gorelick (1993) in Canada and Blight (1995) in South Africa.

Methane Production in Landfills

In 2002, the Australian waste sector (MSW disposal on land, waste water disposal and waste incineration) contributed to approximately 3.2% of the total (total = 550.1 Mt) carbon dioxide equivalent emissions, of which 87.6% was derived from solid waste disposal sites (Scott et al. 2005). Problems associated with landfill gases are:

- Accumulation of gas to explosive quantities (Scott et al. 2005)
- Odour (Hudgins et al. 2002, Sullivan et al. 2004)
- Groundwater acidification (Park and Lee 2002)
- Reduced plant growth after site restoration (Department of the Environment 1986).

Reports of accidents involving explosions and migration of methane from landfills have been reported in recent years (Gendebien et al. 1992, Christensen et al. 1996, Cooper 2008).

Methane, a major component of landfill gas currently counts for 15% to 20% of the greenhouse budget (EPA 2003). Landfill gas is produced from the biological degradation of wastes (Jones and Nedwell, 1993, Giani et al. 2002) and comprises methane (45% to 60% v/v) and carbon dioxide (40% to 60% v/v) (Swarbrick and Dever, 1999). In Australia, methane concentrations range from 50% to 60% (v/v) of the total landfill gas produced (Duffy et al. 1996, Hansen et al. 1998, Yuen 1999) and is comparable with the global emissions (Bogner et al. 1996). However a study conducted by the Department of Climate Change (2006) showed that methane composition of Australian landfill gases was ca. 75% of the total gas released; which may possibly be due to presence of hot tropical conditions in many parts of the country.

Methane production in landfills normally takes place in two stages – the non-methanogenic stage and the methanogenic stage (Farquhar and Rovers 1973, Whalen et al. 1990, Czepiel et al. 1995, Borjesson and Svensson 1997). The non-methanogenic stage is initiated by a hydrolytic process which reduces complex organic matter to smaller soluble compounds such as fatty acids, simple sugars, amino acids and other low molecular weight organic compounds (Imshenetsky 1968). This phase accomplishes further modification of the organic material by capturing energy and forming organic acids, ammonia, water, hydrogen and carbon dioxide (Farquhar and Rovers 1973). The methanogenic stage is initiated by microorganisms under aerobic conditions (Urmann et al. 2007). The microorganisms active in the methanogenic stage are generally bacteria of the genus *Methanobacterium*; common inhabitants of soil, sewage (Alexander 1971) and marine sediments where both oxygen and methane are present (Hanson and Hanson 1996). Odorous gases such as nitrogen and hydrogen sulphide are also produced during this stage (Alexander 1971).

Normally, a steady rate of methane production is reached after 80 to 500 days of waste deposition, and this state is maintained for 10 to 20 years (Moore et al. 1998). The time required for degradation of waste in landfills and the amount of gas formed depends upon a number of factors. These include type and amount of buried waste, water content of the waste, compaction and leachate recycling procedures (Farquhar and Rovers 1973).

Methane Emissions from Landfills

The average annual methane emissions from a municipal solid waste landfill in Canberra were estimated to be 10 kg methane per tonne of waste (Denmead 1995, unpublished data).

Soils are also the only known biological sinks ((Mosier 1998, Hutsch 2001, Reay et al. 2004). Reay et al. (2001) demonstrated the effect of five species (alder, oak, Norway spruce, Scot pine and grassland) on methane oxidation in temperate soils in an experimental plot in Northwest England. The trial plot consisted of (0.2 ha) four pure stands of Scots pine (*Pinus sylvestris* L.), Norway spruce (*Picea abies* L.), alder (*Alnus glutinosa* L.), oak (*Quercus petraea*), together with a grass control. Each of the five treatments was planted in triplicate blocks, with treatment plots being located randomly within each block (Figure 1). No fertilizers were applied either at planting or subsequently. The trial plot showed low methane oxidation rates compared to the Norway spruce, Scots pine and grassland. European beech (*Fagus sylvatica*) exhibited oxidation rates three times higher than that of Norway spruce on the same soil type in Germany (Borken et al. 2003). This difference was due to the variation in monoterpenes in different species (Maurer et al. 2008). Monoterpenes are naturally occurring compounds produced by plants (Amaral et al. 1997). European beech produces and releases these compounds into the soil in very low concentrations (Holzke et al. 2006), whereas Norway spruce produces high concentrations of monoterpenes in needles, twigs, bark and buds (Bufler et al. 1990). Monoterpenes have a tendency to inhibit growth and activities of methanotrophs in the soil (Amaral et al. 1998).

Methane is the main hydrocarbon present in the atmosphere, with an average concentration of 1.7 parts per million (ppm) (Borjesson and Svensson 1997, Humer and Lechner 1999) and "a molecule of methane in the atmosphere has a global warming potential approximately 25 times higher than that of a carbon dioxide molecule (Forster et al. 2007) because of methane's higher infrared activity" (Abichou et al. 2006, IPCC 2007). Bingemer and Crutzen (1987), Richards (1989) and Bogner (2003) estimated 9 to 70 Tg yr^{-1} of methane emission from landfills. The rate of emission of methane (methane flux) varies over orders of magnitude, from less than 0.0004 g m^{-2} d^{-1} to about 4000 g m^{-2} d^{-1} (Bogner et al. 1997), and in some occasions more than 10,000 g m^{-2} d^{-1} was found in the US (Spokas et al. 2006). For example, Jones and Nedwell (1993) and Nozhevnikova et al. (1993) reported 0 to 38.4 g m^{-2} d^{-1} in the UK and Moscow. De Visscher (1999) reported 0.0048 to 4000 g m^{-2} d^{-1} in a laboratory-based soil column study in Belgium, and Tohijima and Wakita (1993) found 200 g

m^{-2} d^{-1} in Tokyo. Methane flux of 45 g m^{-2} d^{-1} in Germany (Jager and Peter, 1995) and 10.5 g m^{-2} d^{-1} in France (Pokryszka et al. 1995) have also been reported. Sengupta et al. (1998) found 0.3 to 0.7 g m^{-2} d^{-1} of methane flux in Indian landfills. Similarly, a laboratory test conducted by Gebert et al. (2009) showed an oxidation rate of 0.2 to 426 g m^{-2} h^{-1}.

Effect of Methane on Plants

Methane is generally non-toxic to plants or other organisms (Chan et al. 1991, Trotter and Cooke 2005). A major effect of methane in soils is methane oxidation which depletes oxygen present in the landfill, increases carbon dioxide levels (Trotter and Cooke 2005) and also raises soil temperatures (Fischer 1999). These changes in soil gas composition may contribute to plant death fatality by enhanced asphyxiation due to lower solubility of gases inside plant cells or by drying of soil (Fischer 1999).

Vegetation Management

Vegetation in and around landfills have to be well maintained, and weeds need to be cleared regularly and the landfills need to be sprayed with herbicide as part of the regular maintenance schedule. Weeding and clearing of vegetation from buffer zone is essential to avoid landfill fires to spread into surrounding properties.

Daily Cover and Final Landfill Capping

To mitigate adverse effects of landfills, the State Governments across Australia haves made it mandatory to cover the waste on a daily basis using soil (daily cover, atleast 300 mm deepthick). The State Governments in Australia has also made it mandatory to cap the landfill using clay once the landfill is completely filled and to maintain the integrity of the clay for at least 30 years to ensure low risk to the community and lower environmental impacts due to leachate and gas generation.

Current Landfill Technologies

Landfills pose threat to the environment due to their very presence and also due to operations involved in managing and burying waste. Landfill operations include, spreading of waste using a dozer, compacting waste and stripping the daily cover. These operations are carried out to maintain the integrity of landfills, increase landfill capacity and reduce environmental impacts. Another landfill operation technology is the bioreactor landfill, which has been successfully implemented in Australia.

Bozkurt et al. (2000) have developed a conceptual model of landfill wastes and their associated physico-chemical properties that lead to the production of leachate. Leachate collection systems (Rowe et al. 1997, Rittman et al. 1998) and landfill liners (Alston et al.

1997, Halse et al. 1990, Benson 2000) have been extensively used in landfill design and construction. A number of other techniques can also be used to reduce impact of leachate, and these are shown in Table 2.

Table 2. Various leachate treatment techniques

Types of Treatment	Reference
Aerated lagoon	Robinson et al. (1992)
Anaerobic lagoon	Blakey et al. (1992)
Rotating biological contactors	Knox (1992)
Physico-chemical treatment	Cossu et al. (1992)
Reverse osmosis	Li and Heine (2007)
Activated sludge technique	Avezzu et al. (1992)

Table 3. Landfill remediation technologies and their drawbacks

Technology	Reference	Drawbacks	Reference
Leachate collection system	Ramke (1989), Rowe et al. (1997), Rittman et al. (1998)	Clogging due to microbial reaction and calcium carbonate deposition	Korfiatis and Demetracopoulos (1986), VanGulck et al. (2003)
Compacted clay liners (CCL)	Anderson (1982), Nirmala et al. (1995)	Pure organic liquid with high acidity increases the hydraulic conductivity	Nirmala et al. (1995), Alston et al. (1997)
Composite liners (High Density Poly-ethylene; HDPE)	Halse et al. (1990)	Sudden potential stress, cracking and unexpected rupture of plastic	Benson (2000)
Composite liners (Poly-Vinyl Chloride; PVC))	Benson (2000)	They contain plasticizers that make the polymer flexible and the plasticizers evaporate over time	Shackelford et al. (2000), Benson (2000)
Geosynthetic Clay Liners (GCL)	Shackelford et al. (2000), Melchoir (1997), Benson (2000)	Bentonite is not chemically compatible and it hydrates when it comes in contact with leachate	EPA (2001), Melchoir (1997), Benson (2000),
Compacted clay covers (CCC)	Drumm et al. (1997), Benson and Othman (1993), Melchoir (1997), Khire et al. (1997)	Desiccation and cracking of clay, short life span and does not allow optimum diffusion of oxygen required for methane oxidation	Drumm et al. (1997), Abichou (2004), Vasudevan et al. (2003)
Composite covers (HDPE)	Melchoir (1997), Floess et al. (1996), Khire et al. (1997)	Expensive. They act as protective barriers to clay caps	Melchoir (1997), Floess et al. (1996), Khire et al. (1997)
Composite covers (PVC)	Stark et al. (2008)	Expensive.as protective barriers to clay caps	Benson (2000)

Gas emission from landfills has led to significant climate change over the last few decades (El-Fadel and Massoud 2000). Significant amounts of money and time have been spent in treating, extracting and trying to reduce methane emission and leachate generation from landfills. Methane recovery, although an expensive affair, is being implemented and adopted in developed countries. Feasibility of this system greatly depends on types of gases, and the amount of methane gas emitted by the landfill (Willumsen 2002). Methane from landfills has been extracted, recovered and used for various purposes including bioenergy generation (De Walle et al. 1978, Borjesson et al. 2000), and these activities have considerably reduced the quantities of methane being released into the atmosphere. Willumsen (2002) reported approximately 25 methane gas recovery plants operational in Australia in 2001, and this number should now have increased. In 2006, 103 MW of electricity was generated from landfill gas in Australia (WMAA 2007). Landfill capping can contribute to reduction in methane emissions. A study by Grossman et al. (2002) examined the natural attenuation of landfill methane through aerobic oxidation in landfill soils. Different types of liners and caps are also being used to reduce water infiltration with a view to reducing methane generation. Numerous scientists, researchers and engineers have been working on technologies that are required to reduce environmental hazards caused by landfills. These technologies and their drawbacks are summarised in Table 3.

Despite evolution of new technologies, there remain several problems in controlling adverse environmental impacts from landfills. Moreover, most landfills in regional areas are unsupervised and charge no gate fee, which puts many of the councils under budget constraints; thereby discouraging them from implementing more advanced technologies which are costly.

Landfill Technologies in Australia

Lining of landfills has been practiced in Australia since early 1990s (Friends of Earth 2000) to prevent the flow of leachate from landfills into groundwater. Geosynthetic clay liners (GCLs) are commonly used in both liquid and solid waste containment in Australia (Phillips and Eberle 1999). Australia instituted the Commonwealth Environmental Protection Agency (CEPA) in 1992. Individual states had also established their own Environmental Protection Agencies/Authorities (EPA) between 1970's and 1990's. At present CEPA has classified landfills according to their types and the types of waste accepted at each site.

Currently, more than 600 operational landfills exist in Australia (Bateman 2005, Johnston 2009). A survey of 230 closed landfills revealed that 70% used clay capping systems, 20% composite and 10% unspecified alternative caps (Bateman 2005). Most sites in Australia had a clay liner and over half of the larger landfills had composite liners. Most landfills had a leachate collection system and many landfill operators treated the leachate before disposal (Bateman 2005). About 40% of the larger landfill sites (>100,000 tonnes yr^{-1}) collected and used methane gas to generate electricity and about 90% of the sites surveyed flared the gases (Bateman 2005). In 2002, there were around 25 methane gas recovery systems installed in landfills throughout Australia (Willumsen 2002), which should have increased by now due to introducing stringent legislations and Carbon Farming Initiative (CFI).

Leachate recirculation is a technique used to encourage saturation of waste, with a view to stimulating the waste degradation processes to achieve quicker stabilisation of the landfill

(Scott et al. 2005). Rather than relying on intermittent rainfall, leachate recirculation has been employed to enhance and maintain waste saturation. Leachate recirculation also benefits in liquid volume reduction via evaporation (spray irrigation) (Percy and Truong, 2005). Concerns have been expressed about elevated pollutant levels around landfills but various studies have found no evidence of elevated pollutant levels in landfills that use leachate recirculation systems (Doedens and Cord 1989, Ehrig 1989, Reinhart and Basel 1996).

Several experiments have shown that methane generation will be enhanced if the acid generation phase is minimised (Komilis et al. 1999); leachate recirculation is known to minimise acid generation. Leachate recirculation has the potential to reduce quantity of leachate being produced. Research into potential application of leachate recirculation has been conducted in Australia (Waste Services 1998, Yuen 1999). Bowman et al. (2002) and Percy and Truong (2005) demonstrated the use of vetiver grass technology and parklands, respectively, for managing landfill leachate. Despite several new technologies such as gas recovery systems, leachate recirculation systems, liners and caps, very little is known about their long-term benefits and sustainability (Scott et al. 2005). Various technologies currently used in landfills to reduce adverse environmental impacts of landfills are shown in Table 4.

Table 4. Landfill remediation technologies practiced in Australia

Technology	Reference
1. Leachate collection	Bateman (2005), EPA (2005)
2. Leachate recirculation	Scott et al. (2005), Yuen (1999)
3. Leachate disposal	EPA (2005), Percy and Truong (2005)
4. Leachate treatment	Bateman (2005), Scott et al. (2005)
5. Methane gas recovery	EPA (2005), Willumsen et al. (2002)
6. Gas flaring	Bateman (2005)
7. Caps	
Clay	EPA (2005), Gourc et al. (2009)
GCL	Bateman (2005)
Alternative caps	Ashwath and Venkatraman (2007), Travor et al. (2009), Yuen et al. (2011)
Caps made of industrial by-products	Ronkainen et al. (2009)
8. Liners	
Clay liners	EPA (2005)
GCL	Bateman (2005)
9. Biofiltration	Dever et al. (2007)
10. Bioreactor	Yuen (1999)

LANDFILL CAPPING

Landfill capping is practiced to isolate landfills from the outside environment, mainly rainwater (Vasudevan et al. 2003). Rain and temperature are the two key elements to be considered in designing a capping system (Berger and Melchoir 2009). Similarly, suitability and type of capping material depends on climatic conditions and topography of the site (Berger and Melchoir 2009). Capping involves placement of clay (usually with materials that are similar to the liner) over filled landfill to minimise percolation of rain or surface water into the waste (Scott et al. 2005). In recent years, conventional capping systems made of compacted clay (Othman et al. 1994), GCLs (Benson, 2000), PVC (Levin and Hammond 1990) and HDPE (Simon and Muller 2004) have been used extensively in both developed and developing countries. Landfill capping systems minimise the entry of water into waste, which is primarily responsible for generation of leachate (Bendz et al. 1997) and methane gas emission (Whalen et al. 1990). Unlike the caps which are usually placed on completion of landfilling, placement of a layer of top soil or local soil is applied daily (daily cover) to reduce odour, stop the waste from being blown away and to prevent vermin from accessing the buried waste (Rohrs and Fourie 2002).

Once MSW landfills reach the end of their life or the permitted maximum height, current regulations require that the waste be covered with low hydraulic conductivity soil layer to minimise entry of water into the waste (Leone et al. 1977). Environmental regulatory authorities in Australia have prescribed the use of clay capping to minimise percolation of water into the waste (EPA 2005). A capping system recommended for landfills in Queensland is shown in Figure 2.

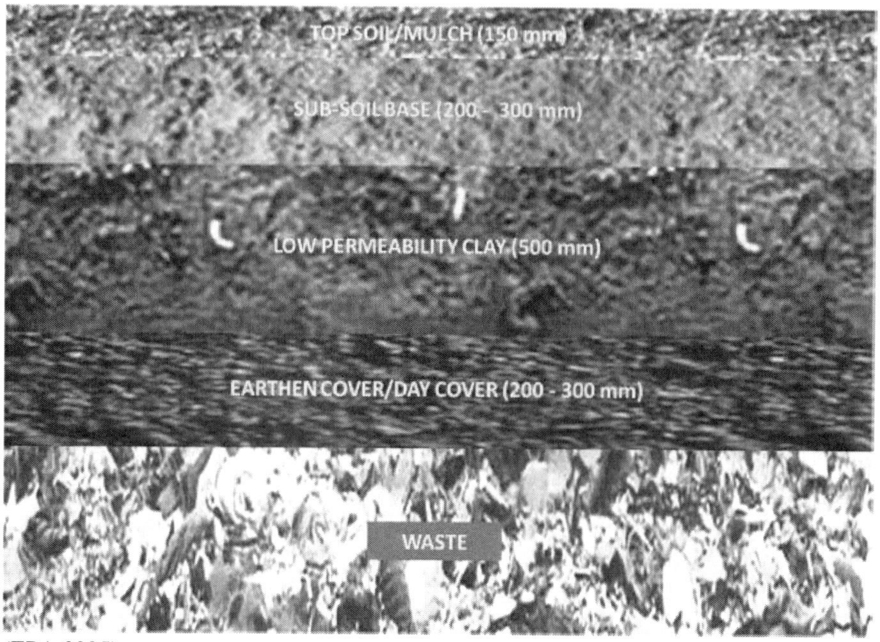

Source: (EPA 2005)

Figure 2. Typical clay capping system in Queensland.

Landfill caps improve aesthetics; prevent waste from blowing across landscape; reduce odour; control insects, rodents and flies; reduce fire hazards and most importantly minimise infiltration of water into the waste (Leone et al. 1977). Although many uses have been proposed for completed landfills, experience has shown that they are generally unsuitable for excavation, crop production or for suburban development. A large number of landfills have been used as play grounds or parks (Pereboom et al. 2009). Recent research studies from Austria (Tintner et al. 2009) and Italy (Delbarba and Mazzata 2009) have shown the possibility of crop production on closed landfills.

The performance of a landfill cap depends on the interactive effects of various processes that regulate a site's water balance (Paige et al. 1996). Negligence in the understanding of these processes by traditional remedial engineering techniques (Nyhan et al. 1990) has led to failure of many remediated landfills (Jacobs et al. 1980, Hakonson et al. 1982). Various techniques such as use of clay capping, GCLs, PVC or HDPE capping (Bowers 2002, Bateman 2005) and phytocapping (Ashwath and Pangahas 2004) have been suggested as effective techniques for landfill capping. Clay caps (Fig 2) have been used in more than 70% of the landfill sites in Australia (Bateman 2005). Amongst the larger sites (>100,000 tonnes yr^{-1}), 25% of the landfills have used GCLs in the capping system, whereas 10% of the larger sites used alternative capping systems (not specified) (Bateman 2005).

Clay caps rely on low saturated hydraulic conductivity of the caps to minimise entry of water into buried waste. However, this is not often achieved due to the formation of micro and macro cracks as the clay caps age and due to drying and wetting cycles associated with seasonal changes in rainfall and relative humidity (Albright and Benson 2001, Albright et al. 2003, Dwyer 2001, Melchoir 1997, Simon and Muller 2004). Clay caps require heavy maintenance and regular environmental monitoring to avoid any occurrences that enhance water infiltration. Clay caps have a limited life span (Vasudevan et al. 2003). Recent studies have shown that clay caps, although popular, often fail over a longer term due to cracking (Albright and Benson 2001, Albright et al. 2004, Benson and Othman 1993, Gourc et al. 2009, Khire et al. 1997, Melchior 1997, Othman et al. 1994) and they do not allow optimal interaction between methane and oxygen, which is necessary for methane oxidation (Abichou 2004).

Geosynthetic Clay Liners comprise a bentonite layer between two Geotextiles layers. These are preferred due to their low cost and space saving (Simon and Muller 2004). In the past, studies on GCLs have reported failure of the capping system due to the freezing/thawing effect and damage by desiccation (Lin and Benson 2000, Melchoir 1997) in temperate regions. GCLs are also susceptible to leakages through holes left behind during construction (Board and Laine 1995, Croizer and Walker 1995). Vegetation such as trees and shrubs that establish voluntarily, or those grown intentionally, will continually compete with grasses for available space sending their roots down to deeper soil layers, thereby both creating and leaving cracks in the capping system following root death (Stonell 1986). HDPE capping systems are very expensive (Simon and Muller 2004) compared to conventional clay caps.

Cossu et al. (1995) recommends the use of bio- covers with high permeability to fasten waste degradation process and environmental impacts of landfills. They also reported that bio- covers have a good field capacity and have the potential to reduce odour from the buried waste. According to Benson (2000), various researchers recommended the use of a PVC geomembrane capping system for sanitary landfills (New York State Department of Environmental Conservation; NYSDEC) due to its flexibility and durability. However, PVC caps often lose their plasticizer, and the corresponding elongation properties reduce by 63% (Stark et al. 2008).

A new technology commonly known 'Phytocapping' was introduced in 1991 by the Idaho National Engineering and Environmental Laboratory (INEEL) for the US Department of Energy (USDoE). This concept was also tested at Lakes Creek Landfill in Rockhampton, Australia. Phytocapping has been trialled throughout US (Ankeny et al. 1997) as the first demonstration project conducted by INEEL in 1991. The trial demonstrated that phytocaps can be as efficient as the clay capping technique in many parts of the US (US EPA 2003).

PHYTOCAPPING – *THE CONCEPT*

The phytocapping concept has been well described by Hauser et al. (2001). In brief, phytocaps have two major components; the plants that act as 'bio-pumps' and 'rain interceptors' and the soil that acts as 'storage' and a 'screen'. The vegetation and soil together minimise percolation of water into the waste (Figure 3). In this system, the soil acts mainly as the 'storage' to absorb and store water. Soil may also act as a 'screen' to restrict methane release into the atmosphere, while also serving as the substrate for plant growth. Plants act as 'bio-pumps' thereby use up stored water and minimise movement of water into the waste (Figure 3) (Venkatraman et al. 2006). Plants also act as 'rain interceptors' by blocking a certain proportion of the rain from reaching the ground surface (via direct evaporation from the tree canopy) (Venkatraman and Ashwath 2006). Trees may also minimise methane emission. (Figure 3) by promoting methane oxidising microorganisms in their root zone.

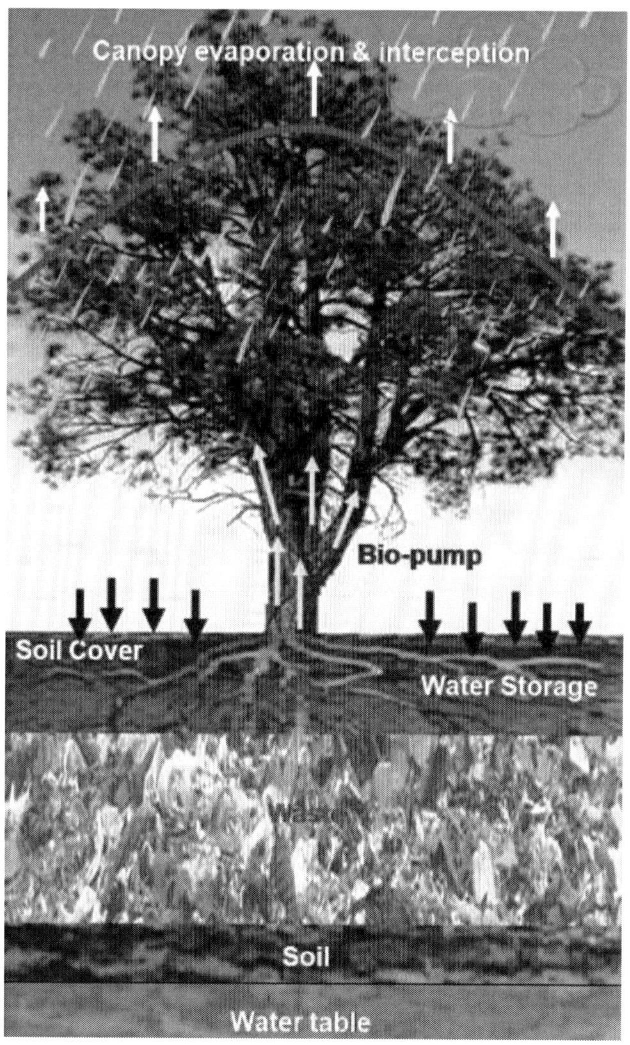

Figure 3. Phytocapping concept.

Sustainability of the phytocap depends upon thickness and properties of the soil used and types of the trees chosen for the phytocapping system (Albright et al. 2003, Dwyer 2001, Hauser et al. 2001). EPA (2005) recommended a standard saturated conductivity of 10^{-8} m s^{-1} for clay caps to reduce the 'bath tub effect' that increases potential of the waste to generate leachate. Phytocapping systems tend to prefer medium textured soils that not only hold reasonable amounts of moisture but also allow good root activity. This system also prefers vegetation that can adapt well to local conditions such that these will remove water rapidly during the wet season and minimise their water uptake during dry periods, with the view of ensuring sustainability of the established phytocaps on landfills. The vegetation should also be tolerant of landfill gases and able to support methanogenic bacteria to enhance methane oxidation. Local soil is preferred to reduce cost of the established phytocaps (Hakonson 1997, Dwyer 2003). Phytocaps has proven to perform better than clay caps at certain sites in the US (Albright et al. 2006, Benson et al. 2002).

Phytocaps are also referred to as phytocovers, evapotranspiration caps (ETC), water balance caps (WBC), alternative earthen final caps (AEC), vegetative landfill caps VLC), soil-plant caps (SPC) and store-and-release caps (SRC) (EPA 2003, Madalinski et al. 2003). The design of phytocaps is based on quantifying water balance components of a given site, which include water storage capacity of the soil, precipitation, surface runoff, evapotranspiration, infiltration (Dwyer 2001, Hauser et al. 2001, Albright et al. 2003, Madalinski et al. 2003) and canopy interception. Greater storage capacity of soil and evapotranspiration properties of plants will reduce potential for percolation through the phytocap (Hakonson 1997, Hauser et al. 2001, Dwyer 2003). Phytocaps are half as expensive as clay caps and these offer other environmental benefits whilst performing equally well as conventional caps (US EPA 2003, Scanlon et al. 2005).

Two types of phytocaps (monolithic barriers and capillary barriers) are used in the US (US EPA 2003). Monolithic barriers, also referred to as monofills, contain a single vegetated soil layer to retain water until it is either transpired or evaporated from the soil surface (Madalinski et al. 2003). Capillary barriers contain a fine-grained soil layer overlying a coarser-grained layer (Wing and Gee 1994). Studies evaluating performance of phytocaps have been conducted at USDoE sites (Nyhan et al.1990, Chichchester and Hauser 1991, Waugh et al. 1994, Anderson et al. 1997, Gee and Ward 1997, Dwyer 2001). More discussion of these capping systems can be found in Koerner and Daniel (1997), Krieth (1994) and Tchobanalogous et al. (1993) and Albright et al. (2010). Following success of the Alternative Capping Assessment Program (ACAP) study in the US, a similar study with a similar concept and design, is has been undertaken in Australia. This is known as the Australian Alternative Capping Assessment Program (A-ACAP) (Yeates et al. 2008, Salt et al. 2011, WMAA 2013).

Soil Cover

In phytocapping systems, soil covers are used both to retain moisture and to support plant and microbial growth. Generally, locally available soils are used in phytocapping projects. It is also possible that the soil may be synthesised using readily available materials such as clean fill, green waste, biosolids and crushed concrete. The soils used in phytocapping, however, must retain adequate quantities of moisture and be free from chemicals that may affect survival and growth of plants and microbes. The properties such as its texture (Auge et al. 2001), water holding capacity (Williams et al. 1983, Schmitz and Sourell 2000), porosity (Johnson et al. 2003, Bird et al. 2005), bulk density (Kalman et al. 1996), hydraulic conductivity (Hopmans and Dane 1986), chemical composition (Lee and Foster 1991) and organic matter content determine the suitability of soil to phytocapping.

Depth of the soil used will have a marked impact on phytocapping, not only from the point of view of its effectiveness, but also from cost of constructing the landfills. It is therefore important to choose the right depth considering the climate (particularly rainfall pattern), nature of the plants used and regulatory requirements.

Hauser et al. (2001) examined the effect on soil depth on plant growth, and concluded that the phytocaps were as effective as clay caps in many cases. A study conducted by Moffat and Houston (1991) at Pitsea Landfill site, Essex, observed that the trees grown in 1500 mm of soil cover grew better than those established in a 200 mm soil cover. Similar conclusions were drawn by Warren et al. (1996) from a four-year trial at a site that had a mean annual rainfall of 2000 mm. They reported a percolation of 300 mm in soil caps having 1500 mm with trees and shrubs, 240 mm in 1500 mm caps that had grass, and 400 mm in 900 mm soil caps that had grass only. The highest percolation in 900 mm cap was found to be due to insufficient depth (Warren et al. 1996) and high soil density (Hauser et al. 2001, Jones 1983). Soil density in the 900 mm cap with grass was reported to be 1.86 Mg m^{-3} (Warren et al. 1996) which is higher than that permissible for normal root growth (1 to 5 Mg m^{-3}) (Grossman et al. 1992, Hauser et al. 2001). Soil densities above 1.7 Mg m^{-3} restrict root growth in most soils (Timlin et al. 1998). Thus it is extremely important to optimise the two main factors, viz soil depth and soil density, in designing phytocapping systems, with the view to improving water holding capacity and promoting root growth.

Inclusion of topsoils in phytocaps promotes plant growth as they contain nutrients, essential microbes and soil seed banks. Likewise, addition of green waste or organic matter will minimise soil evaporation, salinity build up around the root zone, as well as the weed competition (Ashwath 2011). The addition of green waste and top soil can also reduce soil erosion (Madalinski et al. 2003). McGuire et al. (2001) recommended a minimum depth of top soil of 150 mm to support vegetation.

Overall, based on the previous studies and research, it is certain that optimising soil depth is critical to maintain the site water balance of the phytocapping system. Results from the study conducted by Warren et al. (2006) also suggest that selecting the right type of vegetation is important for good performance and long term sustainability of the phytocapping system.

Species

Plants play a key role in maintaining the water balance of the phytocapping system (Weand et al. 1999). However, selecting the right species is very important for a good and sustainable phytocapping system.

In addition to good interception and transpiration properties, plants that are leachate tolerant, drought tolerant and salt tolerant should be selected. For these reasons native plants are preferred (Dwyer et al. 1999), as they adapt to local conditions and are less likely to disturb the ecosystem. Native plants also have the ability to regenerate in case of natural disasters such as storms and bushfires.

Some of the species used in previous studies include wheat grass, clover, rabbit brush, sagebrush, willows, hybrid poplar (Von Der Hude et al. 1999, US EPA 2000, Benson et al. 2002, Licht et al. 2004, Scanlon et al. 2005) bluegrass, alfalfa prickly rose (US EPA 2003), switch grass, bermuda grass, Italian rye grass, black mustard, prickly lettuce, sandburg bluegrass, needle grass, California brome, bluebunch, western wheat grass, blue gramma, green needle grass, creeping wild rye and crested wheat grass (Albright et al. 2004).

In some of the sites, grasses are preferred due to their ability to control erosion and produce extensive fibrous roots (Hauser et al. 2001). Salt cedar (*Tamarix* sp.) has also been used in landfills in the US (Nicholas, 1994) due to its high water uptake capacity of 9.5 mm d^{-1} (Gay and Fritschen 1979) and high drought tolerance (Cleverly et al. 2007). Apart from grasses, poplars and willows have also been extensively used in phytocapping trials in the US due to their local occurrences, good performance record in phytoremediation projects in Europe (Elowson 1999) and North America (US EPA 1998, 2000, 2001) and in phytocapping (EPA 2003, Dwyer S: per. Comm. 2006), tolerance to leachate (Aitchison 2005), efficient in water uptake (up to 4.8 mm d^{-1}) (Hinckley et al.1994, Aaronson 1996, Bassman 2000, Aaronson and Perttu 2001, Negri et al. 2003), good root development (1500 mm deep) (Licht 1990, Licht et al. 2004) and high LAI (Heilman et al.1994, Hinckley et al.1994, Chappell 1997). Past studies have demonstrated the potential of hybrid poplars to reduce the amount of water reaching wastes (Dobson and Moffat 1993, Madison et al.1991) and metabolise pollutants and release these into the atmosphere (Hinchman et al. 1996 Kopp et al. 2001, Reimenschnieder et al. 2001).

About 450 species of *Salix* (willows) are found worldwide; distributed mostly in the northern hemisphere (Argus 1997) with 106 species found in North America (Zomleter 1994). *Salix* species also grows in subtropical and tropical regions in all continents except Antarctica and Australia (Kuzovkina and Quigley 2004). *Salix* species are best known to grow in soils with low nutrient levels (Heijden and Kuyper 2003), have high transpiration rates (Ebbs et al. 2003) and are able to accumulate high levels of heavy metals like cadmium (Klang-Westin and Eriksson 2003), and are tolerant drought and salinity (Kowalchik 2001) and produce large quantity of biomass (Perttu 1993).

In Australia, over 100 native species (including trees, shrubs and grasses) have been tested in various systems (Aswathappa et al. 1986, Ashwath et al. 1993, Ragupapthy et al. 1998) since 2003. However in this study, potential of 21 species to reduce percolation of water into the waste has been tested.

Studies in the US and Australia have tested and endorsed the potential of a wide range of plant species to grow in landfill conditions. Numerous parameters like LAI, Root Length Density (RLD), root depth, transpiration and tree growth rate have been measured (Lee and Schnabel 2000, Benson et al. 2002, Madalinski et al. 2003, Licht et al. 2004, Scanlon et al. 2005, Venkatraman and Ashwath 2007, 2010, Ashwath 2011). Various simulations of site water balance using hydrological models have also confirmed importance of plants in a phytocapping system (Lee and Schnabel 2000, Benson et al. 2004).

Overall, the choice of plant species usually depends on site conditions such as physical properties of the soil material placed over the waste and thickness of the soil cover (Department of the Environment 1984), degree of exposure to environmental factors and the local climatic conditions (Hibberd 1989). Other factors such as rooting depth (Kiniry et al. 1995) and the ability to pump large amounts of water (Robinson et al. 2003) also play a vital role in defining the sustainability of plant species in a phytocapping system.

Findings from the ALCD and ACAP

The Sandia National Laboratories at Kirtland Air Force Base near Albuquerque, New Mexico, tested four alternative capping designs including monolithic ETC, capillary barrier ETC, anisotropic barrier ETC and GCL as part of the ALCD project (Dwyer et al. 2000, Dwyer 2001, Scanlon et al. 2005). Results from a three year study (May 1997 to June 2000) with a total rainfall of 7745 mm showed a seepage flux of 0.19 mm day^{-1} for ETC, 0.16 mm day^{-1} for anisotropic barrier ETC, 0.87 mm day^{-1} for capillary barrier ETC and 1.81 mm day^{-1} for GCL (Dwyer et al. 2000).

The ACAP was established in 1997 by the US Environment Protection Agency (US EPA) with the view to evaluating performance of various capping systems under different agro-climatic conditions in the US (Albright et al. 2003). Eleven pilot scale test sections were established using conventional and alternative capping systems (Scanlon et al. 2005). In 2003, more than 64 landfill sites in the US were found to have incorporated the phytocapping system (Madalinski et al. 2003). This included 56 projects with monolithic capping and 20 projects with capillary barrier capping. The number of landfill sites with phytocaps increased to 93 in 2007. This included 47 demonstration trials and 46 full scale applications (www.cluin.org/products/altcovers). Today there are 223 alternative cover projects (demonstration, full scale or pilot) across US, ET/phytocapping systems have been readily implemented and used as a final closure for MSW landfills in many parts the US (Aitchison 2005) and negotiations are occurring between landfill operators and the EPA in Australia (Yeates et al. 2008). All studies undertaken in the US have considered only the site water balance of the system and not the individual components of trees and soil cover (e.g. species, capping thickness, tree characteristics, methane reduction). A few studies have measured methane emissions.

Results of the ACAP trials suggested that the percolation rates for phytocaps in arid and semi-arid regions with mean rainfall of less than 500 mm can be less than 1 mm yr^{-1} at some sites whilst the percolation rates in humid areas where the mean rainfall ranged between 450 mm yr^{-1} to 1250 mm yr^{-1} were 12 to 128 mm yr^{-1} (Benson et al. 2002). In the same study, at one of the sites in Albany (1280 mm rain), Benson et al. (2002) found that percolation rates decreased drastically with tree growth from 360 mm yr^{-1} to 14 mm yr^{-1}. A similar study on composite covers revealed a percolation rate of 1 mm yr^{-1} in arid and semi-arid regions and 5 mm yr^{-1} in humid regions (Benson 2002).

Benson et al. (2001) demonstrated a field evaluation of phytocaps under different agro-climatic conditions using five different methods for assessing percolation. These include water balance, trend analysis, Darcy's Law calculations, tracer methods and lysimeters. Each of the methods varied significantly in their results. The trend analysis method was found least accurate followed by the water balance method. Percolation rates computed according to Darcy's Laws were reasonably accurate but Benson et al. (2001) found the use of lysimeters to be the most accurate method for estimating percolation rates (200 m^2). The lysimeters gave an accurate result with a precision of 0.00004 mm yr^{-1} to 0.5 mm yr^{-1}. Benson et al. (2001) also suggested that lysimeters of the area >100 m^2 are required as sensitivity of the drainage measurement increases with the size of the lysimeters.

Another study conducted by the US EPA at the Polson landfill in Montana and New Mexico in 1999 (US EPA 2003), used 150 mm of top soil, 450 mm of compacted silt and 600 mm of sandy gravel. The cover was seeded with a mixture of grasses, forbs and shrubs including wheat grass, bluegrass, alfalfa and prickly rose shrubs. Percolation was measured using lysimeters and soil temperature with a heat dissipation unit. Results showed that the phytocapping system had a cumulative percolation of 0.5 mm from 1999 to 2002, in a region that received a total rainfall of 837 mm during the same period. On the other hand, HYDRUS 2D software predicted approximately 0.6 mm percolation in the first year and 0.1 mm over the next 9 years for the same site (US EPA 2003).

Madalinski et al. (2003) reported their findings of a demonstration phytocapping project in the US using wheat grasses, clover, rabbit brush, sagebrush, willows and hybrid poplars grown on three different capping systems. Monolithic cover contained soil cover ranging from 600 mm to 3000 mm, whilst the depth of capillary barrier (with an additional layer of gravel) covered with fine grained soil ranged between 420 mm to 1520 mm, and the coarser grained covers ranged from 152.4 mm to 600 mm. HYDRUS 2D was used to predict site water balance. Similar studies were conducted by the US EPA in landfills at Los Angeles where a 910 mm silty sand/clayey sand layer was established. This trial used UNSAT-H model and predicted a cumulative percolation of 500 mm for phytocaps and 950 mm for conventional cap over 10 years. The study also found that performance of phytocaps improved with maturity of plants grown in this system.

The study from an ALCD project found that percolation varied between 0.05 mm and 0.54 mm depending on the type of phytocaps used at a site that received 267 mm of rainfall in 1997. In 2002, the percolation was reduced to 0 mm. Similarly, the ACAP study that commenced in 2000 at Altamont, California (semi-arid region) reported percolation of 0 mm in the first year in the 225 mm rainfall zone and 5 mm percolation in the 300 mm rainfall zone (US EPA 2003). In 1999, at Polson, 0.05 mm of percolation was recorded 300 mm in the first year at an annual rainfall of 300 mm, and 0.45 mm at 250 mm rainfall in the second year (US EPA 2003). A study conducted by Albright et al. (2004) estimated a seepage flux of 60 mm yr^{-1} in the site receiving an annual rainfall of 760 mm. The ACAP and ALCD projects tested the performance of phytocaps against composite and compacted clay caps (Dwyer et al. 2000, Dwyer 2001 and Scanlon et al. 2005). Results showed that phytocaps worked equally well or better than conventional caps in most occasions (Albright et al. 2004, Benson et al. 2004).

Modelling Site Water Balance

The ALCD and ACAP project in the US used numerical models such as Hydrologic Evaluation of Landfill Performance (HELP), Vadose/W, HYDRUS 1D/2D and UNSTAT- H. Several other modelling studies were undertaken to evaluate the hydrological performance of landfill covers (Chai and Miura 2002, Ho et al. 2004) to predict seepage production (Dho et al. 2002, Ham 2002). Studies in the past have compared the ability of hydrologic models to predict water balance of landfill final caps. Models used in previous studies include UNSAT-H, HYDRUS, Simulation of Heat and Water (SHAW), Vadose/W, Soil Water Balance and Infiltration Model (SWIM), VS2DTI, and HELP (Fayer et al. 1992, Fayer and Gee 1997, Khire et al. 1999, Scanlon et al. 2002, Benson et al. 2004). Of these models, HYDRUS, UNSAT-H, Leachate Estimation and Chemistry Model (LEACHM) and Vadose/W are used most frequently in practical situations (Benson et al. 2004) due to their reliability and ability to incorporate more plant, soil and climatic data and use Richards' equation.

Fayer and Gee (1997) compared water balance data from eight non-vegetated lysimeters located in the semi-arid south-eastern region of Washington State, using UNSAT-H. Soil-water storage was under-predicted during winter months and over-predicted during summer. Khire et al. (1997) compared water-balance predictions made with HELP and UNSAT-H models with lysimeters data for two resistive barrier covers located in Georgia (humid climate) and Washington state (semi-arid climate). The cover profile at both locations consisted of a vegetated surface layer overlying a compacted fine-grained layer. Meteorological, vegetation, and soil data were used, but predictions from UNSAT-H were in better agreement with the measured water balance than those from HELP. Percolation was found to be grossly over-predicted by HELP and slightly under-predicted by UNSAT-H. Errors in predicting snowmelt and frozen ground affected most water-balance parameters and were found to significantly affect runoff predictions during winter months.

Khire et al. (1999) described a comparison between predictions made with UNSAT-H and the field data from a capillary barrier test section consisting of a 150 mm layer of silt overlying a 750 mm layer of sand. The comparison showed that UNSAT-H under-predicted runoff (within 100 mm) and over-predicted percolation (within 50 mm) and the soil-water storage was predicted accurately within 30 mm of measured soil-water storage. Scanlon et al. (2002) compared predictions made with HELP, HYDRUS, SHAW, Vadose/W, SWIM, UNSAT-H, and VS2DTI with data collected from landfill caps in the semi-arid states of Texas, New Mexico, and Idaho, over three years. The cover profile at the Texas site consisted of (from top to bottom) 300 mm of sandy clay blended with 15% gravel, 1700 mm of compacted sandy clay, and 1000 mm of sandy gravel. A 1070 mm thick monolithic cover of silty sand was evaluated at the New Mexico site and a 3000 mm thick monolithic cover of sandy silt was evaluated at the Idaho site. Models employing Richards' Equation predicted the water balance more accurately than the HELP model. Scanlon et al. (2005) suggested that the relationship between the abundance of vegetation, evapotranspiration, and water availability was an important factor affecting accuracy of water-balance predictions, and that most models used today do not account for this interaction.

Benson et al. (2004) compared measured water-balance data from a monolithic cap at a semi-arid site to predictions made with UNSAT-H and Vadose/W. On-site data was incorporated in the model. The study found that more accurate predictions were obtained with Vadose/W than with UNSAT-H. Surface runoff was over-predicted by UNSAT-H, which affected all sub-surface hydraulic processes. In contrast, Vadose/W accurately predicted surface runoff, evapotranspiration, and temporal variations in soil-water storage. Neither model predicted percolation accurately nor succeeded in capturing a key change in the transpiration pattern during the last winter-summer period of the monitoring program. Differences in the method used to simulate precipitation intensity were found to be partly responsible for differences in the accuracy of predicted surface runoff. Simulations conducted to evaluate the importance of the lower boundary condition showed that essentially the same predictions were obtained when the lower boundary was assigned as a unit gradient or seepage face condition.

Chammas et al. (1999) compared HELP and Vadose/W model and concluded that soil cover models gave more accurate predictions and that HELP could only conservatively predict percolation in dry environments. Despite being inaccurate, HELP has been extensively used by various researchers in phytocapping. Interesting conclusions against HELP were made by Fleenor and King (1995) and Khire et al. (1997). Khire et al. (1997) compared HELP and UNSAT-H and concluded that HELP over-estimated and UNSAT-H under-estimated percolation, but that HELP performed better than UNSAT-H. HELP and the Flow Investigation of Landfill Leachate (FILL) model have been extensively used in Australia to predict site water balance of landfill caps and liner systems (Khanbilvardi et al. 1995). However, recent Australian Field Lysimeters and Leachate Yield Tests in Queensland and South Australia (SA) found that the HELP model overestimates infiltration rate and leachate volumes generated in Australian landfills (Bowers 2002).

In 2004, Hauser and Gimon compared Environmental Policy Integrated Climate (EPIC), HELP, HYDRUS and UNSAT-H, and found that EPIC was better than HELP for phytocaps but UNSAT-H and HYDRUS were more accurate and could be used in phytocapping research. Further details and comparison between the various models are given in.

Based on the literature review, it is evident that the phytocapping trials conducted in the US used very few tree species and little or no information has been provided on role of trees in maintaining a desirable site water balance. The ACAP used mixtures of grasses, shrubs, forbs and trees, predominantly willows and hybrid poplars (EPA 2003). None of the previous studies have emphasised the role of plant species in a phytocapping system. Phytocaps can also reduce methane emissions from landfills via methane oxidation. Not much research has been conducted on the role of trees in reducing methane emissions. Phytocapping studies in the past have measured various soil parameters, tree growth, LAI, soil moisture, biomass production and transpiration but have not investigated vital characteristics of plants to intercept rainfall (Crockford and Richardson 1990). Canopy rainfall interception plays a significant role, not only in the hydrological balance of the phytocapping system, but also by helping to optimise soil thickness, which is the most expensive component of the phytocapping system.

Phytocaps can be a preferred future landfill design for many of the landfill sites to achieve stability while maximising leachate volume reduction, minimising methane emissions and encouraging native species (Licht et al. 2004), especially in semi-arid regions. Results from the ACAP study suggested that phytocaps performed better in semi-arid and arid regions than in tropical and temperate regions because of the dominance of summer precipitation (62% to 80%) that corresponds with the periods of high ET (EPA 2003, Dwyer et al. 2005).

KNOWLEDGE GAP

Phytocaps offer several advantages over conventional capping systems. They are more sustainable and provide ecological and aesthetic benefits beyond their transpiration functionality. Phytocaps also provide shelter for many native flora and fauna, thus enhancing the biodiversity value of that region. A series of studies conducted in the US (Albright et al. 2004) has laid a foundation to promote phytocapping as an alternative technique for landfill post closure management. There continues to be issues which need further attention and in-depth study such as plant growth and root development, transpiration, canopy rainfall interception and methane oxidation in phytocaps.

Studies in the past have demonstrated effectiveness of phytocaps in comparison with other clay capping systems using lysimeters, but different components (i.e. trees and soil) of the capping system with regards to their performance, integrity and sustainability are barely reported. Secondly, the ability of phytocaps to reduce methane emission has not been explored. For these reasons, performance of trees in reducing infiltration of water into waste, the role of soil depth, and effects of trees on methane emission, survival and long term sustainability of established trees needs to be defined so that the phytocapping technique can be further developed and used in landfill remediation.

Canopy rainfall interception (rain intercepting effect) is another component that plays a significant role in maintaining the hydrological balance of the phytocapping system. This particular feature of trees not only improves the efficiency of a phytocap but also reduces cost of phytocapping by optimising soil depth, as it reduces amount of rain reaching the soil. Finally, previous trials in the ACAP and the ALCD projects have used very few or no tree species. Thus in-depth study and further understanding of the species to be used in a phytocapping system is warranted.

SUMMARY

Landfills cause environmental hazards due to leachate generation and methane gas emissions (EPA 2002, CSIRO 2001). Efforts have been made to mitigate these impacts through developing landfill technologies such as leachate collection systems (Rittman et al. 1998), landfill gas collection systems (EPA 2006) or construction of landfill caps and liners (Bowers 2002). These technologies are expensive and not practically viable for many of the small and medium sized landfills (landfills accepting <100,000 tonnes yr^{-1}). A minimum of 100,000 tonnes yr^{-1} of waste is required to successfully run a 1 MW engine for electricity conversion (Molloy 2008). Many of the small and medium sized landfills constructed prior to 1990 are not lined. Furthermore, the mandatory requirement to clay cap, revegetate and continuously monitoring the landfill for 30 years after its closure has put a lot of pressure on many of the councils in Australia.

Landfill caps are often used to isolate landfills from the outside environment; chiefly water (Vasudevan et al. 2003). In recent years, conventional capping systems made of compacted clay (Othman et al. 1994), GCLs (Benson 2000), PVC (Levin and Hammond 1990) and HDPE (Simon and Muller 2004) have been used extensively in developed and developing countries. Environmental Protection Authority in the US and Australia have prescribed use of clay capping to minimise percolation of water into buried waste. However clay caps fail over time (Albright et al. 2004) and do not allow for optimal interaction of methane with oxygen, which is essential for methane oxidation (Abichou et al. 2004). GCLs have also reported as constant failure due to the thawing effect and damage by desiccation (Melchoir 1997, Lin and Benson 2000). GCLs are also susceptible to leakages through holes left behind during construction (Board and Laine 1995, Croizer and Walker 1995). HDPE capping systems in turn are very expensive and susceptible to degradation from sunlight as well as chemical and biological degradation (Simon and Muller 2004). Conventional caps require heavy maintenance and constant monitoring to avoid occurrences that enhance water infiltration. PVC caps often lose their plasticizer and the corresponding elongation properties were reduced by 63% (Levin and Hammond 1990).

Hence, a new technology called phytocapping' was trialled at Lakes Creek Landfill, Rockhampton, Australia from 2003 to 2007. Field studies conducted in the US have proved that phytocaps perform equally well or better than clay caps. Phytocaps have been successfully trialled and adopted in the US, but several aspects such as soil depth, plant species to be grown, their transpiration rates, rainfall interception potential, effect of methane on tree growth, effect of plants on methane emission and sustainability of the plant species in harsh landfill conditions are still to be understood to successfully implement phytocaps in Australia.

This is the first study of its kind in Australia and the first study to consider various plant related aspects such as identification of trees to be grown at each site, role of plants in reducing methane emission, sustainability of the established vegetation, their contribution to site water balance via canopy interception and transpiration, and the plant/soil combinations that offer the best site water balance. An additional feature, i.e. role of soil thickness in determining site water balance and sustainability of the established vegetation as well as mineral composition of trees established on phytocaps has also been studied.

Conclusion from Published Research

Some broad conclusions can be drawn from previous studies on landfills and landfill caps:

- Landfills are the easiest and most economic means of disposing waste, and this practice will continue to be used in the future.
- Despite progress made in landfill technology and management, the associated environmental impacts are still a major concern globally and in Australia.
- Landfill capping using compacted clay has been made mandatory in Australia, but clay caps are expensive and often fail to perform over a period of time.
- Properly designed phytocapping - an alternative capping technology has shown to perform equally or better than clay caps in the US and is yet to be proven in Australia.

There is a need for further research on various aspects, such as:

- The type of species to be grown on phytocaps
- Thickness of the soil required for tree growth and water retention
- The role played by canopy rainfall interception
- The effect of phytocaps on methane emission

REFERENCES

Abichou, T., Chanton, J. and Powelson, D. (2006) Field performance of biocells, biocovers and biofilters to mitigate greenhouse gas emission from landfills. Florida Centre for Solid and Hazardous Waste Management, Florida.

Abichou, T., Palueson, D. and Chanton, J. (2004) Bio-reactive cover systems. Florida Centre for Solid and Hazardous Waste Management, Florida.

Aitchison, E. (2005) Tree covers for containment and leachate recirculation. Vol.www. Ecolotree.com, Accessed on 14/11/ 2005

Albrecht, B. A. and Benson, C. H. (2001) Effect of desiccation on compated natural clays, *Journal of Geotechnical and Geoenvironmental Engineering*, 127: 67-76.

Albright, W. G., Benson, C. H., Gee, G. W., Roesler, A. C., Abichou, T., Apiwantrgoon, P., Lyles, B. F. and Rock, S. A. (2004) Field water balance of landfill final covers, *Journal of Environmental Quality*, 33: 2317 - 2332.

Albright, W. H., Benson, C. H., Gee, G. W., Abichou, T., Roesler, A. C. and Rock, S. A. (2003) Examining the alternatives, *Journal of Civil Engineering*, 73: 70-74.

Alexander, M. (1971) *Microbial Ecology*, John Wiley and Sons, Inc, New York, p 511.

Alston, C., Daniel, D. E. and Devroy, D. (1997) Design and construction of sand-bentonite liner for effluent treatment lagoon, *Canadian Geotechnical Journal*, 34: 841-852.

Al-Yaquot, A. F. and Hamoda, M. F. (2003) Evaluation of landfill leachate in arid climate - a case study, *Environmental International*, 29: 593-600.

Anderson, D. (1982) Does landfill leachate make clay liners more permeable, *Civil Engineering*, 52: 66-69.

Anderson, J. E. (1997) Soil plant cover systems for final closure of solid waste landfills in arid regions. In: *Landfill Capping in the Semi-arid West-Problems, Perspectives and Solutions.* (Eds: Reynolds, T. D. and Morris, R. C.) Idaho Falls, pp. 27-38.

Ankeny, M. D., L.M, C., Majumdar, N., Kelsey, J. and Miller, M. (1997) Performance and cost considerations for landfill caps in semi-arid climates. In: *Landfill Capping in the Semi-arid West: Problems, Prospective and Solutions,* Environmental Science and Research Foundation, Idaho Falls.

Argus, G. W. (1997) Infrageneric classification of *Salix* in the new world, *Systematic Botany Monographs*, 52: 121.

Aronson, P. (Ed.) (1996) *Simulated evapotranspiration from a landfill irrigated with leachate,* PhD Thesis, Swedish University of Agricultural Sciences, Uppsala, Sweden.

Aronson, P. and Perttu, K. (2001) Willow vegetation filters for wastewater treatment and soil remediation combined with biomass production, *The Forestry Chronicle*, 77: 293-299.

Ashwath, N. & Pangahas, N, (2004) Phytoremediation trial at Lakes Creek Landfill, Rockhampton. In: *The WMAA Central Queensland Regional Resource Recovery and Waste Management Seminar, CQU*, Gladstone, Australia.

Ashwath, N. and Venkatraman K. (2007) Phytocapping of landfills: a Rockhampton experience. In *WasteQ,* Waste Management Association of Australia, Surfers Paradise, Gold Coast (CD ROM).

Assmuth, T. W. and Strandberg, T. (1993) Groundwater contamination at Finnish landfills, *Water, Air and Soil Pollution*, 69: 179-199.

Auge, R. M., Stodola, A. J. W., Tims, J. E. and Saxton, A. M. (2001) Moisture retention properties of a mycorrhizal soil, *Plant and Soil*, 230: 87-97.

Avezzu, F., Bissolotti, G. and Collivignarelli, C. (1992) Combination of wet oxidation and activated sludge treatment. In: *Landfilling of Waste: Leachate* (Eds, Christensen, T. H., Stegmann, R. and Cossu, R.) Francis and Taylor, pp. 333.

Badv, K. (2000) An overview of the effect of open dump solid waste landfill in Urmia city , Iran, on the the surrounding soil and groundwater resources. Department of Civil Engineering, Urmi University,Urmia.

Bassman, J. H. (2000) Factors affecting water use in poplar culture. In *Hybrid poplars in the Pacific Northwest: Culture, Commerce and Capability* (Ed, Blatner, K. A.) Washington State University Extention, Washington, pp. 69-75.

Bateman, S. (2005) National Landfill Survey. In: *Landfill 2005,* Brisbane, 7-5 September.

Bell, T., D (1999) Australian trees for the rehabilitation of waterlogged and salinity-damaged landscapes, *Australian Journal of Botany*, 47: 697-716.

Bendz, D., Singh, V. P. and Akesson, M. (1997) Accumulation of water and generation of leachate in a young landfill, *Journal of Hydrology*, 203: 1-10.

Benson, C. H. (2000) Liners and covers for waste containment. In: *Proceedings of the Fourth Kansai International Geotechnical Forum,* Kyoto, Japan, pp. 1-40.

Benson, C. H. and Othman, M. (1993) Hydraulic conductivity of compacted clay frozen and thawed, *Journal of Geotechnical Engineering*, 119: 276-294.

Benson, C. H., Abichou, T., Albright, W. G., Gee, G. and Roesler, A. C. (2001) Field evaluation of alternative earthen fian lcovers, *International Journal of Phytoremediation*, 2: 105-127.

Benson, C. H., Albright, W. H., Roesler, A. C. and Abichou, T. (2002) Evaluation of final coverperformance: field data from the Alternative Cover Assessment Program (ACAP). In *Waste Management 2002 Conference,* 24 – 28 February, Tucson, USA, , pp. 1-17.

Benson, C. H., Bohnhoff, G. L., Apiwantrgoon, P., Ogorzalek, A. S., Shackelford, C. and Albright, W. H. (2004) Comparison of model predictions and field data for an ET cover. In *Tailing and Mine Waste*, pp. 137.

Bingemer, H. G. and Crutzen, P. J. (1987) The production of methane from solid wastes, *Journal of Geophysics Research*, 92: 2181-2187.

Bird, N. R. A., Preston, A. R., Randall, E. W., Whalley, W. R. and Whitmore, A. P. (2005) Measurement of the size distribution of water filled pores at different matric potentials by stray field nuclear magnetic resonance, *European Journal of Soil Science*, 56: 135-143.

Blakey, N. C., Cossu, R., Maris, P. F. and Mosey, F. E. (1992) Anaerobic Lagoons and UASB Reactors: Laboratory Experiments. In: *Landfilling of Waste: Leachate* (Eds, Christensen, T. H., Cossu, R. and Stegmann, R.) Taylor and Francis, pp. 245.

Blight, G. E. (1995) Ground water pollution from a landfill in semi arid climate. In: *Proceedings Sardinia 95, Fifth International Landfill Symposium,* Cagliari, Italy, 2-6 October.

Board, M. and Laine, D. (1995) Corralling liner nightmares, *Journal of MSW Management*, 5: 48-51.

Bocanegra, E., Massone, H., Martinez, D., Civit, E. and Farenga, M. (2001) Groundwater contamination: risk management and assessment for landfills in Mar del Plata, Argentina, *Environmental Geology*, 40: 732-741.

Boeckx, P. and Van Cleemput, O. (1996) Methane oxidation in neutral landfill cover soil: influence of temperature, moisture content and N-turnover, *Journal of Environmental Quality*, 25: 178-183.

Bogner, J. and Matthews, E. (2003) Global methane emissions from landfills: new methodology and annual estimates 1980-1996, *Global Biogeochemical Cycles*, 17: 1-18.

Bogner, J., Meadows, M. and Czeipiel, P. (1997) Fluxes of methane between landfills and the atmosphere: and engineered controls, *Soil Use Management*, 13: 268 -277.

Bogner, J., Spokas, K., Burton, E., Sweeney, R. and Corona, V. (1995) Landfills as atmospheric methane sources and sink, *Chemosphere*, 31: 4119-4130.

Bogner, J., Sweeney, R., Coleman, D., Huitric, R. and Ririe, G. (1996) Using isotopic and molecular data to model landfill gas processes, *Waste Management Research*, 14: 367 - 376.

Borjesson, G. and Svensson, B. (1997) Seasonal and diurnal methane emission from a landfill and their regulation by methane oxidation, *Waste Management and Research*, 15: 33-54.

Bowers, N. D. (2002) Specifying landfill environment and management plans and closure plans for arid landfills in Australia. GHD Pvt. Ltd.,Adelaide.

Bowman, M. S., Clune, T. S. and Sutton, B. G. (2002) Sustainable management of landfill leachate by irrigation, *Water, Air and Soil Pollution*, 134: 81-86.

Bozkurt, S., Moreno, L. and Neretnieks, I. (2000) Long term processes in waste deposits, *Science of the Total Environment*, 250: 1101-121.

Chai, J. and Miura, N. (2002) Comparing the performance of landfill liner systems, *Journal of Material Cycles and Waste Wanagement*, 4: 135-142.

Chammas, G. A., Geddis, M. and McCaulou, D. R. (1999) A comparison of two models for simulating the water balance of soil covers under semi arid conditions. In: *National Meeting of the American Society of Surface Mining and Reclamation,* Arizona.

Chan, Y. S. G., M. H. Wong and B. A. Whitton (1991). "Effects of landfill gas on subtropical woody plants." *Journal of Environmental Management* 15: 411-431.

Chappell, J. (1997) Phytoremediation of TCE in groundwater using *Populus*. http://clu.in.org/products/phytotce.htm, viewed on 12/04/2005.

Chichester, F. W. and Hauser, V. L. (1991) Change in chemical properties of constructed mine soils developing under forage grass management, *Journal of soil Science Society*, 55: 451-459.

Christensen, T. H., Cossu, R. and Stegmann, R. (1994) Principles of landfill barrier systems. In: *Landfilling of Wastes: Barriers* (Eds, Christensen, T. H., Cossu, R. and Stegmann, R.) Chapman and Hall, UK, pp 1-10.

Christensen, T. H., Cossu, R. and Stegmann, R. (1996) *Landfilling of waste: Biogas*. E&FN Spon, London, pp 840.

Clister, W., Janechek, A. and Hibbs, S. (1997) Groundwater Impacts Associated with landfill gas migration at municipal solid waste landfill sites: 1-7 (unpublished data).

Cossu, R., Serra, R. and Muntoni, A. (1992) Physico-chemical treatment of leachate. In *Landfilling of Waste: Leachate* (Eds, Christensen, T. H., Stegmann, R. and Cossu, R.) Francis and Taylor, pp. 265.

Crockford, H. and Richardson, D. P. (1990) Partitioning of rainfall in a eucalypt forest and pine plantation in south eastern Australia: ii stemflow and factors affecting stemflow in a dry sclerophyll eucalypt forest and Pinus radiata plantation, *Hydrological Process*, 4: 145-155.

Crozier, F. and Walker, T. (1995) 'CQA + GLLS = TEC: how much does your liner leak?, *Journal of Waste Management*: 2: 24-26.

CSIRO (2001) Australia State of the Environment. CSIRO, Melbourne.

Czeipiel, P., Mosher, B., Crill, P. and Harriss, R. (1996) Quantifying the effect of oxidation on landfill methane emissions, *Journal of Geophysical Research*, 101: 16721 - 16729.

de Visscher, A., Boeckx, T. D. and Van Cleemput, O. (1999) Methane oxidation in simulated landfill cover soil environments, *Environ. Sci. Technol.*, 33: 1854-1859.

Denmead, O. T. (1995) Novel micro technological methods for measuring trace gas flux. In: *Philosophical Transactions of the Royal Society of London (unpublished).*

Department of the Environment (1984) Landfill practices review group, *WLR Technical Note Series, Her Majesty's Stationery Office, London*, pp 65.

Dever, S. A., Swarsbrick, G. E. and Stuetz, R. M. (2007) Passive drainage and biofiltration of landfill gas: Australian field trial, *Waste Management*, 27: 277-286.

Dho, N. Y., Koo, J. K. and Lee, S. R. (2002) Prediction of leachate level in Kimpo metropolitan landfill site by total water balance, *Journal of Environmental Monitoring and Assessment*, 73: 207-219.

Dobson, M. C. and Moffat, A. J. (1994) The potential for woodland establishment on landfill sites, *The Journal of Applied Ecology*, 31: 192 -195.

Doedans, H. and Landwher, C. K. (1989) Leachate recirculation, in sanitary landfilling: process, technology and environmental impact. (Eds, Chritensen, T. H., Cossu, R. and Stegmann, R.) Academin Press, London, pp. 231.

Drumm, E., Boles, D. and Wilson, G. (1997) Desiccation cracks results in preferential flow. In *Geotechnical News* 1

Duffy, B. L., Nelson, P. F. and Williams, D. J. (1996) Composition of trace hydrocarbon and chlorinated hydrocarbon compounds in landfill gas. In: *Proceedings from the 3rd National Hazardous and Solid Waste Convention* Melbourne, Australia, pp. 494.

Dunnet, S. C. (2004) Current issues at the south Fremantle landfill site, Western Australia, *Journal of Rural and Remote Environmental Health*, 3: 40-51.

Dwyer, S. F. (2001) Finding a better cover, *Journal of Civil Engineering*, 71: 58-63.

Dwyer, S. F. (2003) Water balance measurements and computer simulations of landfill covers. In: *PhD Discussion,* University of New Mexico, Albuquerque.

Dwyer, S. F., Reavis, B. and Newman, G. (2000) Alternative Landfill Cover Demonstration. Sandia National Laboratory, Albuquerque.

Dwyer, S. F., Stormont, J. C. and Anderson, C. E. (1999) Mixed Waste landfill Design Report. Sandia National Laboratories.

Ebbs, S., Bushey, J., Poston, S., Kosma, D., Samiotakis, M. and Dzombak, D. (2003) Transport and metabolism of free cyanide and iron cyanide complexes by willow, *Plant Cell and Environment*, 26: 1467-1478.

Ehrig, H. J. (1989) Leachate quality in sanitary landfilling: process, technology and environmental impact (Eds, Christensen, T. H., Cossu, R. and Stegmann, R.). Academic Press, London, pp. 213.

El-Fadel, E. and Massoud, M. (2000) Emissions from landfills: a methodology comparative assessment, *Environmental Technology*, 21: 965-978.

El-Fadel, M., Findikakis, A. N. and Leckie, J. O. (1997) Environmental aspects of solid waste landfilling, *Journal of Environmental Management*, Vol 50, pp 1 - 25.

Elowson, S. (1999) Willow as a vegetation filter for cleaning of polluted drainage water from Agricultural Land, *Biomass and Bioenergy*, 16: 281-290.

Ensley, B. D. (2000) Rational use of Phytoremediation. In: *Phytoremediation of Toxic Metals-using Plants to Clean up the Environment.*(Eds, Raskin, I. and Ensley, B. D.) John Wiley and sons, Inc., New York, pp. 3-12.

EPA (1998) A citizens Guide to phytoremediation. US EPA, pp. 1-6.

EPA (2000) Introduction to Phytoremediation. Office of Research and Development, EPA, Washington D.C (Report).

EPA (2001) Geosynthetic clay liners used in municipal solid waste landfills.

EPA (2001) Phytoremediation bibliography- Online. US EPA, Washington D.C.

EPA (2002), www.epa.gov/region02/superfund/npl/0201290c.html. Viewed on 13/08/2006

EPA (2003) Evapotranspiration landfill cover systems fact sheet. EPA, pp. 1-12.

EPA (2005) Guideline: landfill siting, design, operation and rehabilitation. Environmental Protection Agency, Brisbane, Queensland.

EPA (2006) The state of waste and recycling in Queensland. Environment Protection Agency, Brisbane, Queensland.

Farquhar, G. J. and Rovers, F. A. (1973) Gas production during refuse decomposition, *Water, Air and Soil Pollution*, 2: 483-495.

Fatta, D., Papadopoulos, A. and Loizidou, M. (1999) A study of the landfill leachate and its impact on the groundwater quality of the greater area, *Environmental Geochemistry and Health*, 21: 175-190.

Fayer, M. J. and Gee, G. W. (1997) Hydrologic model tests for landfill covers using field data. Landfill capping in the semi arid west: Problems, Perspectives and Solutions. Environmental Science and Research Foundation, Idaho Falls, USA.

Fayer, M. J., Rockhold, M. L. and Campbell, M. D. (1992) Hydrologic modelling of protective barriers: Comparison of field data and simulation results, *Journal of Soil Science Society*, 56: 690-700.

Fischer, C. (1999) Gas emissions from landfills: An overview of issues and research needs. ECONS SA, Environmental Engineering, Stockholm, pp. 1-52.

Fleenor, W. E. and King, I.P (1995) Identifying limitations on use of the HELP model. In *Landfill Closures, Environmental Protection and Recovery*, (Eds, Dunn, R. J. and Sing, U. P.) *American Society of Civil Engineers Geotechnical Special Publication*, New York, 53: 121-138.

Floess, C. H., Smith, R. F. J. and Hitchcock, R. H. (1996) Capping with fibre clay, *International Journal of Rock Mechanics and Mining Sciences and Geomechanics*, 33: 327A-327A(1).

Flyhammer, P. (1995) Leachate quality and environmental effects at active Swedish municipal landfill. In: *Fifth International Landfill Symposium, Cagliari*, Italy, 2-6 October.

Friends of Earth. (2000) Friends of Earth Citizens Guide to Municipal Landfills. In *Friends of Earth*, www.foe.org/ptp/manual.ktm, Viewed on 12/10/2005

Gailey, R. M. and Gorelick, S. M. (1993) Design of optimal, reliable plume capture schemes: Application to Glouschter landfill ground water contamination problem, *Groundwater*, 31: 107-114.

Gay, L. W. and Fritschen, L. J. (1979) Energy budget analysis of water use by Salt cedar, *Water Resources Research*, 15: 1589-1592.

Gee, G. and Ward, A. (1997) Still in quest of the perfect cap, Landfill capping in semi-arid west: problems, perspective and solutions. (Eds, Reynolds, T. D. and Morris, R. C.) Idaho, pp. 145-165.

Gendebein, A., Pauwels, M., Ledrut-Damanet, M. J., Nyns, E. J., Willumsem, H. C., Butson, J., Fabry, R. and Ferrero, G. J. (1992) Potential landfill gas damages to vegetation. In *Landfill gas from environment to energy,* Luxembourg: Commission of the European Communities, pp. 35-46.

Giani, L., Bredenkamp, J. and Eden, I. (2002) Temporal and spatial variability of the methane dynamics of landfill cover soils, *Journal of Plant Nutrition and Soil Science*, 165: 205-210.

Grossman, E. L., Cifuentes, L. A. and Cozzarelli, I. M. (2002) Anaerobic Methane Oxidation in a landfill-leachate plume, *Environmental Science and Technology*, 36: 2436-2442.

Grossman, R. B., Benham, E. C., Harms, D. S. and Sinclair, H. R. J. (1992) Physical root restriction prediction in a mine spoil reclamation protocol. In *National Symposium on Prime Farmland Reclamation,* University of Illinois, Urbana, pp. 191 -196.

Hakonson, T. E., L. J. Lane, J. G. Stegar and G. L. DePooter (1982). Some interactive factors affecting trench cover integrity on low level waste sites. *Conference on low level waste disposal: site characterisation and monitoring* Arlington, VA, USA, Los Alamos National Lab, NM, USA.

Halse, Y., Lord, A. and koerner, R. (1990) Ductile-to-brittle transition time in polyethylene geomembrane sheet. In: *Geosynthetic Testing for Waste Containment Application*, pp. 95-109.

Ham, J. M. (2002) Uncertainty analysis of the water balance technique for measuring seepage from animal waste lagoons, *Journal of Environmental Quality*, 31: 1370-1379.

Hansen, J., Sato, M., Lacis, A., Ruedy, R., Tegen, I. and Mathews, E. (1998) Climate forcings in the industrial area era. In: *Proceedings of the National Academy of Science*, 90: 12753-12758.

Hanson, R. S. and Hanson, T. E. (1996) Methanotropic bacteria, *Microbiological Review*, 60: 439-471.

Hauser, P. E., Weand, B. L. and Gill, M. D. (2001) Natural covers for landfills and buried waste, *Journal of Environmental Engineering*, 127: 768-775.

Heijden, E. W. v. d. and Kuyper, T. W. (2003) Ecological strategies of ecomycorhizal fungi of *Salix repens*: Root manipulation versus root replacement, *Oikis*, 103: 668-680.

Heilman, P. E., Ekuan, G. and Fogle, D. (1994) Above and below ground biomass and fine roots of 4 year old hybrid *Populus trichocarpa* x *Populus deltoides* and parental species in short rotation culture, *Canadian Journal of Forest Research*, 24: 1186-1192.

Hibberd, B. G. (1989) Urban Forestry Practice, *Forestry commission Handbook*, 5.

Hinchman, R. R., Negri, C. M. and Gatliff, E. G. (1996) Phytoremediation: Using green plants to cleanup contaminated soil, groundwater and wastewater. Argonne National Laboratory, Illinois, pp. 1-10.

Hinckley, T. M., Brooks, J. R., Cermak, J., Ceulemans, R., Kucera, J., Meinzer, F. C. and Roberts, D. A. (1994) Water Flux in a hybrid poplar stand, *Tree Physiology*, 14: 1005-1018.

Ho, C. K., Arnold, B. W., Cochron, J. R., Taira, R. Y. and Pelton, M. A. (2004) A probabilistic model and software tool for evaluating the long term performance of landfill covers, *Environmental Modelling and Software*, 19: 63-88.

Hopmans, J. W. and Dane, J. H. (1986) Temperature dependence of soil hydraulic properties, *Soil Science Society America. Journal*, 50: 4-9.

Hudgins, M. and Leon, G. (2000) Innovative landfill gas and odour control using an anaerobic landfill system. In: *Odour and VOC Emissions 2000* Water Environment Federation, Alexandria, Cincinnati, OD - USA, pp. 619-641.

Humer, M. and Lechner, P. (1999) Methane oxidation in compost cover layers on landfills. In *Proceedings of the Seventh International Waste Management and Landfill Symposium*, Cagliari, Italy, 4-8th October.

Hutsch, B. (2001) Methane oxidation, nitrification and counts of methanotropic bacteria in soils from a long term fertilisation experiment, *Journal of Plant Nutrition and Soil Science*, 164: 21-28.

Imshenetsky, A. A. (Ed.) (1968) *The Ecology and Soil Bacteria,* University of Liverpool, University of Toronto Press, Toronto, Canada.

Jacobs, D. G., Epler, J. S. and R.R, R. (1980) Identification of technical problems encountered in the shallow land burial of low level radioactive wastes. Oak Ridge National Lab.,Oak Ridge.

Jager, J. and Peter, J. (1995) Messung de oberfl enemission cvon deponigas, *Stuttgrater Berichte Abfallwirtschaft*, 19: 337-345.

Janechek, A. and Prosser, R. (1995) Landfill gas collection and groundwater protection. In *Eigth International Madison Waste Conference,* Department of Engineering Professional Development, University of Wisconsin, Madison, pp. 9.

Jeevanrao, K. and Shataram, M. V. (2003) Soil and water pollution due to open landfills. In *Workshop on Sustainable Landfill Management,* Chennai, India, pp. 27-38.

Johnson, A., Roy, I. M., Matthews, G. P. and Patel, D. (2003) An improved simulation of void structure, water retention and hydraulic conductivity in soil with the Pore-cor-three-dimensional network, *European Journal of Soil Science*, 54: 477-489.

Jones, H. A. and Nedwell, D. B. (1993) Methane emission and methane oxidation in landfill cover soil, *FEMS Microbiological Ecology*, 102: 185-195.

Kalman, R., Sandor, K., Van Genuchten, M. and Per Erik, J. (1996) Estimation of water retention characteristics from the bulk density and particle size distribution of Swedish soils, *Soil Science*, 161: 832-845.

Kayuzhni, S., Epov, A., Sormunen, K., Kettunen, R., Rintala, J., Privalenko, V., Pender, S., Colleran, E. and Nozhevnikova, A. N. (2003) Evaluation of the current status of operating and closed landfills in Russia, Finland and Ireland with regards to water pollution and methane emission. In: *Anaerobic Digestion of Solid Wastes III* (Eds, Wilderer, P. and Moletta, R.) Munich.

Khire, M. V., Benson, C. H. and Bosscher, P. J. (1997) Water balance modelling of earthen final covers, *Journal of Geotechnical and Geoenvironmental Engineering*, 123: 744-754.

Khire, M. V., Benson, C. H. and Bosscher, P. J. (1999) Field data from a capillary barrier and model predictions with UNSAT-H, *Journal of Geotechnical and Geoenvironmental Engineering*, 125: 518-527.

Kiniry, J. R., Major, D. J., Izaurralde, R. C., Williams, J. R., Gassman, P. W., Morrison, M., Bergentine, R. and Zentner, R. P. (1995) EPIC model parametrs for cereal, oilseed and forage crops in northern great plains region, *Canadian Journal of Plant Science*, 75: 679-688.

Klang-Westin, E. and Eriksson, J. (2003) Potential of *Salix* as phytoextractor for Cd on moderately contaminated soils, *Plant Soil*, 249: 127-137.

Knox, K. (1992) Rotating biological contactors. In: *Landfilling of Waste: Leachate* (Eds, Christensen, T. H., Cossu, R. and Stegmann, R.) Francis and Taylor, pp. 211.

Koerner, R. M. and Daniel, D. E. (1997) *Final Covers for Solid Waste Landfills and Abandoned Dumps,* Thomas Telford, ASCE Press, pp. 253.

Komilis, D. P., Ham, R. K. and Stegmann, R. (1999) The effect of landfill design and operation on waste degradation behaviour: a review, *Waste Management and Research*, 17: 20 - 26.

Kopp, R. F., Abrahamson, L. P., White, E. H., Volk, T. A., Nowak, D. J. and Fillhart, R. C. (2001) Willow biomass production during ten successive annual harvests, *Biomass and Bioenergy*, 20: 1-7.

Korfiatis, G. P. and Demetracopoulos, A. C. (1986) Flow characteristics of landfill leachate collection systems and liners, *Journal of Environmental Engineering*, 112: 538-550.

Kowalchik,B.L. (2001), http//www.reo.gov-/col/wetland_classification, Viewed on 23/03/2005.

Kreith, F. (1994) *Handbook of Solid Waste Management,* MCGraw Hill, New York.

Kuvozkina, Y. A. and Quigly, M. F. (2004) Willows beyond wetlands: uses of *Salix* L. species for environmental projects, *Journal of Water, Air and Soil Pollution*, 162: 183-204.

Lee, W. J. and W. E. Schnabel (2005). *Implementation of a pilot scale evapotranspiration landfill cover for use in cold regions.* Proceedings of World Water and Environmental Recources, Anchorage, Alaska, USA, American Society of Civil Engineers.

Leone, I. A., Flower, F. B., Arthur, J. J. and E.F, G. (1977) Damage to woody species by anaerobic landfill gases, *Journal of Arboriculture*, 3: 221-225.

Levin, S. B. and Hammod, M. D. (1990) Application of PVC in a 'Top Cap' application. In *Geosynthetic testing for waste containment applications*(Ed, Robert, M. K.) American Society for Testing and Materials, Philadelphia.

Li, F. and Heine, W. (2007) Purification of landfill leachate with open channel reverseo osmosis membrane module. In *Landfill and Transfer Station 2007,* 19 -21 November, Melbourne, Australia.

Licht, L. (1990) Deep rooted poplar trees grown in the riparian zone for biomass production and non-point source pollution control. University of Iowa, IA.

Licht, L., Aitchison, E. and Rock, S. A. (2004) Evapotranspiration tree caps: research prototype results, full scale histories and possible future directions. In *SWANA Landfill Symposium,* Monterey, CA, pp. 1-13.

Lin, L. and Benson, C. H. (2000) Effect of went-dry cycling on swelling and hydraulic conductivity of GCL's, *Journal of Geotechnical and Geoenvironmental Engineering*, 126: 40-49.

Madalinski, K. L., Gratton, D. N. and Weisman, R. J. (2003) Evapotranspiration covers: An Innovative approach to remediated and close contaminated sites, *Remediation (online publication)*: 55-67.

Madison, M. F., Licht, L., Ricks, F. B. and Grabhorn, H. (1991) Landfill cap closure utilising a tree ecosystem, *American Society of Agricultural Engineers*, paper no. 912137.

McGuire, P. E., England, J. A. and Andraski, B. J. (2001) An evapotranspiration cover for containment at a semiarid landfill site. In *International Containment and Remediation Technology Conference,* Orlando, Florida.

Melchior, S. (1997) "In situ studies on the performance of landfill caps". In *Proc. Int. Containment technol. Conference*St. Petersburg, pp. 365-373.

Moore, S., Freund, P., Riemer, P. and Smith, A. (1998) Abatement of methane emissions. In *IEA greenhouse gas R&D programme,* CRE group LtD, UK, pp. 37.

Mosier, A. R. (1998) Soil processes and global change, *Biology and Fertility of Soils*, 27: 221-229.

Neufeld, D. A. (2000) An ecosystem approach to planning for groundwater: the case of Waterloo Region, Ontario, Canada, *Hydrogeology Journal*, 8: 239-250.

Nirmala, g., Lewis, B. and Finno, R. (1995) Microstructural changes in sand-bentonite soils when exposed to aniline, *Journal of Geotechnical Engineering*, 121: 119-125.

Nozhevnikova, A. N., Lifskitz, A. B., Lebedev, V. S. and Zavarirzin, G. A. (1993) Emission of methane into the atmosphere from th landfills in the former USSR, *Chemosphere*, 26: 401-417.

Nyhan, J. W., Hakonson, T. E. and Drennon, B. J. (1990) Water balance study of two landfill cover designs for semi arid regions, *Journal of Environmental Quality*, 19: 281-288.

Othman, M., Benson, C. H., Chamberlain, E. and Zimmie, T. (1994) Laboratory testing to evaluate changes in hydraulic conductivity caused by freeze thaw: state of the art. In *Hydraulic conductivity and waste containment transport in soils* (Eds, Trautwein, S. and Daniel, D. E.), pp. 227-254.

Paige, G. B., Stone, J. J., Lane, L. J., Yakowitz, D. S. and Hakonson, T. E. (1996) Evaluation of a prototype decision support system for selecting trench cap design, *Journal of Environmental Quality*, 25: 127-135.

Park, H. I. and Lee, S. R. (2002) Long term settlement behaviour of MSW landfills with various fill ages, *Waste Management and Research*, 20: 259-268.

Percy, I. and Truong, P. (2005) Landfill Leachate Disposal with Irrigated Vetiver Grass. In *Landfill 2005,* Brisbane, Australia, 7-9th September.

Perttu, K. (1993) Biomass production and nutrient removal from municipal wastes using willow vegetation filters, *Journal of Sustainable Forestry*, 1: 57-70.

Phillips, P. and Eberle, M. (1999) The use of Geosynthetic Clay Liners (GCL's in containment applications - an Australian perspective.

Pokryszka, Z., Tauziede, C. and Cassini, P. (1995) Development and validation of a mehtod for measuring biogas emission using a dynamic chamber. In *Fifth International Landfill Symposium*, Vol. 3 Sardinia, Italy, pp. 495-506.

Ragle, N., Kissel, J., Ongrerth, J. E. and Dewalle, F. B. (1995) Composition and variability of leachate from recent and aged areas within municipal landfill, *Water Environmental Research*, 67: 238-243.

Ramke, H. G. (1989) *Leachate collection systems,* Academic Press, New York.

Reid, J. B. and Huck, M. G. (1990) Diurnal variation of crop hydraulic resistance: a new analysis, *Agronomy. Journal*, 82: 827-834.

Reinhart, D. R. (1996) Full scale experiences with leachate recirculation landfills: case studies, *Waste Management and Research*, 14: 347.

Richards, K. (1989) Landfill gas: Working with Gaia, *Biodeterioration abstracts*, 3: 317-331.

Riemenschneider, D. E., Berguson, W. E., Dickmann, D. I., Hall, R. B., Isebrands, J. G., Mohn, C. A., Stanosz, G. R. and Tuskan, G. A. (2001) Poplar breeding and testing strategies in the north- central U.S : Demonstration of potential yield and consideration of future research needs., *The Forestry Chronicle*, 77: 245-253.

Rittmann, B. E., Fleming, I. R. and Rowe, R. K. (1998) Leachate chemistry: its implication for clogging, *North American Water and Environment Congress and Destructive Water*: 28-33.

Robinson, B., Green, S., Mills, T., Clothier, B., Laplane, R., van der Velde, M., Fung, L., Deurer, M., Hurst, S., Thayalakumaran, T. and Van Den Dijssel, C. (2003) Phytoremediation: using plants as biopumps to improve degraded environments, *Australia Journal of Soil Research*, 41: 599-611.

Robinson, H. D., Barr, M. J. and Last, S. D. (1992) Leachate collection, treatment and disposal, *Journal of the Institution of Water and Environmental Management*, 6: 321-332.

Rohrs, L. H. and Fourie, A. B. (2002) The myth of sustainable landfills in South Africa. School of Civil Engineering, University of Witwatersrand.

Rowe, R. K., Fleming, I. R., Armstrong, M. D., Cooke, A. J., Cullimore, D. R., Bennett, P., Longstaffe, F. J. and Rittman, D. R. (1997) Recent advances in understanding the clogging of leachate collection systems. In *Proceedings of the Sixth International Landfill Symposium, Lining, drainage and Stability in Landfills*Sadrinia, pp. 383-392.

Sanchez, L. D. M., Dorronsoro, J. L., Fernadez-Serrano, M. E. and Herraez, I. (1993) Study of the leachates generated by two different landfills located in the municipality of Madrid. In *Fourth International Landfill Symposium,* Cagliari, Italy.

Scanlon, B. R., Marty, C., Reedy, R. C., Porro, I., Simunek, J. and Flerchinger, G. N. (2002) Intercode comparisons for simulating water balance of surfacial sedimanets in semi arid regions, *Water Resources Research*, 38: 1323 - 1338.

Scanlon, B. R., Reedy, R. C., Keese, K. E. and Dwyer, S. F. (2005) Evaluation of Evapotranspiration covers for waste containment in arid and semiarid regions in south western USA, *Vadose Zone Journal*, 4: 55-71.

Schmitz, M. and Sourell, H. (2000) Variability in soil moisture measurements, *Irrigation Science*, 19: 147-151.

Scott, J., Beydoun, D., Amal, R., Low, G. and Cattle, J. (2005) Landfill management, leachate generation and leach testing of solid wastes in Australia and overseas, *Critical Reviews in Environmental Science and Technology*, 35: 239-332.

Sengupta, S., Trehan, J. and Koul, S. (1998) Determination of methane emission from municipal solid waste by gas chromatography, *Fresenius Environmental Bulletin*, 7: 413-416.

Shackelford, C., Benson, C. H., Katsumi, T., Edil, T. and Lin, L. (2000) Evaluating hydraulic conductivity of GCL's permeated with non standard liquids, *Journal of Geotextiles and Geomembrane*, 18: 133-161.

Shouse, P. J., Russell, W. B., Burden, D. S., Selim, H. H., Sisson, J. B. and Van Genuchten, M. (1995) Spatial variability of soil water retention functions in a silt loam soil, *Soil Science*, 159: 1-12.

Simon, F. and Muller, W. W. (2004) Standard and alternative landfill capping, *Environmental Science and Policy*, 7: 277-290.

Simunek, J., Van Genuchten, M. T. and Sejna, M. (2005) The Hydrus1D software package for simulating the one dimensional movement of water, heat and multiple solutes in variably saturated media. University of California - Riverside, California, USA, pp. 240.

Spokas, K., Bogner, J., Chanton, J., Morcet, M., Aran, C., Graff, C., Moreau le Govan, Y. and Hebe, I. (2006) Methane mass balance at three landfill sites: what is the efficiency of capture by gas collection systems, *Waste Management*, 26: 516-525.

Stonell, K. (1986) Design parameters for containment sites. In *Landfill water management symposium proceedings.* (Ed, Harwell) Harwell, UK, Harwell, UK, pp. 30-38.

Sullivan, Patrick, S. and Raymond, H. (2004) Air quality and odour impacts from landfill related emissions. In *Odours and Emissions 2004*Bellevue, WA - USA, pp. 531-557.

Swarsbrick, G. and Dever, S. (2004) Landfill biofiltration trails: Preliminary design and analysis. In *Proceedings from Enviro 2004,* Melbourne, Australia.

Tchobanoglous, G., Theisen, H. and Vigil, S. (1993) *Integrated solid waste management: engineering principles and management issues,,* McGraw Hill, Singapore.

Themelis, N. J. and Ulloa, P. A. (2007) Methane generation in landfills, *Renewable Energy*, 32: 1243-1257.

Thorton, P. (1994) Do landfills have a future in Australia? In *Proceedings from Enviro 2002,* Melbourne, Australia.

Timlin, D. J., Ahuja, L. R. and Heathman, G. C. (1998) Preferential transport of a bromide tracer applied in a pulse method of ponded water, *Journal of Environmental Quality*, 27: 505-514.

Tohjima, J. and Wakita, H. (1993) Estimation of methane discharge from a plume: a case of landfill, *Geophysics Research Letters*, 20: 2067-2070.

Trotter, H. D. and H. A. Cooke (2005). "Influence of landfill gas on the microdisctribution of grass establishment through natural colonisation." *Environmental Management* 35: 303 - 310.

van Moorst, H. (1996) Some preliminary research findings, report for Werribee Residents Against Toxics Dumps.

VanGulck, J. F., Rowe, K., R, Rittmann, B. E. and Cooke, A. J. (2003) Predicting biogeochemical calcium precipitation in landfill leachate collection systems, *Biodegradation*, 14: 331-346.

Vasudevan, N. K., Vedachalam, S. and Sridhar, D. (2003) Study on the various methods of landfill remediation. In: *Workshop on Sustainable Landfill Management,* Chennai, India, pp. 309-315.

Venkatraman, K, Ashwath, N and Su, N (2008) Environmental Performance of a Phytocapped landfill in a semi arid region. In: *Proceedings of 2008 National Landfill Managers and Operators Conference*, Gold Coast, Australia, pp. 28 – 33.

Venkatraman, K. and Ashwath, N. (2006) Canopy interception contributes up to 10% reduction in effective rainfall in a phytocapped landfill In *2nd International Conference on Environmental Science and Technology,* Houston, Texas, USA.

Venkatraman, K. and Ashwath, N. (2006) Phytocapping: can it mitigate methane emissions from municipal landfills? In: *2nd International Conference on Environmental Science and Technology,* Houston, Texas, USA.

Venkatraman, K. and Ashwath, N. (2007) Phytocapping: a potential alternative technique to clay caps. In *Landfill 2007,* 19 -21 November, Melbourne, Australia,.

Venkatraman, K. and Ashwath, N. (2007) Phytocapping: an Alternative technique for reducing leachate and methane generation from municipal landfills. *The Environmentalist*, 27: 155 – 164.

Venkatraman, K. and Ashwath, N. (2008) Phytocapping of Landfills – An Australian Experience, In: *5th International Phytotechnologies Conference*, Nanjing, China.

Venkatraman, K. and Ashwath, N. (2009) 'Can phytocapping technique reduce methane emission from municipal landfills?' *International Journal of Environmental Technology and Management*, Vol. 10, No. 1, pp.44–55.

Venkatraman, K., Ashwath, N. and Sharma, A. (2006) Water use by Australian native species grown on a phytocapped landfill site. In: *2nd International Conference on Environmental Science and Technology.* Houston, Texas, USA.

Von Der Hude, N., Melchoir, S. and Mockel, S. (1999) Construction of a capillary barrier in the cover of the Breinermoor landfill. In: *Seventh International Waste Management and Landfill Symposium,* CISA, Caligari, Italy, pp. 1-12.

Warren, R. W., Hakonson, T. E. and Bostik, K. V. (1996) choosing the most effective hazardous waste landfill cover., *Remediation (online publication).*

Waste Services, N. S. W. (1999) Annual Report 98/99. Waste Services NSW.

Waugh, W. J., Thiede, M. E., Bates, D. J., Caldwell, L. L., Gee, G. W. and Kemp, C. J. (1994) Plant cover and water balance in gravel ad-mixtures at an arid waste-burial site., *Journal of Environmental Quality*, 23: 676-685.

Weand, B. L., Horin, J. D., Hauser, V. L., Gimon, D. M., Gill, M. D., Mehta, M. and Casagrande, D. J. (1999) Landfill covers for use at Air Force installations. AFCEE/ERT Texas.

Whalen, S. C., Reeburgh, W. S. and Sandbeck, K. A. (1990) Rapid methane oxidation in a landfill cover soil, *Applied Environmental Microbiology*, 56: 3405 - 3411.

Williams, J., Prebble, R. E., Williams, W. T. and Hignett, C. T. (1983) The influence of texture, structure and clay mineralogy on the soil moisture characteristics, *Australian Journal of Soil Research*, 21: 15-32.

Willumsen, H. (2002) Energy recovery from landfill gas in Denmark and worldwide. www.lei.lt/Opet/pdf/Willumsen.pdf, Viewed on 23/12/2007

Wing, R. N. and Gee, G. W. (1994) Quest for a perfect cap, *Civil Engineering*, 64: 38-41.

WMAA (2007) Responses to the discussion paper on a possible design for a national greenhouse gas emissions trading scheme. Waste Management Association of Australia.

Xu, X., Rudolph, V. and Greenfield, P. F. (1999) Australian Urban landfills: Management and Economics, *Waste Management and Research*, 17: 171.

Yeates, R., Stuart, R., Jones, M. (2008) The Australian Alternative Covers Assessment Program. In: *Proceedings of the 5th International Phytotechnologies Conference*, Nanjing, China, pp 126 - 137

Yuen, S. T. S. (1999) Bioreacteor landfills promoted by leachate recirculation: A full scale study. In: *Department of Civil and Environmental Engineering,* University of Melbourne, Melbourne.

Zheng, C., Bennett, G. D. and Andrews, C. B. (1991) Analysis of ground water remedial alternatives as a superfund site, *Groundwater*, 29: 838-848.

Zomleter, W. B. (1994) Guide to flowering plant families. In: *The University of North Carolina Press,* Chapel Hill and London, pp. 430.

In: Landfills and Recycling Centers
Editor: Carla H. Jackson
ISBN: 978-1-63463-792-3

Chapter 6

Health and Ecological Risk of Methane and Volatile Organic Compounds Emission from Solid Waste Disposal: A Case Study in Thailand

Sansarith Thianpopirug, Sudarat Kucamau, Wilai Chiemchaisri and Chart Chiemchaisri*
Department of Environmental Engineering, Faculty of Engineering, Kasetsart University, Chatuchak, Bangkok, Thailand

Abstract

The objective of this research was to evaluate the emission and distribution of methane and volatile organic compounds (VOCs) from solid waste disposal site in Thailand. Gaseous emissions from solid waste piles with and without soil cover were determined. Highest average methane emission rate was found to be 53.51 $g/m^2.d$ from solid waste pile with cover soil during rainy season. This rate was significant higher than 20.87 $g/m^2.d$ under dry season and 5.23 $g/m^2.d$ and 3.99 $g/m^2.d$ from open waste pile under rainy and dry conditions respectively. The emission of BTEX compounds was found highest from the waste pile with cover soil under dry condition at $7.11x10^{-2}$, $9.76x10^{-2}$, $4.8x10^{-2}$ and $7.62x10^{-2}$ $g/m^2.d$. The source of VOCs was identified in fresh waste samples. The dispersion of methane and BTEX compounds in ambient air was modeled. The toxicity of BTEX compounds to workers and nearby communities was at low level as determined by hazard index. However, the carcinogenicity risk from benzene to those people were found to be $2.45x10^{-5}$ and $2.85x10^{-5}$, respectively. Geno-toxicity study using plant comet assay (Pothos) suggested that there was a relationship between DNA damage after 96 hours exposure and VOC emission rate from solid waste disposal area.

Keywords: Gas emission, landfill, methane, solid wastes, volatile organic compounds

---

* E-mail: corresponding author: fengccc@ku.ac.th.

1. Introduction

About 65% of municipal solid wastes collected in Thailand are disposed by unsanitary method [1] which creates several environmental problems. The waste disposal activities also released gaseous pollutants, mainly methane and carbon dioxide. In addition to those greenhouse gases, volatile organic compounds (VOCs), though present at lower concentration, could pose severe hazard to human and the environment. In Thailand, it was estimated that about 0.28 million tons of methane was released from solid waste disposal sites to the atmosphere each year [2]. Globally, about 20-70 million tons of this greenhouse gas was emitted from waste disposal activities [3]. Non-methane organic compounds emission was estimated at 13,000 tons [4]. In the past decade, many countries have paid the attention on methane and VOCs emission from solid waste disposal site for adoption of mitigation measures. In developed countries, USA, municipal landfills in operation need to comply with regulations to reduce methane and VOCs emission from the diversion of organic and hazardous wastes from municipal landfills. Nevertheless, the study in developing countries on this issue is still very limited.

This study is focusing on the field investigation of methane and VOCs (Benzene, Toluene, Ethyl-benzene and Xylene or BTEX compounds) emission from solid waste disposal site in Thailand. The scope of study includes determination of methane and VOCs emission rate from waste pile with and without cover soil, contamination of BTEX compounds in solid waste components, dispersive transport of methane and BTEX compounds off the site and their effect to workers and nearby community health risk. Furthermore, ecological risk of gaseous emission to plants was evaluated using plant comet assay.

2. Methodology

2.1. Site Characteristic

The study was conducted at a solid waste disposal site in Thailand (Figure 1). The site covers an area of about 9.28 ha and it has been in operation since 1982. Total amount of waste deposited at the site was more than 2 million tons. In 2007, this site has 3 closed waste cells and 2 waste dumping piles in operation. The quantity of solid wastes received at the site was about 800 tons/day. The wastes were disposed in open dumping manner forming a mountain of about 7-10 meters above ground. Cover soil was occasionally placed on the waste pile at every 6 months interval. Leachate formed during raining period at the site was gravitationally drained from the waste pile through open channel and pumped into an open storage pond.

The surrounding topographic feature is flat terrain with maximum elevation difference of less than 2 meters. The site is surrounding by rice fields. Residential areas were located about 1 kilometer adjacent to the site in northern, northeastern, eastern and southeastern directions. There are two nearby canals where the water is utilized for agricultural and domestic consumption purposes.

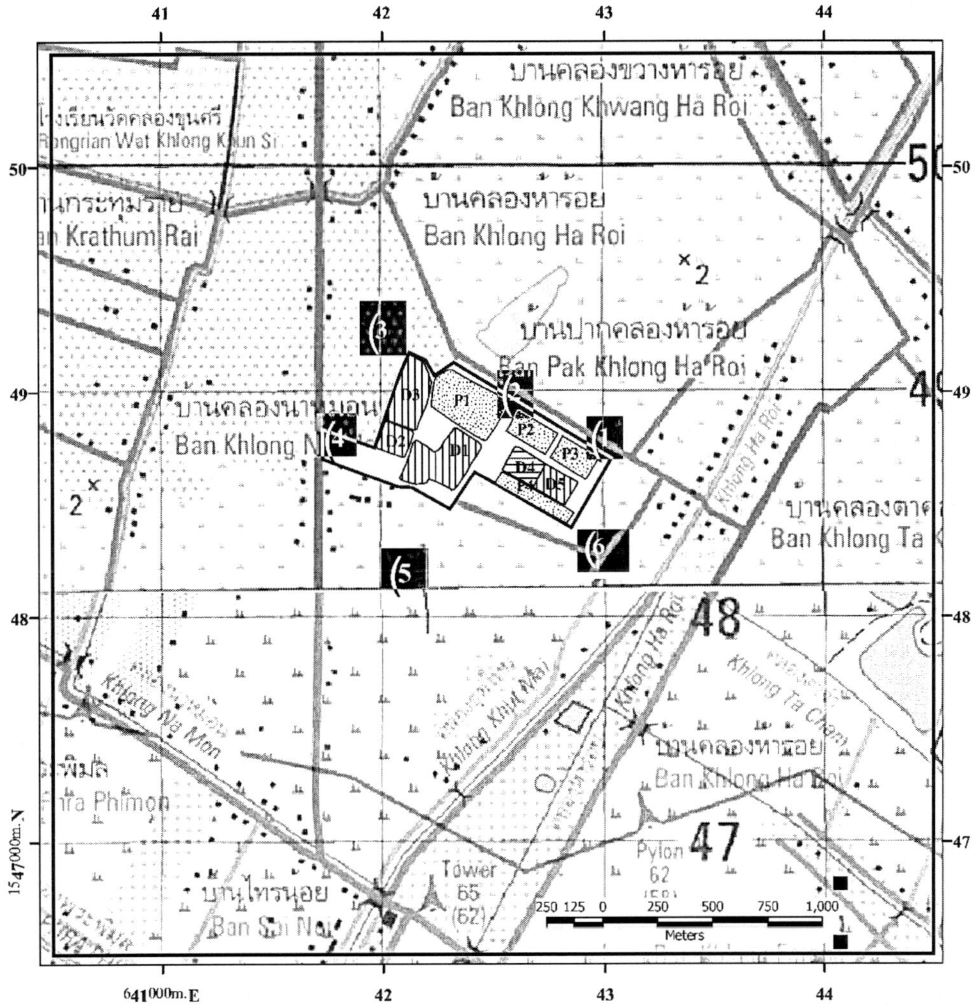

Figure 1. Geographical location of solid waste disposal site.

2.2. Collection of Samples and Analysis

Surface isolation flux chamber operated under static mode [5] was used for the determination of methane and VOCs emission rate from the surface of the two operating waste piles (1 year old). The gas emission rate was measured at surface of waste pile with and without cover soil. The schematic diagram of closed flux chamber used for the determination of gas emission rate is shown in Figure 2. The chamber was made from stainless steel with 0.049 m^3 volume and 0.188 m^2 cover area.

After the chamber had been placed on the surface of the waste pile (either with or without cover soil), gas samples were collected at the starting time and consecutively at every 20 minute interval for 1 hour. The methane concentration in collected samples was then analyzed using gas chromatography (Agilent 6890-ECD detector).

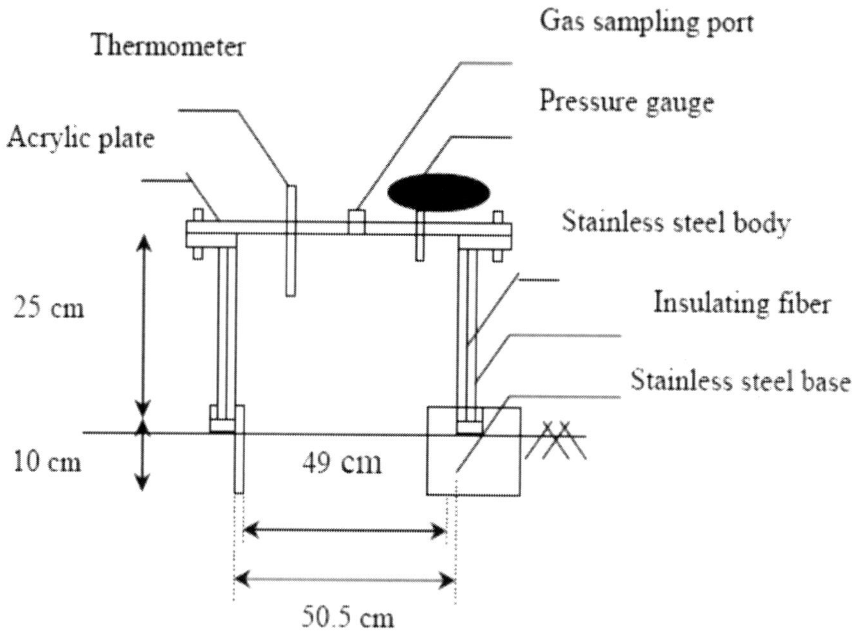

Figure 2. Schematic of closed flux chamber.

The closed flux chamber was operated by allowing gas from surface of waste pile to accumulate inside the chamber and the increment of gas concentration was determined for a period of time. The increasing rate in gas concentration with time was used to determine the gas flux as described in equation 1.

$$F = \frac{\rho V \Delta C}{A \Delta t} \qquad (1)$$

where F is the flux of gas (g/m^2.h), ρ is the density of the gas (g/m^3), V is the volume of chamber (m^3), A is the surface of chamber enclosed area (m^2), ΔC is the change of gas concentration, and Δt is the time interval which samples were taken (h).

For VOC determination, the gas samples were withdrawn from the chamber after 1 hour. The optimum sampling period was referred from our previous study [6]. The collected gas sample was passing through the adsorbent tube (SKC-activated coconut shell charcoal-20/40mesh) at 30 ml/min flow rate for 20 minutes for VOC adsorption. The adsorbed VOCs in the adsorbent tube were then extracted by using carbon disulfide (CS_2) and analyzed by gas chromatography (Shimadzu GC14B-FID detector).

The emission rate from closed flux chamber samples were interpolated into spatial average emission by inverse distance weight (IDW) interpolation method. The VOC concentrations in ambient air were determined at upwind and downwind sampling stations (indicated as points No.1, 2, 5 and 6 in Figure 1). The ambient air sampling stations (point no.1 and 2) were located on downwind direction during the sampling period whereas point No.5 and No.6 were the sampling stations towards the residential area. The station No. 5 is located near residential village and point No. 6 is located at an agricultural village.

2.3. Pollutant Dispersion Model-AERMOD

The transportation of VOCs was determined by AERMOD model. The AMS/EPA regulatory model (AERMOD) is a steady-state gaussian plume modeling tool for assessing potential environmental and health effects of continuous emissions to the atmosphere. The AERMOD can be used to assess pollutant concentration downwind from a source, including point, area, volume and open pitch source type [7]. For the application of the AERMOD at solid waste disposal site, area sources characterization was used to simulate emission that initially disperse in two dimensions with little or no plume rise, such as ground-level or low-level emissions from this study site. It should be noted that the approach that AERMOD uses to address plume meander had not been implemented for area sources. As a result, concentration predictions for area sources may be overestimated under very light wind condition. In general, this is not expected to be a problem for meteorological data collected using standard wind instruments that threshold are generally too high and a worst case scenario was applied. In order to avoid over estimates for area sources during light wind condition, simulation using pseudo-point sources is used.

The assessment of methane and VOCs emission from two solid waste dumping piles currently in operation was based on direct field measurement, whereas the emission rates from other three closed waste cells were obtained from previous study [5, 6]. These emission rates were then input into AERMOD model. They were also varied with the presence or absence of cover soil. Seasonal effect was determined by varying precipitation and evaporation in each month using actual meteorological data. In this study year, the wet season started from July until October and other months were classified as dry season. Other required meteorological data were wind speed, wind direction, dry bulb temperature, cloud cover and ceiling height. They were obtained from a weather station in Bangkok which is the nearest site with available information. The surface meteorological data were then rearranged into format suitable for input (SCRAM) before transfer into preprocessing process for the calculation of planetary boundary layer. The upper air sounding data or rawinsonde was also obtained from Bangkok's weather station. The cartesian grid receptor networks were defined in coverage domain of 4x4 km^2, each receptors were 100 meter interval. Thus, 1681 receptors were calculated for predictable concentration. In AERMOD model, three control parameters related to the surface characteristics were the surface roughness, albedo and the bowen ratio. These parameters were used to calibrate model by matching data from model and ambient air concentration sampling. These control parameters were then corrected using the values suggested by the model developer [7].

2.4. Health Risk

There are two types of assessment on health effects: non-carcinogenicity and carcinogenicity [8]. In this study, the non-carcinogenic health risk from dispersed VOCs exposed to the workers at the site and population in nearby communities was determined by hazard index. The VOC concentrations were derived from annual average concentrations from AERMOD model. Hazard quotient (HQ) or Hazard index (HI) was used to indicate level of non-carcinogenic risk by comparing the intake rate with reference dose (Rf_d, mg/kg.d). The critical ratio is defined as 1.0 [8]. Inhalation intake was derived from Equation 2 and inhalation reference dose was retrieved from IRIS [9].

$$I = \frac{CRt_g f_g D_t}{W_B t_{avg}} \quad (2)$$

$$HI = I / Rf_d \quad (3)$$

where I is the inhalation intake (mg/kg.d), C is the gas concentration (mg/m^3), R is the rate of inhalation (average 20 m^3/day for adult), t_g is the time of exposure (hours/day), f_g is the frequency of exposure (days/year), D_t is the duration of exposure (year), W_B is the body weight (70 kg average), and t_{avg} is the total days of exposure (days; equal to D_tx365). Hazard index (HI) was determined from Equation 3 using intake rate (mg/kg.d) and reference dose (Rf_d, mg/kg.d) of compounds. The same amounts of concentrations of VOC compounds in atmosphere via inhalation were applied in calculation for both effects. For carcinogenic risk, the duration of exposure for carcinogenicity is lifetime exposure (70 years). Equation 4 is used in order to assess the cancer risk [8]. Because only benzene in this case is classified as carcinogen [13] whereas the others (ethylbenzene, toluene and xylene) are identified as non-carcinogens for human due to inadequate up to date experimental data [13] . A slope factor (SF, $(mg/kg.d)^{-1}$) of benzene 2.9×10^{-2} $(mg/kg.d)^{-1}$ [8] was applied for assessment of carcinogenicity in this study.

$$\text{Carcinogenic Risk} = I \times SF \quad (4)$$

2.5. Ecological Risk

During the last decade, the Comet assay has been extensively applied to plants (leaves, shoot, and roots) to detect DNA damage arising due to more chemicals in polluted areas [14]. Gichner and Plewa (1998) [15] developed a sensitive method for isolation of nuclei from leaf tissue of *Nicotiana tabacum*. The method resulted in high resolution and constant, low tail moment values for negative controls, and hence could be incorporated for in situ plant environmental monitoring. Pothos *(Epipremnum aureum)*, is a climbing herbaceous plant. Fully developed leaves are large and waxy. In term of biomonitor, it was used successfully via Comet assay in evaluation of genotoxicity caused by air pollutants from transportation [16]. It was selected as the monitoring species for evaluation geno-toxicity in order to reveal relationship between the DNA damage in Pothos and the polluted air caused from landfill gas emission. Pothos used in this study was obtained from a local breeder and transported immediately to the laboratory. Seedlings of -15days age, Pothos were cultured in 30x30x50 cm^3 pots area (compacted in clean coconut husk). At one month of age, a set of 20 healthy plants which having 20 cm heights in average was used in the experiment. Determination of total VOC emission was perform from 8 am to 2 pm during wet season (September to November) and dry season (February to April) monthly basis. The approach or parameter used must be clearly defined [17]. The percentage of DNA damage can be calculated as illustrated in (5).

$$\% \text{ of DNA damage} = \frac{\sum_{i=1}^{n}(T/(T+H))}{n} X100 \quad (5)$$

When T is the tail area of comet image, H is the head area of comet image, n is the number of comet images [16].

Statistical Analytical for comet assay was performed using the following procedures. Frequency distribution of DNA migration ratio showed 100% of data in 3-D visual effect. The percentages of DNA damages with the exposure time were plotted where exponential equation analysis was employed to evaluate the correlation between the percentages of DNA damage observed in test species and the exposure time using SPSS (version.16) statistical for windows. The relationships between physicochemical parameters (COD, BOD, ammonia, pH and conductivity) and DNA damages of the test plants using Pearson correlation and the association between effective concentration (EC) and physicochemical parameters was performed using multiple linear regressions. Correlation coefficient (x) was classified according to the following criteria; [x] < 0.5 = poor relationship, 0.5<[x]< 0.8 = moderate relationship, [x]>0.8 = strong relationship. For the statistical analysis, significance was evaluated at $P < 0.01$ and $P < 0.05$.

3. RESULTS

3.1. Methane and VOCs Emission from Waste Pile

Surface emission rate of methane (M) and VOCs (BTEX compounds) from solid waste pile without cover soil was determined as shown in Table 1. For comparison, the emission from waste pile with cover soil was presented in Table 2. It was found that average methane emission rate from open waste pile ranged between 2.47 and 7.51 $g/m^2.d$ during dry period with an average of 3.99 $g/m^2.d$. During wet period, the emission rate was 0.51-13.07 $g/m^2.d$ (avg. 5.23 $g/m^2.d$). These emission rates were considerably lower than that of waste pile with cover soil (20.87 and 53.51 $g/m^2.d$ during dry and wet period respectively) as shown in Figure 3. The results suggest that anaerobic condition and moisture content in solid waste pile were the key factors governing methane emission from solid waste disposal site. The moisture content in cover soil affected the microbial activities, and microorganisms could be inactive when the moisture content dropped below 5% [10].

Table 1. Methane and VOCs emission from waste pile without cover soil in dry and rainy seasons

	Emission rate ($g/m^2.d$)			
	Dry season		Rainy season	
	Range	Avg.	Range	Avg.
M	2.47-7.51	3.99	0.51-13.07	5.23
B	0.0107-0.0279	0.0214	0.0044-0.0203	0.0103
T	0.0067-0.0329	0.0194	0.0065-0.0203	0.0140
E	0.0080-0.0115	0.0101	0.0050-0.0105	0.0088
X	0.0098-0.0197	0.0151	0.0047-0.0122	0.0092

Note: M: methane, B: benzene, T: toluene, E: ethyl-benzene, X: xylene.

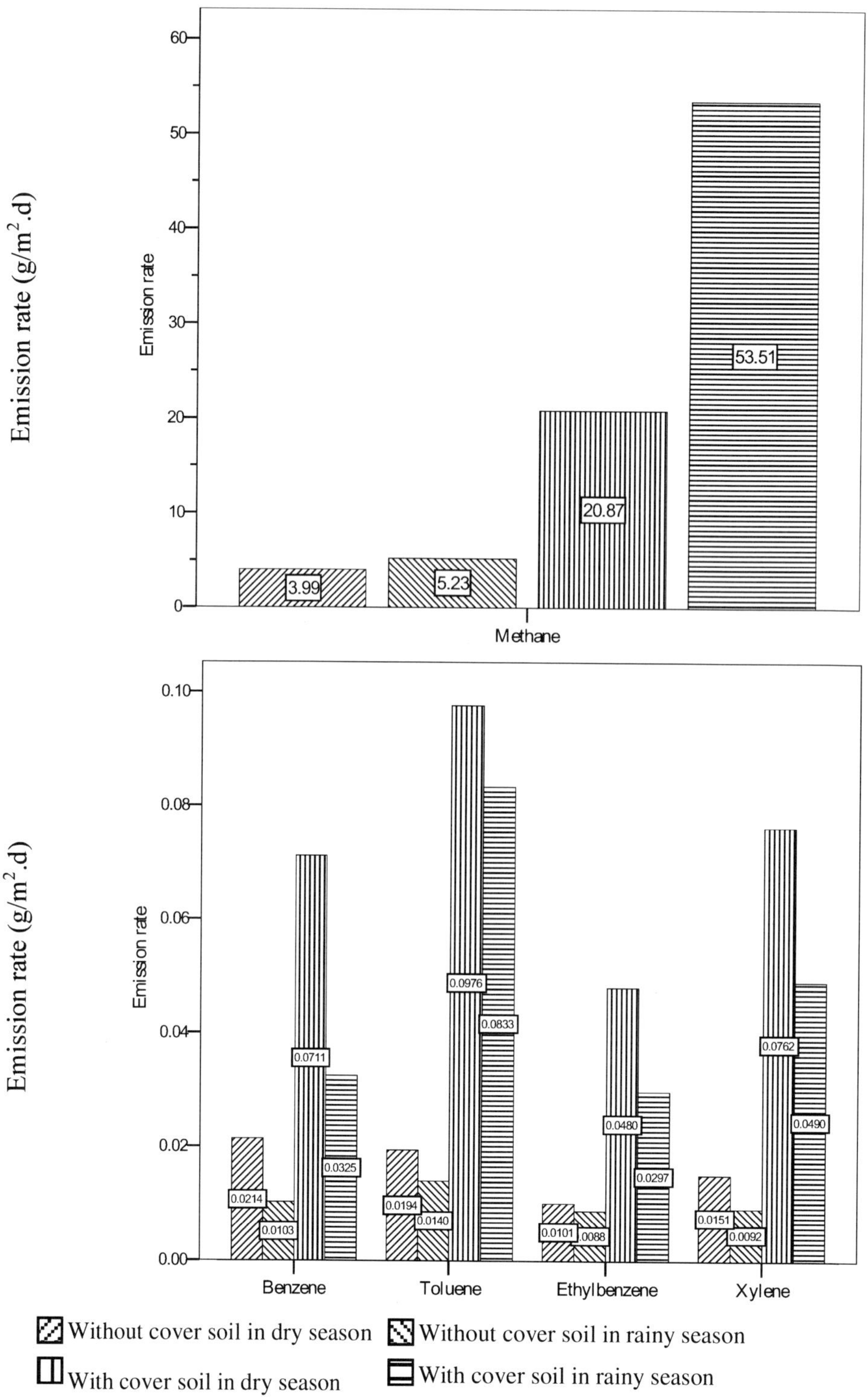

Figure 3. Comparison in methane and VOC emission from waste pile with and without cover soil during dry and rainy period.

Table 2. Methane and VOCs emission from waste pile with cover soil in dry and rainy seasons

	Emission rate (g/m^2.d)			
	Dry season		Rainy season	
	Range	Avg.	Range	Avg.
M	3.20-53.93	20.87	20.01-86.70	53.51
B	0.0248-0.1401	0.0711	0.0046-0.0942	0.0325
T	0.0135-0.1921	0.0976	0.0191-0.1930	0.0833
E	0.0211-0.0794	0.0480	0.0090-0.0669	0.0297
X	0.0387-0.1070	0.0762	0.0205-0.0740	0.0490

Note: M: methane, B: benzene, T: toluene, E: ethyl-benzene, X: xylene.

The emission rates of BTEX compounds from open solid waste pile during dry period were 2.14x10^{-2}, 1.94x10^{-2}, 1.01x10^{-2} and 1.51x10^{-2} g/m^2.d, higher than 1.03x10^{-2}, 1.40x10^{-2}, 0.88x10^{-2} and 0.92x10^{-2} g/m^2.d during wet period. Higher emission rates were detected from solid waste pile with cover soil during dry period, i.e. 7.11x10^{-2}, 9.76x10^{-2}, 4.80x10^{-2} and 7.62x10^{-2} g/m^2.d, whereas they were 3.25x10^{-2}, 8.33x10^{-2}, 2.97x10^{-2} and 4.90x10^{-2} g/m^2.d during wet period. The temperature in the waste pile was found to be the major factor affecting VOCs emission from solid waste pile [6]. During the higher temperature period especially in summer with dry condition, higher emission rates were detected as compared to the rainy season [11, 12].

In our previous study [6], the sources of VOCs in fresh waste samples were investigated. The emission VOCs from mixed and separated waste component was determined at 40 and 80°C simulating ambient and disposed waste conditions. The contamination of toluene and benzene were detected at 6.46 and 0.47 mg/kg. The elevated temperature increased VOCs emission from waste. The waste components contributing majority of VOCs were plastic and yard wastes. Toluene was found at highest concentration in mixed wastes.

3.2. Distribution of VOCs in Ambient air

BTEX level in ambient air was determined by direct measurement at different sampling stations locating surround the site. Maximum BTEX concentrations detected were 0.71, 0.59, 0.17and 0.43 □g/m^3. These results were used to calibrate the control parameters (the surface roughness, albedo and bowen ratio) in AERMOD model. Moderate agreement (R=0.55) between the coefficients from the field measurement and model was obtained.

Annual average concentrations of BTEX compounds within the site and at nearby communities, as suggested by AERMOD model are shown in Table 3. They were averaged at 0.20, 0.28, 0.13, 0.20 μg/m^3 and 0.03, 0.05, 0.02, 0.03 μg/m^3 respectively. Figure 4 presents the box plot showing minimum and maximum concentration range, 25th percentile, 75th percentile, median, and outlier which presenting extra higher concentration in ambient air outside the receptor area. Considering in nearby communities, the annual average concentration of BTEX is not exceeding 0.1 μg/m^3, but higher concentration could be found over 0.1-0.5 μg/m^3 in some other areas. Figure 5 shows the affected area from the dispersion of benzene surrounding the solid waste disposal site.

Table 3. Annual average BTEX concentrations in ambient air as suggested by AERMOD model

	Ambient air concentration (μg/m^3)			
	Waste disposal site		Nearby communities	
	Range	Avg.	Range	Avg.
B	0.06-0.40	0.20	0.003-0.28	0.03
T	0.07-0.65	0.28	0.004-0.41	0.05
E	0.03-0.28	0.13	0.002-0.18	0.02
X	0.05-0.44	0.20	0.003-0.28	0.03

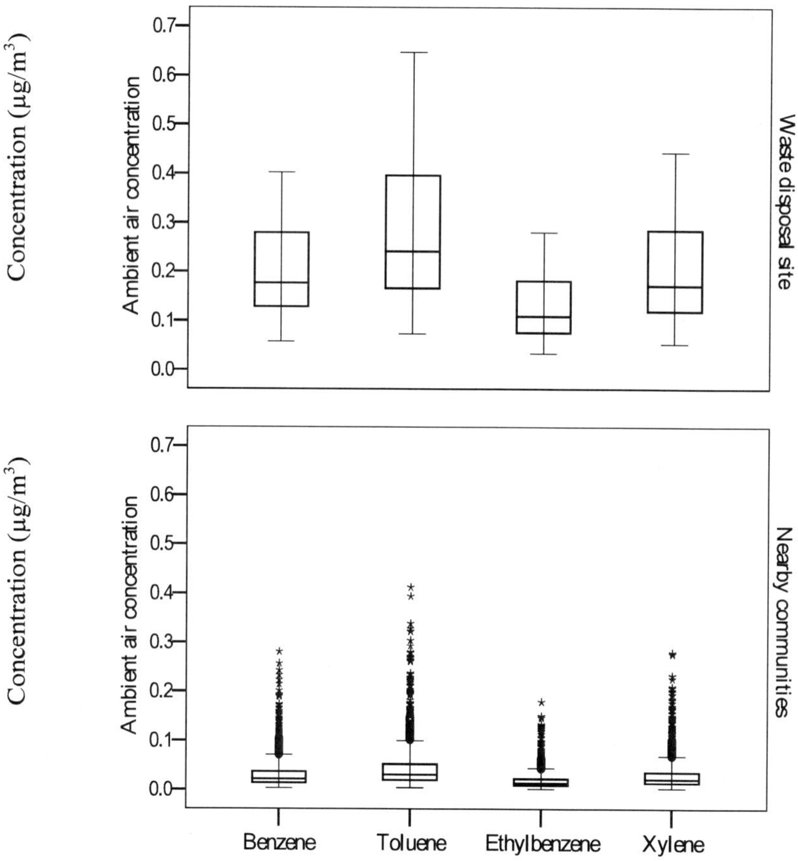

Figure 4. Box plot of annual average concentration of benzene.

In Figure 5, the dispersive transport of benzene was mainly into the north and southeast direction from the site. As a result, the affected areas from this dispersion include the agricultural village and residential village (station No.6 and 5 in Figure 1 respectively).

The annual average BTEX concentrations (Figure 5) were derived from constant input from the defined sources in 2007. In order to study the seasonal and wind effect, wind-rose diagram of the study area as shown in Figure 6 was used. From the diagram, wind direction covered most except northern direction. In addition to wind speed and direction, the temperature and cloud cover were also varied in each month giving different micro-climate condition. The emission rates from waste piles were also found varied monthly depend on seasonal condition. Therefore, monthly average concentration was also determined giving the possible impact from the source at any direction and period of the year.

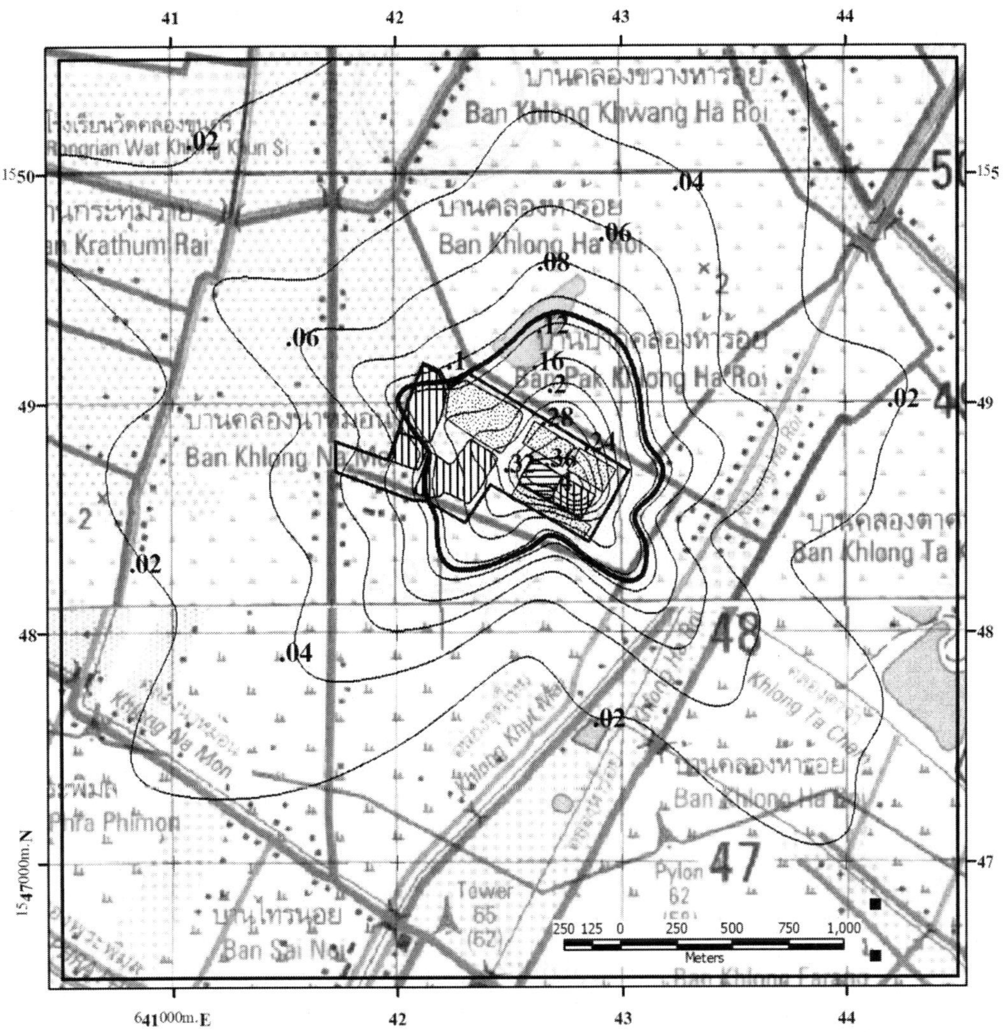

Figure 5. Annual average concentration of benzene from AERMOD model.

Highest monthly average concentration of BTEX compounds to the workers within the site and nearby communities, as suggested by AERMOD model, were 0.47, 0.64, 0.30, 0.47 $\mu g/m^3$ and 0.10, 0.13, 0.06, 0.10 $\mu g/m^3$ respectively. The results in Table 4 and Figure 7 also show the variation of maximum monthly average value within the disposal site and nearby communities.

Figure 8 shows the dispersion pattern of benzene by using monthly average concentration in each month of the year. The concentrations were normalized by the difference between 95th percentile and 5th percentile of annual average as reference value and then shown as 0.5 and 1.0 level in map for each month.

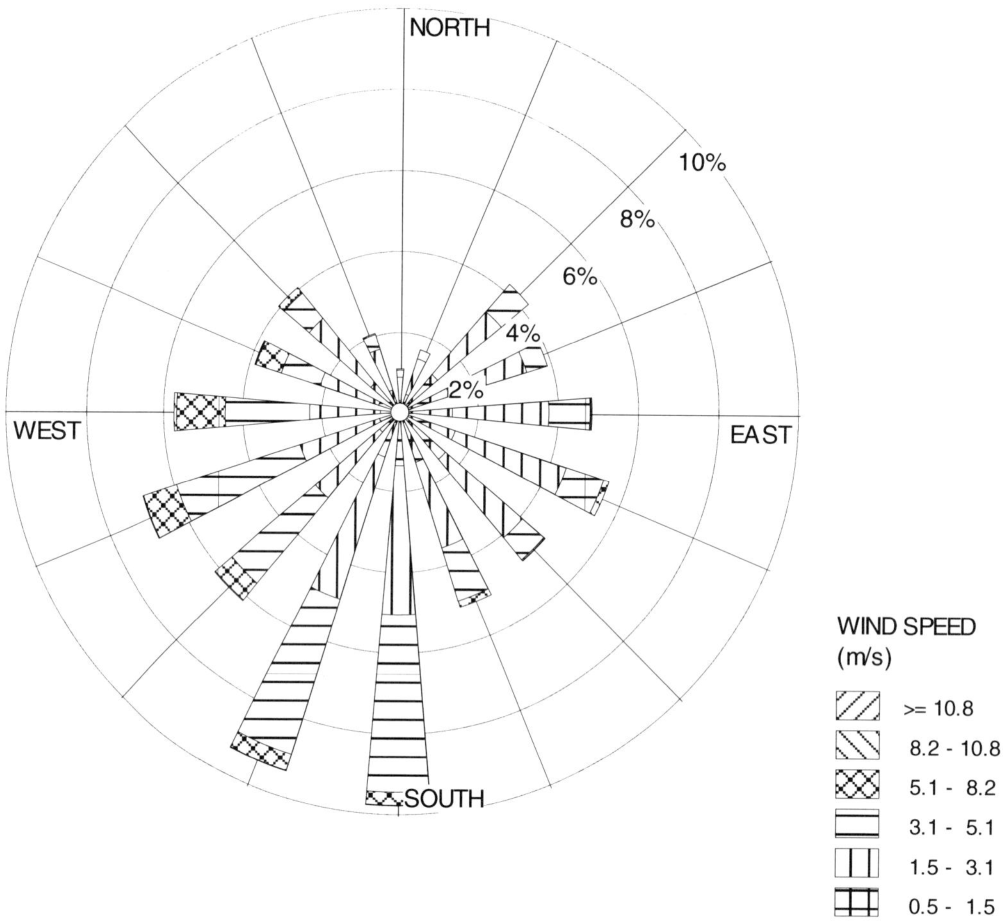

Figure 6. Wind rose diagram in year 2007 at the weather station in Bangkok.

Table 4. Highest monthly average ambient air concentration of BTEX compounds

	Ambient air concentration (μg/m^3)			
	Waste disposal site		Nearby communities	
	Range	Avg.	Range	Avg.
B	0.14-0.87	0.47	0.009-0.76	0.10
T	0.16-1.50	0.64	0.010-0.96	0.13
E	0.08-0.56	0.30	0.005-0.48	0.06
X	0.12-0.89	0.47	0.009-0.75	0.10

From the model, highest concentrations were predicted during winter period (November and December) as a result of higher emission rate and wind effect. During summer (May and June), high dispersion was also observed.

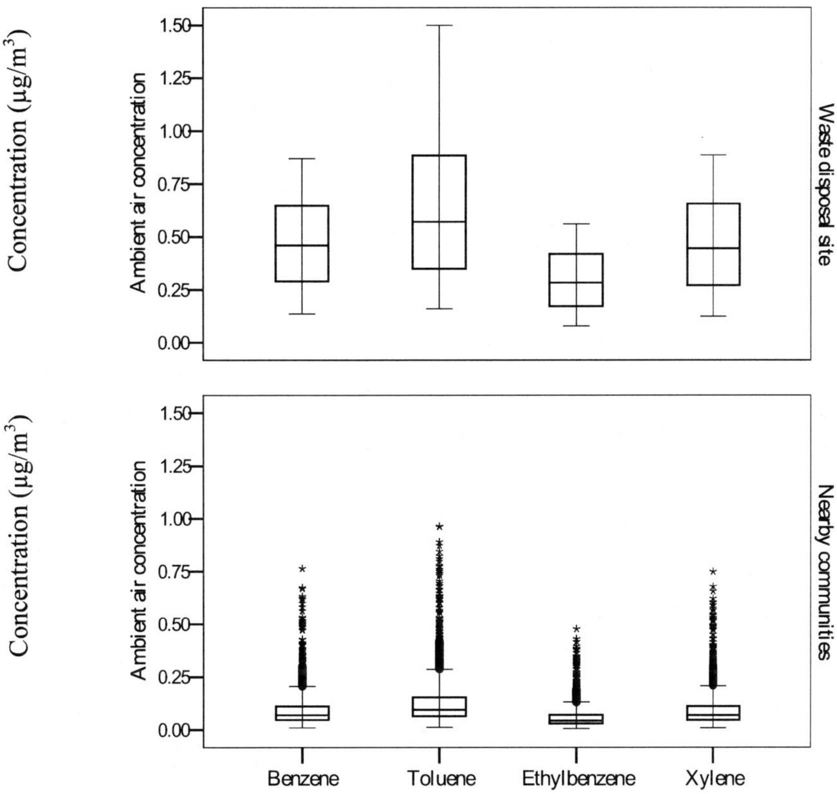

Figure 7. Box plot of highest monthly average concentration of benzene.

3.3. Health Risk to Workers and Nearby Communities

Evaluation of health effects from VOCs to the workers was estimated using maximum 8 hours weight average (TWA) concentration of BTEX compounds obtained from AERMOD model and compared them to standard values, i.e. TLV (ACGIH), REL (NIOSH) and PEL (OSHA). In most cases, the VOC concentrations in ambient air were comparably lower than the standards suggesting low risk levels to the workers.

The non-carcinogenic risk of total BTEX in term of hazard index to the workers who are working 8 hours/day 5days/week and 48 weeks/year for 45 years, were found to be between 0.02-0.15 which is less than the critical value of 1.0 (Figure 9). The health toxicity from solid waste activity to the people living in surrounding communities (24 hours/day 7days/week 50weeks/year for 70 years) were ranged between 0.01-0.40, depending on the direction and distance from the site. In case of nearby communities, the BTEX concentrations for lifetime exposure which give hazard index equal to 1.0 were 0.87, 141.94, 29.40 and 2.94 $\mu g/m^3$ whereas the maximum annual concentrations of BTEX in 2007 from the AERMOD model were only 0.4028, 0.6473, 0.2800 and 0.4438 $\mu g/m^3$. These results suggested that the dispersive transport of BTEX to nearby communities were at low level. Figure 10 shows the risk map of VOC exposure at the site and nearby communities.

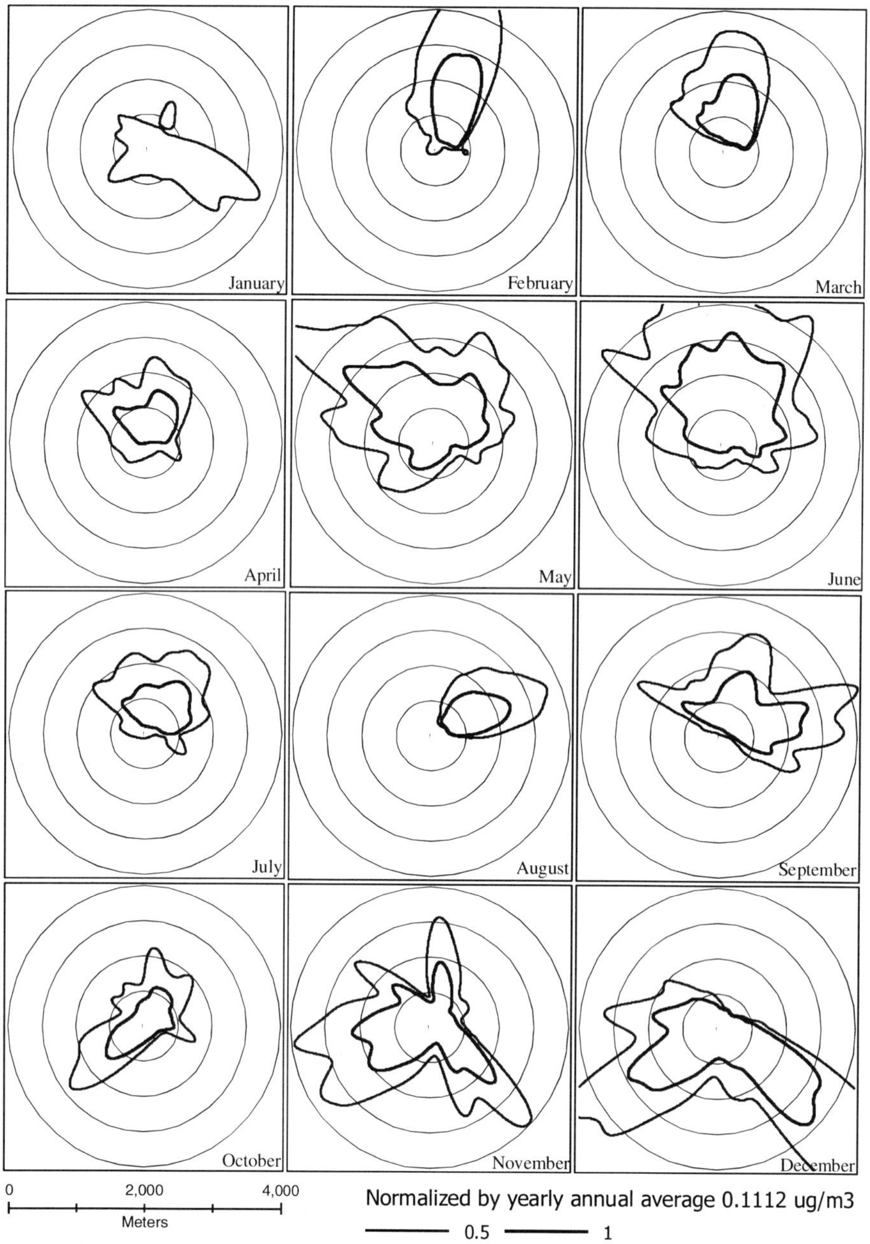

Figure 8. Normalized monthly average concentration of benzene from AERMOD.

For carcinogenic risk of benzene exposure to the workers over their lifetime (70 years, 200 days/year and 8 hours/day), the risk from exposure to average benzene concentration of 0.47 $\mu g/m^3$ (Table 4) was $2.45x10^{-5}$. In case of people living in nearby community, the carcinogenic risk of benzene (0.1 $\mu g/m^3$) is $2.85x10^{-5}$ which is slightly different from that of the workers due to all day exposure estimation. As the acceptable risk is set at $1x10^{-6}$, there is a potential risk to workers and people living in nearby communities of the disposal site for VOC (benzene) exposure.

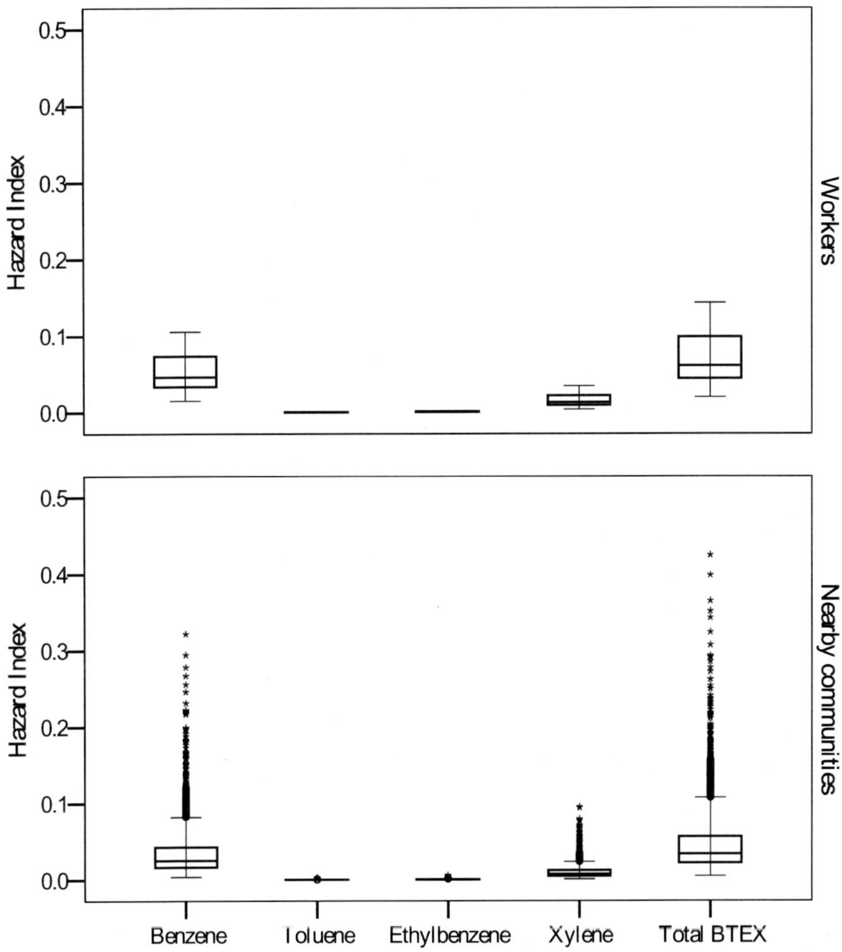

Figure 9. Box plot of health risk to on site workers and nearby communities.

3.5. Ecological Risk

In order to find correlations between gas emissions and geno-toxicity (DNA damage) in different settings and seasons, the experiments were conducted at the uncovered (open dump) and covered (landfill) part of the same solid waste disposal site during dry and rainy seasons. The highest concentration was benzene which was detected at 100% frequency of the samples. Beside benzene, there were other compoounds emited from the uncovered and covered part of waste disposal site, e.g. thiofuran, chloropropene, toluene, ethylpentane , tetramethylbutane, dimethylcyclohexane, dimethyloctane, tetrachloroethylene, dimethylpentane ,ethylbenzen, xylene, styrene, isopropylbenzene ,ethyltoluene, butylcyclo-hexane, trimethylbenzene, cymene, trithiapentane ,butylbenzene ,diethylbenzene, cyclo-propylbenzene, propyl hexanoate ,ethyl heptanoate, methylacetophenone etc. The emissions of total volatile organic compounds (VOCs) during rainy season in open-dump site (84.3±13.3 mg/m^2.d) were not different ($p>0.05$) from those in dry season (83.9±15.1 mg/m^2.d). Similarly at the landfill part, those in rainy season (31.2±3.8 mg/m^2.d) were not different ($p>0.05$) from those in dry season (30.7±3.8 $mg/m^2.d$). However comparison between two part of the disposal site, VOCs emissions in open-dump were signiticantly higher than in landfill area ($p<0.05$). According to the literature, there was 38 VOCs emission during winter and 60 VOCs in summer in which aromatic hydrocarbon was the major VOCs in Chinese landfill [12].

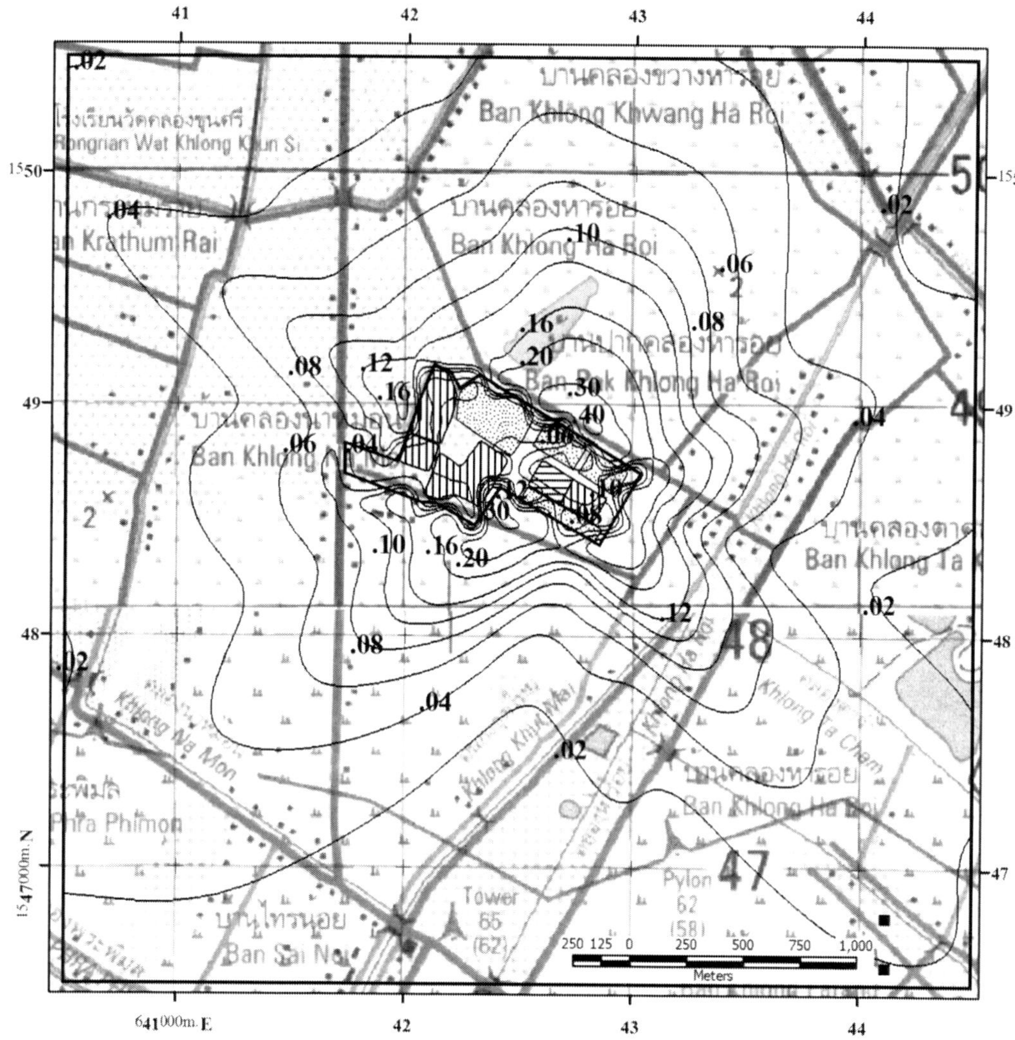

Figure 10. Risk map of VOC exposure.

Genotoxicity to plant species was evaluated by comet assay of Pothos under variation of VOCs emission rates (dry/wet seasons; open dump/landfill area). It suggests that under atmosphere of open dump and landfill area, there were high DNA damages of Pothos. Nevertheless, DNA damage can naturally occur from denature of aged cells too [18]. In this case, DNA damage of the control Pothos (none VOCs exposure) was 20%. For 96 hours exposure at waste disposal site, significant higher damages of DNA were found for open dumping site during dry/wet seasons than that of landfill area ($p<0.05$). A plot of DNA migration and tail moment is shown in Figure 11. Referencing to the difference between VOC emission rates from uncovered (open-dump) and covered (landfill) area of the same waste disposal site, it can conclude that there is relationship between DNA damage of Pothos after 96 hours of exposure and the VOC emission rate from waste disposal area.

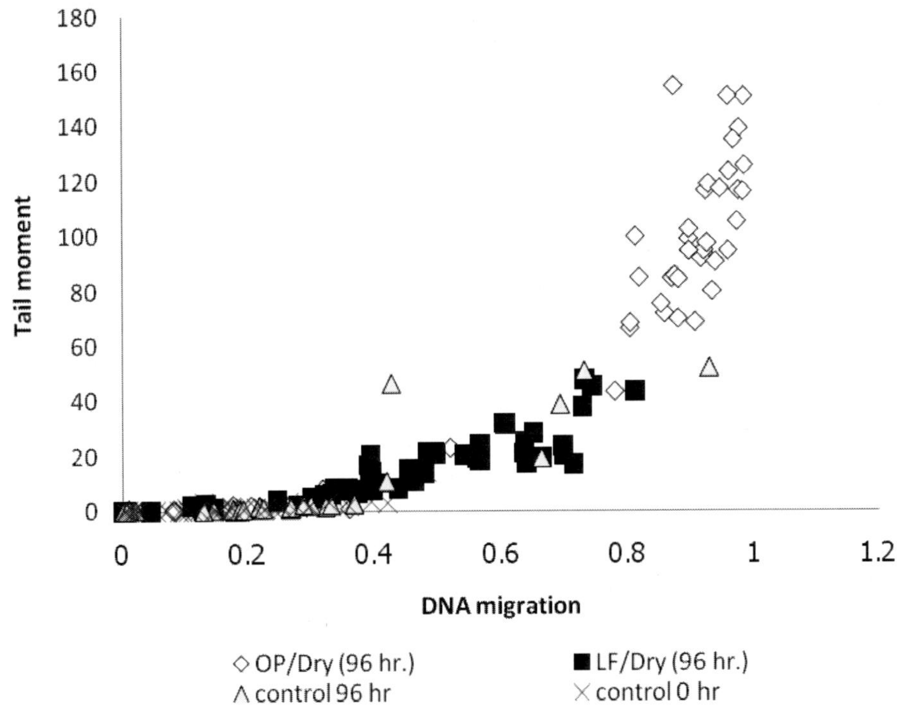

Figrue 11. DNA migration of Pothos in different conditions.

CONCLUSION

Methane and VOCs emission and distribution from solid waste disposal activities at a municipal solid waste disposal site in Thailand were studied. Methane and VOCs emission rate was found highest from solid waste pile with cover soil during wet season and dry season respectively. VOCs were found contaminated in various waste components. The distribution of VOCs in ambient air within the site and surrounding communities were determined by direct measurement and AERMOD model. The concentration of VOCs distribution varies seasonally. Nevertheless, toxicity effects to the workers within the site and nearby communities were at low level as determined by hazard index. However, carcinogenicity from benzene to those people were 2.45×10^{-5} and 2.85×10^{-5}, respectively. From geno-toxicity study using plant comet assay, there was correlation between Pothos DNA damages after 96 hours exposure and VOCs emission rate from the waste disposal site.

ACKNOWLEDGMENTS

This research was financially supported by Swedish International Development Cooperation Agency (SIDA) through Asian Regional Research Program on Environmental Technology (ARRPET).

REFERENCES

[1] Asian Institute of Technology. 2004. Municipal Solid Waste Management in Asia. *Asian Regional Research Program on Environmental Technology (ARRPET),* ISBN: 974-417-258-1.

[2] Thailand Environment Institute. 1996. Climate Change Local Solution for Global Problems: Thailand's Role in a Global Context. *Proceeding of Thailand Environment Institute Annual Conference.* Bangkok, Thailand.

[3] IPCC. 1996. *Climate Change 1992: Supplementary Report to the IPCC Scientific Assessment. Inter-governmental Panel on Climate Change.* University of Cambridge Press, Melbourne, Australia.

[4] Eklund B., Anderson E.P., Walker B.L. and Burrows D.B. 1998. Characterization of Landfill Gas Comparison at the Fresh Kills Municipal Solid Waste Landfill. *Environ. Sci. Tech.,* 32: 2233-2237.

[5] Chiemchaisri C., Chiemchaisri W. and Sawat A. 2006. Mitigation of Methane Emission from Solid Waste Disposal Site in the Tropics by Vegetated Cover Soil. *Asian J. Water Environ. Pollution,* 3(2): 29-33.

[6] Chiemchaisri C., Chiemchaisri W., Boocha M., Chomsurin C. and Urase T. 2006. Evaluation of Volatile Organic Compounds Emission from Solid Waste Disposal Site in Thailand. *Proceeding of 12th International Joint Seminar on Regional Deposition Processes in the Atmosphere*, Peking University, P.R. China, November 13-15, 2006.

[7] U.S. Environmental Protection Agency. 2004. *Description of Model Formulation.* Protection agency Office of Air Quality Planning and Standards, North Calrolina, USA.

[8] Louvar, J.F. 1998. *Health and Environmental Risk Analysis Fundamentals and Application.* Prentice Hall PTR, New Jersy, USA.

[9] Integrated Risk Information System. 2007. *Compare IRIS Values Results.* U.S. EPA, USA.

[10] Glinski, J., and Stepniewski, W. 1986. *Soil Aeration and its role for plants.* CRC Press, Florida, USA.

[11] Kim, K.H.and Kim, M.Y. 2002. The distributions of BTEX compounds in the Ambient Atmosphere of the Nan-Ji-Do Abandoned Landfill Site in Seoul. *Atmospheric Environment,* 36 (2002): 2433-2446.

[12] Zou, S.C., Lee, S.C., Chan, C.Y., Ho, K.F., Wang, X.M., Chan, L.Y. and Zhang, Z.X. Characterization of Ambient Volatile Organic Compounds at a Landfill Site in Guangzhou, South China. *Chemosphere,* 51 (2003): 1015-1022.

[13] http://www.epa.gov/iris/subst/0270.htm

[14] Singh, N.P., McCoy, M.T., Tice, R.R., Schneider, E.L., 1988. A Simple Technique for the Quantitation of Low Levels of DNA Damage in Individual Cells. *Exp. Cell Res.,* 175: 184–91.

[15] Gichner, T, Plewa, M.J., 1998. Induction of Somatic DNA Damage as Measured by Single Cell Gel Electrophoresis and Point Mutation in Leaves of Tobacco Plants. *Mutat. Res.,* (Fundam. Mol. Mech. Mutat.) 401:143–152.

[16] Sriussadaporn, C., Yamamoto, K., Fukushi, K., and Simazaki, D. 2003. Comparison of DNA Damage Detected by Plant Comet Assay in Roadside and Non-Roadside Environment. *Mut. Res.,* 514:31-44.

[17] Hartmann, A., Agurell, E., Beevers, C., Brendler-Schwaab, S., Burlinson, B., Clay, P., Collins, E., Smith, A., Speit, G., Thybaud, V. and Tice, R.R. 2003. *Recommendations for conducting the in vivo alkaline Comet assay.* Mutagenesis Vol.18, No.1:45–51.

[18] Grant, W.F. 1999. Higher Plant Assays for the Detection of Chromosomal Aberrations and Gene Mutations—a Brief Historical Background on Their Use for Screening and Monitoring Environmental Chemicals. *Mutat Res.,* 426:107–12.

In: Landfills and Recycling Centers
Editor: Carla H. Jackson

ISBN: 978-1-63463-792-3

Chapter 7

MODELS FOR EVALUATING PREVENTION OF WASTE ON A MACROECONOMIC LEVEL

Predrag Korica[1] and Andreja Žgajnar Gotvajn[2]
[1]Croatian Environment Agency, Zagreb, Croatia
[2]University of Ljubljana, Faculty of Chemistry and Chemical Technology, Ljubljana, Slovenia

ABSTRACT

The aim of this work is to give review of methods for measuring prevention of waste used by authors in their attempts to assess the impacts which prevention of waste activities can have on the environment, economy and society. The most evident problem with the measuring of waste prevention on a macroeconomical level is that without the clear measurements prevention activities cannot be chosen, proscribed, implemented or evaluated.

Proscribing as target just the reduction of amounts of produced waste without the consideration of other variables that influence the generation of waste wouldn't give a right description of the results of waste prevention activities. Mostly used principles for the evaluation of prevention of waste include models for the conducting of life-cycle assessment (LCA), monitoring approaches of the certain locality (measuring the produced waste before and after the implemented activity), and econometric models. These models still don't give the clear answer to the question on how to measure prevention of waste; however, they make the effects of implemented measures of waste prevention more clear to decision makers so that problems like transferring of an environmental pressure to other side can be avoided. General approach in the European Union for measuring of prevention of waste on a macroeconomic level is to observe the connections which economic variables (e.g. GDP, domestic material consumption etc.) with the amounts of produced waste. In this paper, by using the statistical data available, connections between the generation of waste and economic variables are evaluated in other to assess their potential to be used as waste prevention indicators on a macroeconomic level.

Keywords: LCA, macroeconomical indicators, prevention of waste

1. INTRODUCTION

Waste can be defined as any substance or object which the holder discards or intends or is required to discard. Waste management roughly can be divided into several strategies: prevention, preparing for the re-use, recycling, and other types of recovery (e.g. energy recovery) and disposal (e.g. incineration, landfilling etc.), first being the most desirable and last being the most undesirable, respectively (Directive 2008/98/EC). This is usually called "Waste Hierarchy" and it's usually shown as a pyramid where on the top the waste prevention has been placed, and on the bottom is disposal (Suzuki et al., 2010). The waste hierarchy is presented as an inverted pyramid because the essential thrust of policy is to take action first and foremost on preventing the generation of waste (UNEP, 2013). The waste hierarchy pyramid according to the Waste Framework Directive is shown in the Figure 1 (Directive 2008/98/EC). Generally speaking, energy recovery represents that some of the energy potential of the waste is recovered while recycling results in partial recovery of materials, indicating loss of material or energy, respectively. Some studies show that the differences between regarding environmental impacts of energy recovery and materials are negligible, although still perform better than landfilling (Eriksson et al., 2005). Disposal such as landfilling does not only lead to severe environmental degradation but also results in loss of natural resources (Yedla, 2005).

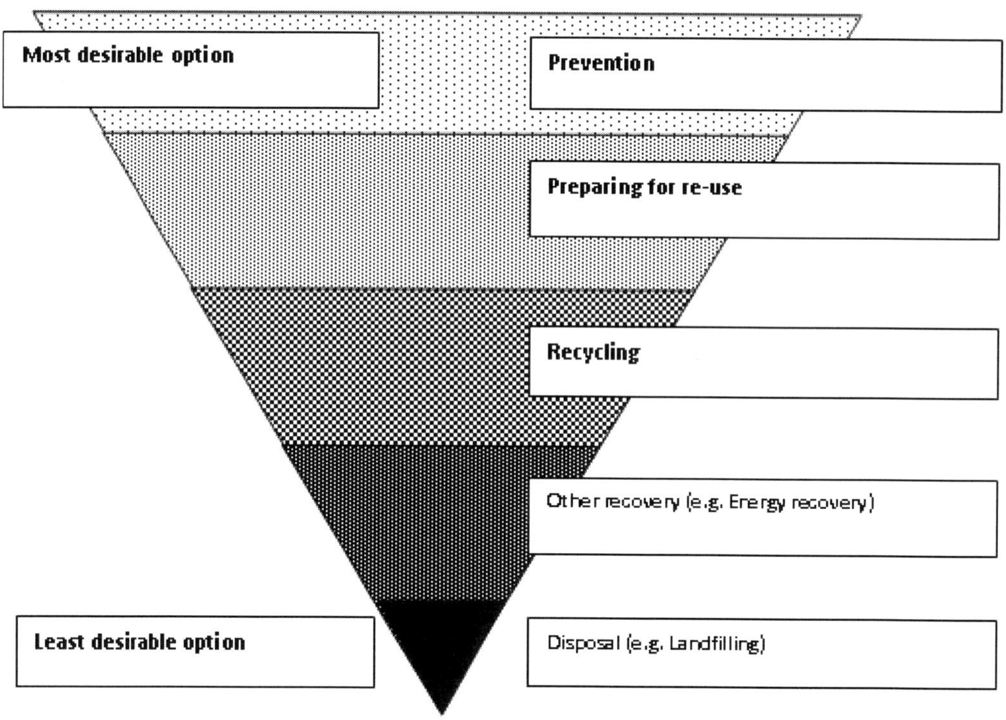

Figure 1. Waste Hierarchy (Directive 2008/98/EC).

Measures of waste prevention eliminate the need for above mentioned measures and thus lower the impact which waste has on the environment. Waste prevention measures can be any form of reducing the quantities of materials used in a process or any form of reducing the quantity of harmful materials which may be contained in a product. Prevention and minimisation also include processes or activities which avoid, reduce or eliminate waste at its source or result in re-use or recycling.

For organisations prevention measures include processes like for e.g. conserve supplies and inventory; eliminate, reduce and reuse products and packaging; use more durable, reusable, repairable and less toxic products and packaging; reduce food and yard waste, including through on-site composting etc. For the citizens waste prevention can also include buying products with less environmental harm (e.g. the least amount of packaging; only the amount of a product that is needed, less harmful products); reusing or repairing items etc. (Zorpas and Lasaridi, 2013).

Having in mind highly recognized impacts on the human health and the environment, European legislation covering waste sector is one of the strictest among its legislation. The first EU legislation covering the generation and treatment of waste was introduced in 1975, and there are now over 20 groups of legislation (e.g. waste batteries, on landfilling of waste, on incineration of waste etc.) in place on waste management. These laws have been designed to bring about a major change away from the widespread dumping of waste in landfills to the imaginative exploitation of waste as a resource that can be re-used, recycled or treated to produce energy (Jackson and Watkins, 2012).

European Commission recognized this and has proscribed through the Waste Framework Directive the waste prevention on the top of the waste management hierarchy and demands from each of the Members State to develop their own waste prevention plans (Directive 2008/98/EC). However, since than waste prevention provisions have been poorly implemented in the national legislations. It can be also noticed that today's waste management systems which are applied in the EU and the rest of the world comprise minimally of the measures for the prevention and minimisation of waste. Although, the waste prevention measures which reduce the quantity of harmful materials which may be contained in a product can be measured (e.g. hazardous substances in the product before and after the measure implementation), it is not clear enough how to measure amounts of prevented waste. Most of scientific and expert public describes the problem with: "How to measure something that does not exist?" implying the need to estimate the amounts of changes reduced by waste prevention (Wilts, 2012; Sharp et al., 2010; Zorpas and Lasaridi, 2013).

2. Measuring Waste Prevention

The issue which has raised undoubtedly a lot of heat in the waste management circles in the recent years is certainly the problem of the evaluation of waste prevention. It is clear that policies and legislative instruments cannot be adopted nor implemented without setting clear targets which have to be achieved. Some other issues regarding the monitoring of waste prevention have also been identified (OECD, 2004):

- Waste prevention can only be measured indirectly as something that could have happened (e.g. as an amount of waste that could have been generated in the absence of the certain waste prevention measure).

- Most of the waste prevention measures are directed toward the upstream processes (e.g. towards influencing the consumer and producer behaviours).
- Efects of some actions of waste prevention take time to appear.
- Waste prevention directs the attention towards which waste and what amount of waste should be managed so some waste prevention measures include also other waste management options (e.g. designing of the product to be reused or more easily recycled).

The most widely used group of indicators of prevention of waste in European Union is "decoupling of growth of BDP from the growth of the amounts of generated waste" (Mazzanti and Zoboli, 2008; Gentil et al., 2010, Sjöström et al., 2010).This indicator is based on the theory that if economic gain increases without the increases in the amounts of generated waste prevention is successful. Many authors have used econometric models in order to analyse these connections and to estimate future amounts of waste to be generated in the EU member states, by making correlations between the generated amounts of waste, incomes and economic activity (Östblom et al., 2010; Bruvoll et al., 2003; Böhringer and Löschel, 2006, Skovgaard et al., 2005; Skovgaard et al., 2007; Wang et al., 2011; Abrate and Ferraris, 2010; Mazzanti, 2008). Recently, some authors also have used the Computable General Equilibrium (CGE) models and econometric models for the analyses of the relation between economic activity and waste generation, having in mind the above mentioned thesis on the connection of the waste amounts with the growth of GDP (Östblom et al., 2010; Bruvoll et al., 2003; Böhringer and Löschel, 2006). Of course, some EU member states have suggested different types of indicators like the total amounts of generated waste or number of implemented measures of waste prevention or implemented Environmental protection systems (e.g. EMAS or ISO 14000) (OECD, 2004). This type of indicators cannot show whether the decrease in the amounts of waste has been successful because of the implemented measures or because of the some socioeconomic reason, neither they can show what are the results of the implemented measures, respectively. These indicators can be helpful for the assessment and comparison of waste prevention activities on an international level, but they are inappropriate for evaluating the effectiveness of specific waste prevention measures on waste quantities (Salhofer et al., 2008).

Growing awareness on the potential adverse effects on the environment which causes production of products contributed to the development of many methods in order to protect the environment. In recent years a significant shift has taken place from waste disposal to management of all resources during the whole life cycle of the product which at the end of its life will become waste. It is believed that the environmental optimisation of energy and material recovery from waste activities can only be made possible by the environmental assessment of waste management from a system perspective (Giugliano et al., 2011). There are many tools for conducting these kinds of assessments, one that is most commonly used is life-cycle assessment (LCA). Examining the resources which come or leave the system of the production of certain product as well as the potential impacts on the environment of that product through its life cycle is called "the life-cycle approach" (Horne et al., 2009). LCA helps expanding the perspective beyond the waste management system and because of this, LCA has gained in acceptance as a tool for waste management planning and policy-making (Ekvall et al., 2007). The application of LCA allows making evidence-based conclusions on the potential environmental benefits of certain waste management strategies because waste is also a valuable resource. Life cycle analysis and other similar tools for environmental systems analysis can contribute to the basis for making decisions which have to be made now in other to protect the environment, not by making it complete but by making it more comprehensive (Ekvall et al., 2007). So, one of the most important purposes of waste LCA models is to communicate the outcomes to develop evidence-based waste management policy. The international standard "ISO 14040 Life Cycle Assessment – LCA" was developed as a helping tool, which describes the basic principles, framework methods and minimal requirements for conducting this assessment (ISO 14040:2006). Methodology of LCA is more and more applied for evaluations the waste prevention measures because it can yield optimal expected results in the relation to available data (Azapagic, 1999; Gentil et al., 2010; Jolliet et al., 2003; Katajajuuri et al., 2011; Sjöström et al.; 2010; Solano, 1999).

3. Life-Cycle Assessment Approach

Over the years, valuable experiences have been made in the field of assessing waste management systems, some being made independent from each other's, some learning from each other's so today there are many validated datasets available and the comprehensiveness of these datasets and quality of the data is still rising (Gentil et al., 2010). Assessment of the life cycle is an objective process for the evaluation of the burden put on the environment by certain product, process or activity, and is being used for the identification and quantification of the consumption of energy and materials, but also of the waste being discharged in the environment along of the whole system. In this way, the influence like consumption of the energy, exploitation of material resources and emissions can be measured, so results can create the opportunities for the improvements in environmental protection (Winters-Lynch, 1993). LCA is used to examine all of the environmental burdens of some product or activity through the all of the major stages: raw material acquisition and processing, transportation, manufacturing, usage and maintenance and waste management (Sefic Williams, 2009). By focusing on the general environmental impacts, LCA can help to avoid that some impacts are transferred from one place to the other. E.g. comparing packaging of juices in glass or plastic bottles, continuous usage of the glass bottles will not decrease the complete impact on the environment. Environmental burden followed usage of the carbohydrates which are needed for the production of the plastic bottles will be replaced by the burden of energy needed and wastewaters generated during the transportation, collection, cleaning and re-usage of the glass bottles (Hoff, 2008). Since the public opinion is very important in the environmental protection, LCA should cover the informative instruments used for communicating with the public (e.g. environmental protection clams, eco-labelling, and TV commercials). LCA results have played a key role in procurement decisions, for example, as companies have sought to assess the relative performance of competing suppliers. This activity, in many cases, has been driven by a recognition, that while public opinion may move in great surge tides, the underlying trend on most environmental issues is still moving steadily upwards (EEA, 1997). Life cycle analysis should be quick and accurate enough to be able to be used by product technologists, designers and manufacturers in making early design and manufacturing decisions. It should also provide information in clear, simple, accessible and independent ways, with the potential also to provide useful information for environmental managers, policy makers and consumers (Horne et al., 2009). Through the design of the product by applying the environmental criteria, the impact on the environment is lowered through all of the stages in the life assessment. This process must revise the whole system of the product which consists of 3 main elements: product, process and distribution (Jeganova, 2004). LCA is the strategy of environmental production where the „extended producer responsibility" is applied through tree mayor instruments: regulatory, economic of informative instruments. The combination of these instruments determinates the exact shape of the extended responsibility of the producer (Lindhqvist, 2000).

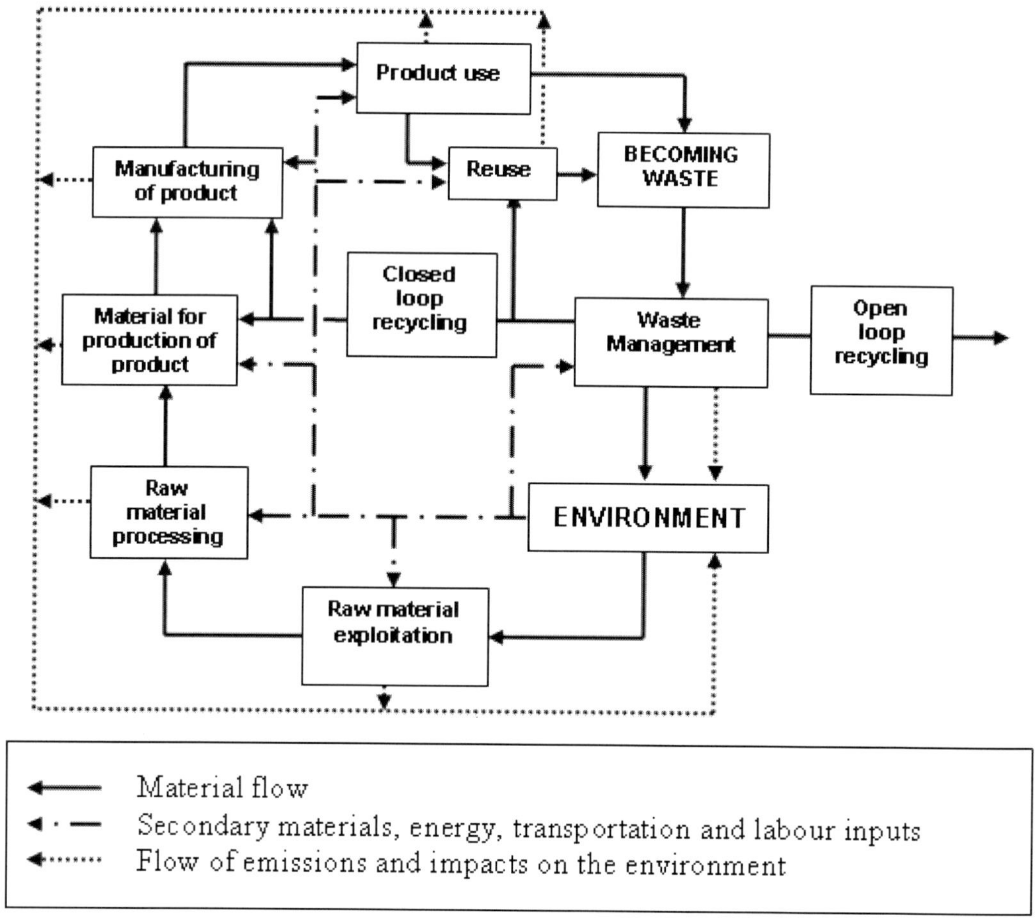

Figure 2. Hypothetical material life cycle system.

The general diagram of the life cycle of a waste management system is shown in Figure 2. The life cycle starts start with exploitation of materials from the environment which are then processed and from them a product is made. Product is consumed directly or through certain service after which it is either discarded or is reused (once or more times) after which it also becomes waste. Waste can be managed by preparing for the reuse, recycling or material energy recovery operations or disposed of (landfilled or incinerated). Waste can be recycled into the same material for the same product (Closed loop recycling) or can be used in another product system (Open loop recycling).

Energy recovered can be used in the same way as the recovered material. All the processes in a life cycle of materials have certain impacts on the environment and have emissions. Furthermore, all of the processes need secondary materials (e.g. materials built in the buildings), energy, transportation and labour inputs which are obtained from environment. While there are constant improvements in the field of the standardisation, international and suggested standards in the ISO 14 040 series are generally accepted and give a consensus on the framework for the life-cycle analyses (Rebitzer *at et al.*, 2004). The first standards ISO 14 040 – 43 where replaced in the year 2006 with 2 revised standards: with new framework standard 14 040 (ISO 14040:2006) and with the standard ISO 14044 (ISO 14044:2006) which defines the demands (Klöpffer, 2012).

It was attempted not to do changes in the content of "old" standards but rather to improve their presentation (Finkbeiner et al., 2006). Standards ISO 14 040 and ISO 14 044 are joined in one set so if LCA study is conducted according to the principles proscribed by ISO 14 040, detail rules and demands proscribed with ISO 14 044 must be taken into the account. Thus, if the study is conducted according to the ISO, it is not allowed to conduct the LCA analysis only according to the 14 040, without obeying to the demands proscribed by 14 044 (Klöpffer, 2012). The working method for LCA is structured along a framework according to the world-wide consensus which divides the entire LCA procedure into four distinct phases (Guinée, 2004):

- Goal and scope definition;
- Inventory analysis;
- Impact assessment and
- Interpretation.

Implementation of the LCA is guided by the definition of the goal of the study, by the method of the data collecting, by the analysis of the data and by the method of the communicating the results. Since LCA is relatively new discipline, this type of assessment will continue to develop. Much more time is needed so that practical experiences (results of many conducted LCA assessments, documentation on their interpretation, their explanations and their implementations) can be incorporated in its methodology (ISO 14040:2006). Depth of assessment is limited by a variety of factors which include purpose, boundaries of the assessment and the level of specificity needed and data availability. The practitioner has to weigh his available resources with other limiting factors (e.g. subjectivity in the assessment, need for assumptions, availability of crucial data etc.).

Authors generally use two concepts of conducting the LCA: an attributional LCA and a consequential LCA. Attributional LCA is used to describe the product/activity system with all of its exchanges with the environment (Rebitzer et al., 2004). Usually, it is used for comparisons of the two product systems or for evaluation of the waste management systems (Coleman et al., 2003; Cleary, 2012). Consequential LCA analyses the casual connections of different impacts on the environment connected with different decisions (Cleary, 2012).

3.1. Life-Cycle Assessment and Waste Prevention

Life cycle analysis is a useful tool for the estimation of the impact which waste prevention has on the society, environment and economy, but as a tool it stills needs further development.

Life cycle assessments of municipal solid waste management systems are undertaken to "optimise the infrastructure system for managing a given amount and composition of waste" (Coleman et al., 2003). There is an strong need for the improvements of the methodology for life-cycle assessment towards being more easily applicable for the creating of the legislative since the regulatory bodies have to concentrate on the single measure of prevention (e.g. the type of material used for beer packaging). The mostly needed improvements need to be done in the field of assessing the environmental benefits of waste prevention activities because the traditional LCA has excluded waste prevention activities form its system boundaries (Cleary, 2010). Some argue that consequential LCA is the only one appropriate for this task of evaluating waste prevention, having in mind that waste would be generated if it wasn't for the implemented waste management measures. Rather, since in the waste management hierarchy waste prevention is considered equal to the other waste management options, the attributional LCA could also be applied (Ekvall et al., 2007; Cleary, 2012). Cleary also claims that it would not be appropriate to consider waste prevention through reduced consumption as a functionally equivalent method of managing waste because waste treatments do not affect the composition and magnitude of product services supplied to the population by waste-generating product systems (Cleary, 2012). However, the efficiency of waste treatment does influence the composition (e.g. end-of-life vehicles) and identification of products which is difficult to manage as a waste, can yield in the strict regulations and environmental taxes which lowers the consumption (e.g. plastic bags). Gentil et al. used the LCA approach to assess the environmental consequences of applying measures of waste prevention (reuse of the beverage packaging, vegetable waste in municipal waste, meat waste and refusing unsolicited mail) on the waste management systems. Two hypothetical systems where used in the assessment: one depended mainly on engineered landfilling for the disposal and one which consist of a higher level of recycling and energy recovery. Since their study was concentrated on the waste management system, the authors removed the upstream processes (the processes before the product becomes waste), however only partially since the in some cases the upstream processes are also effected by waste management prevention (e.g. reuse of bottles has as consequence lower production of bottles). Authors have concluded that prevention of waste has the largest effects on the emissions in the environment if it is applied on the waste management system with low ratio of recovery (Gentil et al., 2011).

In most of the conducted LCA studies it is very difficult to draw conclusions based on the results because so many published LCAs do not make clear what are their system boundaries and assumptions (Cleary, 2012). Although the LCA methodology is used by many authors to compare different waste management options, like recycling vs energy recovery, waste management is very often not included in these comparisons (Ekvall et al., 2007).

The U.S. Environmental Protection Agency developed "**WA**ste **R**eduction **M**odel" (WARM) to help with the evaluation of waste management practices in the regard of their green house gases (GHG) emissions. WARM calculates and totals the relative GHG emission and energy impacts of baseline and alternative materials management practices–source reduction, recycling, combustion, composting, and landfilling using emission factors that

EPA has developed based on a materials life-cycle approach. This model reduces the GHG emissions by (USEPA, 2014):

- Avoiding the "upstream" GHGs emitted in the raw material acquisition, manufacture or production and transport of the source-reduced material.
- Increasing the amount of carbon stored in forests (when wood and paper products are source reduced).
- Avoiding the downstream GHG emissions from waste management.

Practitioner using the WARM first runs the calculator using a baseline scenario of landfilling, recycling, combustion or composting (depending on the likely fate of the material or product if it is not reused), and an alternate scenario of source reduction. For e.g. wood packaging in the baseline scenario is designated for recycling and now it will be reused. The mix of virgin and recycled materials used in the manufacture of the material from which an item is being made is chosen second. The GHG emissions reduction result is multiplied by the number of times the material is reused. The reuse number should equal one less than the number of total uses to account for the production of the initial material. Energy use can be similarly calculated by replacing the GHG emission factors with energy use factors. In the case of material substitution, WARM can calculate the emissions only if the user has the same material in the baseline and alternate scenarios, not in the case of source reduction of one type of material by substituting another material. However, methodology is given what to do in this case. User can calculate, using WARM, the green house emissions from original material, from substitute material and can calculate the number of tons of substitute material used per ton of original material. Then by subtracting the emissions that would have been generated to produce the baseline material from the emissions generated by producing an equivalent amount of the substitute materials, total emissions benefits can be calculated. The U.S. Environmental Protection Agency is continuously working on the improvement of this model to address the complexity of some waste streams like for e.g. paper sludge at paper mills and is updating the calculator model (USEPA, 2014).

In order to evaluate the waste prevention, Cleary developed his own model of attributional LCA which he named "*Waste Management and Prevention* (WasteMAP) life cycle assessment model". In this model, waste prevention is considered a functional equivalent to other waste management options. In his study, author studied the effects which the prevention of waste packaging from wines and spirits supplied to residents of the City of Toronto, where he developed two scenarios: original product system and the alternative one. The indicators which the author used are volume, estimated sales in reference area, per cent of wine/spirits contained in each packaging size (in regard to all types of packaging), and per cent of each container size by type of container (in regard to all types of packaging - aseptic carton, Glass bottle, PET bottle). In the alternative hypothetical scenario author used refillable glass bottles, aseptic cartons, PET bottles in lightweight single use bottles (assumed to have a 20% lower mass than conventional single use container). In order to collect the needed data, author used the SimaPro LCA software, ReCiPe method for the life cycle impact assessment, collected data through the questioners sent to the producers, conducted the field and laboratory research, used the statistical data available from producers and producers unions and made assumptions which author explained with logical reasoning.

SimaPro is the one of the most widely used LCA software. The main difference from the ISO 14040 standard is that it offers standardisation as well as certain level of flexibility that offers users ability to do variations of model. For example, ISO 14044 does not allow weighting across impact categories for public comparisons between products. SimaPro does support weighting which means that it is responsibility of the user to use weighting in a proper way. A similar example can be made for issues such as allocation rules, system boundaries etc. Note has been made that if user deviates form from standards, user cannot claim that LCA was conducted according to the international standards and it will be more difficult to convince others of the reliability of results of such study (Goedkoop et al., 2013). SimaPro comes with a complete implementation of the database Ecoinvent. In late 2000, the project Ecoinvent database was officially launched by several Swiss Federal Offices offices and research institutes. They joined effort to harmonise and update life cycle inventory data for its use in life cycle assessment in order to make consistent, coherent and transparent LCA datasets. In late 2003, and mid 2004, the first versions of the Ecoinvent data sets have been published and made available in a central database accessible via the Internet. The Ecoinvent database can give user information in the way that it can use them both in the attributional and consequential life cycle inventory analysis. One of the most valuable assets of this database: variability and parameter uncertainty of unit processes inputs and outputs (e.g. data uncertainties, space and time variations etc.) are quantified (Frischknecht et al., 2004).

ReCiPe is a life cycle impact assessment (LCIA) method which comprises harmonised category indicators at the midpoint and the endpoint level. A number of methods used for LCIA convert the emissions of hazardous substances and extractions of natural resources into impact category indicators at the midpoint level (such as acidification, climate change and ecotoxicity), while others employ impact category indicators at the endpoint level (such as damage to human health and damage to ecosystem quality). Life cycle assessments commonly assess damage to human health using the concept of 'disability-adjusted life years' (DALY). The DALY of a disease is derived from human health statistics on life years both lost and disabled. Values for disability-adjusted life years have been reported for a wide range of diseases, including various cancer types, vector-borne diseases and non-communicable diseases. DALY is equal to the sum of years of life lost and years of life disabled, the latter of which using a subjective valuation of health disabilities on quality of life. The damage to ecosystem diversity endpoint indicator assumes that species diversity is a suitable indicator of ecosystem quality. This indicator is measured in terms of the potentially disappeared fraction of species integrated over area (or volume, in aquatic and marine habitats) and time. The damage to resource availability indicator quantifies its impact as a marginal increase in resource extraction costs due to the resource consumption attributed to the life cycle (Cleary, 2012; Goedkoop et al., 2009).

In order to quantitatively estimate the potential of waste prevention of waste paper from advertising mail, waste beverage packaging, waste one-way diapers, waste from the food and beverage industry and retail; and waste from big events Salhofer et al. used a ratio model (if the detailed data on consumption and waste generation of certain product was available) or used participation figures from the data reported from other regions. The waste prevention measures studied were formally affirming or refusing advertising mail, substituting one-way packaging by refillable ones, sponsoring of reusable diapers, implementing food banks and substitution of dishes and cutlery by reusable ones and by ones made of biodegradable plastics (Salhofer et al., 2008)..

4. ECONOMETRIC MODEL APPROACH

Generation of waste is strongly connected with the economic variables such as spending, consumption and production. Many authors use the econometric models to analyse the connections between economic variables and the generation of waste (Östblom et al., 2010; Bruvoll et al., 2003; Greyson et al., 2007, Moutavtchi et al., 2010). Econometrics is the unified study of economic models, mathematical statistics, and economic data (Hansen, 2014). All three, statistics, economic theory and mathematics are necessary, but not by itself a sufficient, condition for a real understanding of the quantitative relations in modern economic life. It is the unification of all three that is powerful and that constitutes econometrics (Frisch, 1933). Models are used as a help in determining the causality relationships between correlated variables (Creel, 2014).Thus, econometric models are used so future amounts of waste could be determined by making the correlations between the generated amounts of waste and population growth, population personal income and economic activities (Wang et al., 2011, Skovgaard et al., 2005). LCA models also can be combined with the economic models in which the system boundary includes the entire economy (Hendrickson and Horvath, 1998). Recently, some authors have used Computable General Equilibrium (CGE) models in their attempts to predict the development of the situation after certain waste management measure has been implemented (Sjöström and Östblom, 2010; Bosselo et al., 2012). CGE models present one of the bridges between the economic theory and practice and in the same time present one of the most efficient tools which economic theory in this moment can offer, and are used to analyse the effects of the economic politics measures, influence of which is determined by the simulation of market mechanism (Šohinger, 2002). CGE models build upon general equilibrium theory that combines behavioural assumptions on rational economic agents with the analysis of equilibrium conditions. They provide counterfactual ex ante comparisons, assessing the outcomes with a reform in place with what would have happened had it not been undertaken (Böhringer and Löschel, 2006). Böhringer and Löschel divided CGE models into three general categories (Böhringer and Löschel, 2006):

- sectorial details versus economy-wide scope,
- econometrically estimated models based mostly on time series data or on parameters of functional forms and
- micro–/macro-founded approaches and simple accounting frameworks.

The authors noted that operational versions of energy–economy–environment CGE models have a good coverage of economic indicators, but are week in environmental indicators because of the complex science background (Böhringer and Löschel, 2006). CGE model was used by Sjöström and Östblom to determinate the future trends in the generation of waste in the Swedish economy and to determinate the possibilities for decoupling of amounts of generated waste from GDP. The authors used the Baseline scenario (assumed that the intensities connected with consumption of materials will continue to grow, while the intensities connected with production and households will stay unchanged) and the decoupling scenario (decreased all of the intensities in order to accomplish total decoupling in the amounts of generated waste (in all waste streams and all together)). In this study, CGE model was used to project the future quantities of waste in Sweden based on the economic data (Sjöström and Östblom, 2010).

There is empirical evidence that some pollutants follow an inverse-U-shaped pattern relative to countries income. This type of relationship is called "Environmental Kuznets Curve", EKC. Kuznets was the first economist that studied the particular relationship between economic growth and unemployment. He showed that unemployment increases until some level of income after which level it begins to decrease. The EKC literature is connected to the delinking (decoupling) theory as indicator of environmental effectiveness with respect the economic activity (Abrate and Ferraris, 2010). In case of waste generation, this thesis proved correct for the developing countries being industrialised (Khajuria et al., 2012). Although these observations can lead to conclusion that the growth of income can insure the further development of environmental protection, there are many other factors which influence the environmental pollution in the opposite directions, and also, there are no quarantines that these trends will continue in the future (Bruvoll et al., 2003, Bartoszczuk et al., 2002). Results of lot of studies conducted in order to find evidence on EKC hypotheses on the connection of fortunes of certain country and generation of waste do not support this thesis (Dinda, 2004). However, if the research is modified with different factors (e.g. as altitude and population density) or if type of waste is specified (e.g. waste which has not be separated and recycled or waste being sent to landfills) some studies found evidence to support EKC thesis (Abrate and Ferraris, 2010; Mazzanti and Zoboli, 2008).

5. Measuring the Waste Prevention on the National Level

Europe's approach to waste has evolved significantly in the last two decades by putting the waste prevention on the top of waste hierarchy (Figure 1). Europe has succeeded in shifting waste management up the waste hierarchy in recent years, with less waste landfilled as a result of reduced generation of some wastes, and increased recycling and energy recovery. The issue that remains to be tackled is waste prevention for which the clear targets and methods of measuring have to be developed. EU member States usually use as a general approach to measuring waste prevention in EU the development of indicators which show changes in the ratio of waste generation to economic production or consumer spending. When it comes to indicating the reduction in the amount of produced waste, mostly used indicators are the decoupling of the production of different waste streams for which data in time series is available (e.g. on municipal waste, end-of-life vehicles etc.) from the costs of private consumption and the productivity of resources – ratio of the GDP and domestic material consumption. Also, one of the suitable indicators for this purpose is the minimisation of the packaging waste – ratio of the total amounts of packaging waste and total amount of packed products. The indicators using ratios with economic variables can be shown graphically, but in order to be used as clear targets still have to be quantified somehow. Authors of the study requested by European Commission on the production and prevention of biowaste solved the issue of calculating the decoupling of generation of waste from economic variables by dividing theirs linear regression slopes during the 5 year period (indexed on first year of the period), as shown in the following Equation (1) (European Commission, 2011):

$$D_{y\text{-}5\rightarrow y} = b(EP)_{y\text{-}5\rightarrow y} \,/\, b(DF)_{y\text{-}5\rightarrow y} \qquad (1)$$

where $D_{y\text{-}5\rightarrow y}$ is the decoupling factor for a time interval of five years from y-5 to y, and $b(EP)_{y\text{-}5\rightarrow y}$ is the slope of the linear regression of the waste generation (environmental pressure) over the last five years, EP (environmental pressure) is expressed as an index with y-5 = 100 and $b(DF)_{y\text{-}5\rightarrow y}$ is the slope of the linear regression of the private consumption expenditure (driving force) over the last five years. DF (driving force) is expressed as an index with y-5 = 100.

In the second study requested by European Commission in the year 2012 in which the analytical view of the achievements in the waste management within the EU member states, the authors argued that the result obtained by the Equation (1) can clearly show did the decoupling really taken place, so they have modified it to the Equation (2) which is as follows (BiPRO, 2012):

$$D_{y\text{-}5\rightarrow y} = b(DF)_{y\text{-}5\rightarrow y} - b(DP)_{y\text{-}5\rightarrow y} \qquad (2)$$

If the decoupling factor D is larger than 0 the decoupling is accomplished; if the decoupling factor is equal to 0 coupling is present; and if decoupling factor is smaller than 0 the reverse decoupling is present.

In this study, authors made a calculation for the period 2000 till the year 2010 in the way that they calculated the 5 year periods and compared the changes in the result amounts. The results showed that the level of decoupling is generally decreasing, except for some Member States (Belgium, Ireland, Italy, Malta, Netherlands, Austria, Slovakia and Sweden), and that decrease in the decoupling can be seen because of the world wide recession in the last periods (2005 - 2009 and 2006 – 2010). The most interesting finding was that some Member States show also the reverse decoupling – the GDP is decreasing and the amounts of waste are still rising (BiPRO, 2012).

Ireland in its „National Strategy to 2020 incorporating Ireland's National Waste Prevention Programme" suggests using as indicators comparison of the generation of waste with economic activities. Ireland considers this method to be the closest approximation to measuring quantitative waste prevention. Two of suggested indicators are "Societal wealth generation and waste" and "Industrial economic productivity and waste generation". First indicator is shown graphically on the chart as ratio of GDP and generation of municipal waste, meaning that if the ratio increases prevention of waste shows results. The same can be used for the late indicator, respectively (IEPA, 2014).

In its waste prevention plan, Germany made a remark that the data which have been derived from the statistics on the mass flows can give some indication on how the waste prevention is successful, but they cannot give causal connection to the activities implemented to confirm their success. This demands statistical data keeping of the every single activity or/and every single mass flow which are connected with a single measure. If the applied measures are extrapolated on the certain waste flows, it is possible to calculate the mass of the prevented waste during certain period.

As indicators which can measure the success of the waste prevention the following is suggested (FMENCNS, 2013):

- Reuse of electrical and electronic equipment – ratio of the discarded equipment which has been reused and the amount of the discarded equipment in each of the categories, as well as changes over the years and changes in regard to certain year.
- Packaging reuse – changes in the ratio of packaging which is reused in the different types of packaging in regard with total amounts of packaging.
- Waste intensity – connection of the amounts of generated waste in individual sectors (e.g. industry, construction etc.) and the real net product and a number of employees in this sectors, as well as changes over the years and changes in regard to certain year.
- Raw material productivity – ratio between GDP and direct consumption of materials.

In the German waste prevention plan, authors give remark that making the connection of the data on the waste intensity with economic data can lead to misapprehension because, different from total GDP, the effects of continuous structural changes are being masked (FMENCNS, 2013). In the England's waste prevention programme, indicators are planed which show the measures of household expenditure or income which can be used to show waste arisings per unit household economic activity. For the commercial, industrial and construction and demolition waste they will use arisings per unit of Gross Value Added (GVA) in constant price (volume) terms. In its plan, England made a remark that measuring waste only by the weight does not reflect completely all of the environmental or financial burdens of generated waste and can result in the bias towards action on heavier wastes (DEFRA, 2013). Like England, waste prevention programmes of Wales and Scotland suggest as indicators generated waste from households per unit of household consumption, and generation of commercial, industrial and construction and demolition waste per unit of Gross Value Added (Welsh Government, 2013, Scottish Government, 2013). Holland in its waste prevention plan also makes a note that it is not always easy to measure accomplishment of all of the targets and that it will use the following criteria in the selection of the indicators (Loesje, 2013):

- Is the information available or not?
- Can it be expected that needed information will be available in the future?
- Is this indicator valid for the measuring waste prevention or not?
- Are the some disrupting factors present and is the correct interpretation possible based on the available data?

Slovakia in its waste prevention plan suggests using as indicator generation of waste per unit of GDP, and that the additional indicators will be developed according to the needs (MŽRPSR, 2014). Spanish waste prevention plan suggest as indicators the ratio of generated amounts of waste in regard to GDP (total waste, municipal waste, construction waste in regard to GDP and hazardous waste in regard to GVA) (MAAMA, 2013). Lithuania in its waste prevention plan also suggests using the amounts of generated waste in every sector pre unit of GDP, however, suggest also to monitor (LRAM, 2013):

- Collected amounts of municipal waste and other specific waste streams to compare them with the amount which could have been prepared for the reuse.

- Amount spent on the domestic market for packaging (tons and kilograms per capita) and the number of packaging units ready to be reused (in tons, circulating during the year on the domestic market in percentages).
- Collected Waste Electrical and Electronic Equipment (WEEE in tons or kilograms per capita) and from it the amount prepared for reuse (in tons per year from the EE waste generated in percentages).

As suggestions on how to measure waste prevention, study conducted in New York gives couple of the ways of calculating how much of waste has been prevented in the waste prevention actions:

- Calculating the rate of the reduction of source by first calculating the projections of the generation of waste in the state by using the model of linear regression with the historical data on generated waste per capita. After that, from that amount the real amount of produced waste with the real amount of recycled waste in that year should be subtracted to get the projected waste reduced at the source (SAIC, 2000).
- By comparing the consumption of the consumer with the generated amounts of waste. The study conducted by US Environmental protection agency on the statistical data has shown that the generation of waste is in the close correlation with "personal consumer expenses" more than it is with population trends. To determinate the projected amounts of generated waste, the following equations can be used (SAIC, 2000):

$$AWG_{5\to y} / CS_{5\to y} = PWG_y / CS_y \tag{3}$$

where $AWG_{5\to y}$ stands for the Actual Waste Generation in the year 5 years before the selected year, $CS_{5\to y}$ Consumer Spending in the year 5 years before the selected year, PWG_y Projected Waste Generation in the selected year and CS_y Consumer Spending in the selected year.

$$SR_y = PWG_y - AWG_y \tag{4}$$

where SR_y is the Source Reduction in the selected year and AWGy is Actual Waste Generation in the selected year.

Principle similar to this was suggested also in the study conducted in the New South Wales, i.e. by using the data on the population or GDP, amounts of waste that could have been produced in certain year can be projected based on the rates of generation of waste in previous years. The difference between the projected and real amount gives the picture on whether the generation of waste has been accomplished (DECNSW, 2004).

- Resources productivity ratios give a measure of how well are resources spent which are needed to produce a product or service. E.g. product sales (e.g. sold objects, sales, claims, budget, produced products or costumers served) divided with expenses or amounts of raw materials (e.g. supplies, raw materials, labour, energy, water etc.). If raw materials are conserved, the ratio of sales to raw materials will increase. The drawback of described method is that this is applicable only if the composition of the product remains unchanged (e.g. if lighter weight materials are substituted, material costs could increase and the ratio decrease, even though the amount of waste prevented would decrease) (SAIC, 2000).

- Assessing the amount of waste which can be reduced. This methodology is relaying on the existing detail data on the generated waste, waste streams, potential participants and technological limitations in order to calculate program potential. Equation is as follows (SAIC, 2000):

$$\boxed{\text{General Waste Category}} \times \boxed{\text{Applicability Factor}} \times \boxed{\text{Feasibility Factor}} \times \boxed{\text{Technology Factor}} = \boxed{\text{Program Potential}} \quad (6)$$

For example, the amount of waste from General Waste Category is given (e.g. collected amount of mixed municipal waste in the Republic of Croatia is 1287927 t), Applicability Factor allows to estimate the total waste stream generated by a certain sector (e.g. from organised collections it is 75%), Feasibility Factor identifies the percent of the waste stream that could be reduced through source reduction (e.g. 10% of households is willing to participate in the programme of reduction of waste by home composting), Technology Factor calculates for reduction because of the technical restrains (it is possible only to compost 67% of mixed municipal waste since this is the percentage of biodegradable waste in this waste category). So for this example the calculation is:

1287927 t x 0.75 x 0.1 x 0.67 = 64718.33 t or 5.9% of possible reduction of waste.

Also in regard to effects which the prevented amounts of waste would have on the environment if they had not been prevented, there are some other quantitative indicators suggested like for e.g. the scale of greenhouse gas emission reduction attributable to specific waste avoidance programs, landfill space saved, avoided soil and water contamination, avoided air emissions or avoided human exposure and effects (DECNSW, 2004). These are valuable indicators that give information on the effects of waste prevention measures, but they can be calculated only if the amounts of prevented waste are calculated first. In regard to effects which the prevented amounts of waste would have on the economy and social structure if they had not been prevented, there are many indicators available depending on the interesting parties (e.g. business and industry, government, consumers and households) and which effects are of their interests (e.g. attitude and behaviour of certain groups or institutional collaboration), which include (DECNSW, 2004):

- Household savings from adjusted fees for municipal waste management;
- Short-term loss of income from reduced waste available for recycling;
- Market and trade implications (e.g. sales revenues from reduced waste);
- Level of willingness to take actions to reduce waste;
- Quality of infrastructure e.g. to facilitate on-site composting or for product reuse and
- Perceptions of government performance against objectives etc.

EEA launched a study "The European environment – state and outlook 2015 SOER", which is currently in the progress, where authors suggested using as indicators changes in waste efficiency — the ratio of waste generation to economic production or consumer spending. Thus three indicators where suggested:

- Waste generation in manufacturing and gross added value (GVA) in manufacturing,
- Waste generation in services and gross added value (GVA) in services and
- Municipal waste generation and household expenditure.

The authors mentioned that because of the data gaps and differences in national methodologies for calculating waste in other productive sectors (e.g. agriculture, mining and construction), it is not possible to draw reliable conclusions on efficiency trends in those sectors (EEA, 2014). The main issue with using the national statistics lies in the small number of years for which the data is currently available (only 7 in this case), in the methods and sources of collecting the data (e.g. the data on waste is from waste management facilities who have rigorous levels of control, while the data on GVA is from business companies reported on voluntary basis; even the methods of data collecting from country to country are very different). Looking at specific data, the main issue is that the data is collected not on the enough specific level (e.g. the waste from manufacturing is not specific enough to be used for the assessing the waste from the production of PVC bottles for detergents and the expenditure from households). This all indicates that there is a need to find better economic and social indicators in the official national statistics which have better influence on variables on waste before the conclusions on waste prevention in certain country can be made.

6. Case Study

The objective of the case study presented was to review the possible indicators which can be applied to assess the achievements of waste prevention measures in European countries, having in mind the availability and the applicability of data. As the models for the case study the Republic of Croatia and the Republic of Slovenia were chosen. Data which were analysed were collected by searching the databases maintained by Eurostat, Croatian Environmental Agency, Statistical Office of the Republic of Slovenia and the Croatian Bureau of Statistics. Four indicators of the prevention of waste were chosen:

- Relation between the generation of the municipal waste and the household expenditure.
- Relation between the generation of the waste from manufacturing sectors and the Gross value added (GVA) from manufacturing sectors.
- Relation between the generation of the waste from services sectors and the GVA from services sectors.
- Relation between GDP and Domestic Material Consumption.

6.1. Collecting and Preparing of the Data

- The data on the Croatian generation of municipal waste for the period from the year 2000 till 2012 was collected from the Croatian Environment Agency. The official data on the municipal waste had a break in time series in the year 2011 since in the data from this year on the waste from services similar to the municipal waste was also accounted in the sum. These amounts were subtracted from the amounts for the years 2011 and 2012 so time series would have continuum. The data on the Slovenian generation of municipal waste for the period from the year 2000 till 2012 was collected from the Statistical Office of the Republic of Slovenia. The data on the household expenditure for the period from the year 2000 till 2012 for both the Croatia and the Slovenia were collected by searching the Eurostat's database.
- The data on generation of the waste from the manufacturing sectors and the GVA from the manufacturing sectors for the period from the year 2002 till 2012 was collected by searching the Eurostat's database for both of the countries. As the manufacturing sectors the data from the Mining and quarrying, Manufacturing of the products and the Construction was taken. Since the statistical data on the waste is collected by Eurostat on the biyearly bases (even years), the data for the odd years was calculated using the method of interpolation. The data on the generation of waste for the Republic of Croatia for the year 2006 was missing so the data from the Croatian Bureau of Statistics was used. Also Eurostat datasets on the waste do not contain the data for the year 2002 so the data for this year was used also from the Croatian Bureau of Statistics and the Statistical Office of the Republic of Slovenia, for the Croatia and the Slovenia, respectively. This data differs a bit from the data in the Eurostat database since Eurostat uses the NACE[1] 2010 classification of economic activities, and national statistical office published the data according to the NACE 2007 classification. The data on the GVA from manufacturing sectors was only available for the period from the year 2002 to 2010, and for the years 2011 and 2012 only data on the total GVA was available (all sectors together). The GVA from manufacturing sectors for those years was calculated from the total GVA by the ratio of GVA which these sectors had in the total GVA for the year 2010.
- The data on the generation of the waste from the services sectors and the GVA from the services sectors for the period from the year 2002 till 2012 was collected by searching the Eurostat's database for both the Croatia and the Slovenia. As the services sectors the data from the Wholesale and retail trade, Hotels and restaurants, Transport, Financial intermediation, Real estate, Public administration and community services and Activities of households was taken. The missing data was collected and calculated in the same way as for the data from the manufacturing sectors.
- The data on the Domestic Material Consumption and the national GDP for the period from the year 2001 till 2013 was collected by searching the Eurostat's database for both of the countries.

[1] NACE is derived from the French title "Nomenclature générale des Activités économiques dans les Communautés Européennes" (Statistical classification of economic activities in the European Communities).

6.2. Methodology

All of the four above mentioned indicators are calculated and presented in three different ways:

- As the ratio between the driving force (household expenditure, GVA and GDP) and the environmental pressure (generation of municipal waste, generation of the waste from the manufacturing and services sectors and domestic material consumption). The indicator states that if the ratio increases the waste prevention is being achieved because the driving force is increasing and the environmental pressure is decreasing. The results for the certain time period are then shown in the Figures 3, 5, 7 and 9.
- By subtracting the linear regression slopes of the driving force variable with the environmental pressure variable during the 5 year period (indexed on first year of the period) as shown above in the Equation (2). For each of the 5 year period in the time series decoupling indicator is calculated and the results are shown in the Figures 4, 6, 8 and 10.
- By calculating the rate of prevented waste by first calculating the projection of the environmental pressure by using the model of linear regression analysis with the historical data for the last year in the time series. After that, from the projected amount the real amount of the produced environmental pressure is subtracted to get the amount of prevented environmental pressure. The result (the amount of prevented waste) can be given as the percentage of the amount of the waste that could have been generated or as the percentage of the actual amount of the waste that was generated in that year, if the measures of waste prevention have not been implemented.

6.3. Results

Relation between the Generation of the Municipal Waste and the Household Expenditure

The results of the indicator "Ratio of the household expenditure and the generation of the municipal waste" are shown in the Figure 3. The ratio is slightly increasing for the Republic of Croatia, and significantly increasing for the Republic of Slovenia.

Figure 4 shows the results of the subtracting of the linear regression slopes of the household expenditures and the generation of the municipal waste during the 5 year period for the period 2000 till 2012 (total of 9 five year periods). The decoupling factor for both countries has been decreasing till the end of the period 2003 till 2007, after which for both countries decoupling factor increases indicating waste prevention.

The rate of prevented waste was calculated by projecting the amount of generated municipal waste for the year 2012, as the last year in the time series, by using the model of linear regression analyses of the historical data of the household expenditure and the generated municipal waste. The results for the Republic of Croatia show that projected amount for the year 2012 of generated municipal waste is 10.1% larger than the actual amount of generated municipal waste in that year, indicating prevention of waste of 9.2% in relation to the projected amount of generated municipal waste for that year. The results for the Republic of Slovenia show that projected amount for the year 2012 of generated municipal waste is 38.7% larger than the actual amount of generated municipal waste in that year, indicating prevention of waste of 27.9% in relation to the projected amount of generated municipal waste for that year.

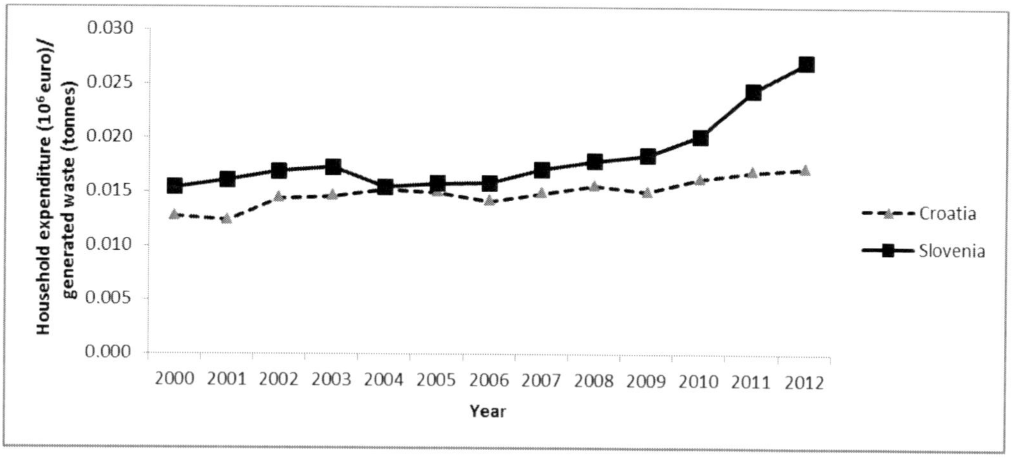

Figure 3. Ratio of the household expenditure and the generation of the municipal waste.

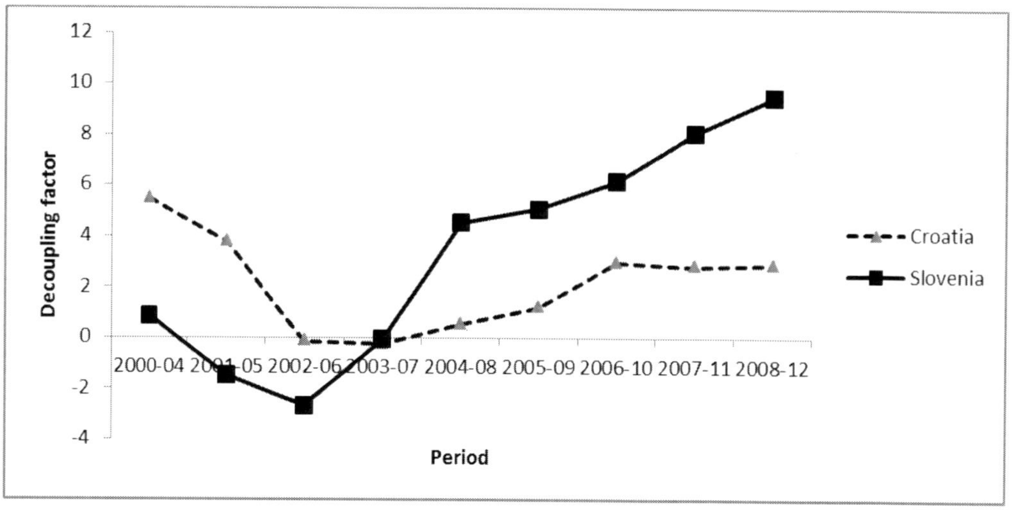

Figure 4. Decoupling factors of the household expenditures and the generation of the municipal waste.

Relation between the Generation of the Waste from Manufacturing Sectors and the GVA from Manufacturing Sectors

The results of the indicator "Ratio of the GVA from manufacturing sectors and the generation of the waste from manufacturing sectors" during the time period of 2002 till 2012 are shown in the Figure 5. The ratio is significantly increasing for the Republic of Croatia as the amounts of generated waste from manufacturing decrease faster than the GVA from manufacturing. For the Republic of Slovenia ratio is almost steady during the observed time period, except for the last year where the result show a slightly larger increase in relation to the previous year.

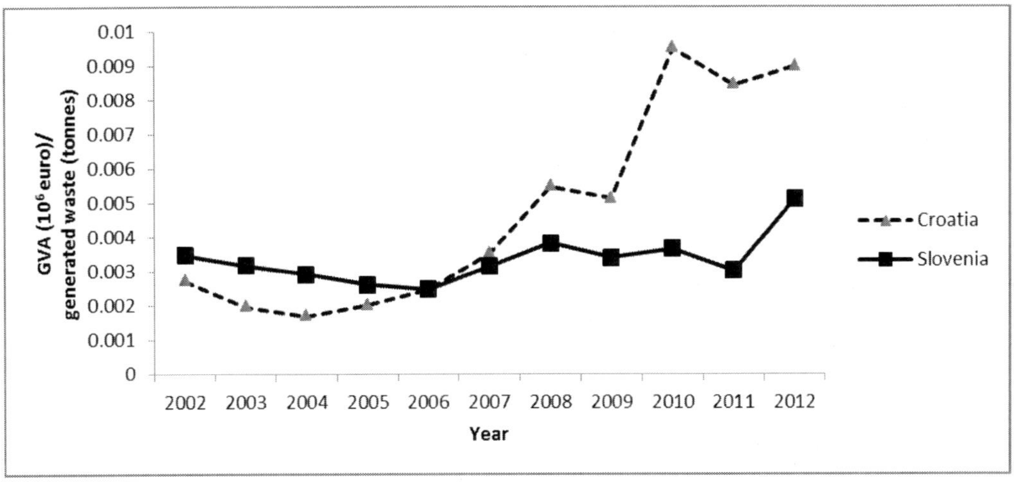

Figure 5. Ratio of the GVA from manufacturing sectors and the generation of the waste from manufacturing sectors.

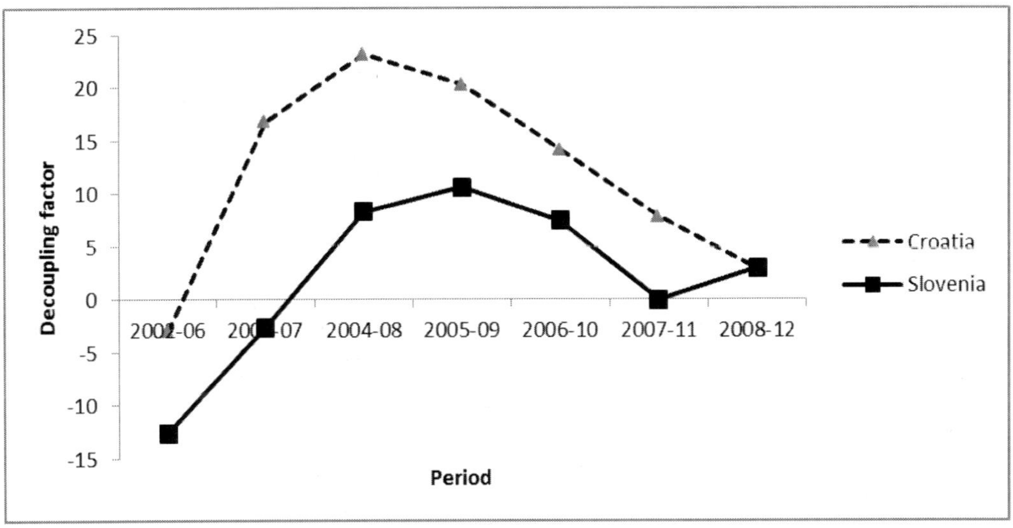

Figure 6. Decoupling factors of the GVA from manufacturing sectors and the generation of the waste from manufacturing sectors.

Figure 6 shows the results of the dividing of the linear regression slopes of the GVA from manufacturing sectors and the generation of the waste from manufacturing sectors, during the 5 year period for the period 2002 till 2012 (total of 7 five year periods). The decoupling factor for the Republic of Croatia has been decreasingfalling in the last 4 five year periods, after the increase in the first three years. The decoupling factor for the Republic of Slovenia showed increase during the first 4 five year periods, after which a decrease is shown in the periods 2006 – 2010 and 2007 – 2011, while in the last five year period an increase is shown.

The waste prevention results can be seen before the 2008 (the start of the worldwide recession) when the amounts of waste from manufacturing sectors declined, and the GVA continued to increase. After 2008 the linear regression slopes for generated waste from manufacturing declined faster than the slopes of GVA for both countries, which led to the low decoupling factors at the end of the observed period. The rate of waste prevention was calculated by projecting the amount of generated waste from manufacturing sectors for the year 2012, as the last year in the time series, by using the model of linear regression of the historical data on the GVA from manufacturing sectors and the generation of the waste from manufacturing sectors.

The results for the Republic of Croatia show that projected amount for the year 2012 of generated municipal waste is 182.84% larger than the actual amount of generated municipal waste in that year, indicating prevention of waste of 64.64% in relation to the projected amount of generated municipal waste for that year.

The results for the Republic of Slovenia show that projected amount for the year 2012 of generated municipal waste is 34.83 % smaller than the actual amount of generated municipal waste in that year, indicating that 53.45% more of waste was generated in the 2012 than the projected amount of generated municipal waste for that year.

Relation between the Generation of the Waste from Services Sectors and The GVA from Services Sectors

The results of the indicator "Ratio of the GVA from services sectors and the generation of the waste from services sectors" during the time period of 2002 till 2012 are shown in the Figure 7. The ratio has significantly increased for the Republic of Croatia in the last two years of the time series after the period of no significant changes. The ratio for the Republic of Slovenia is decreasing from the year 2008, after the period of significant increase.

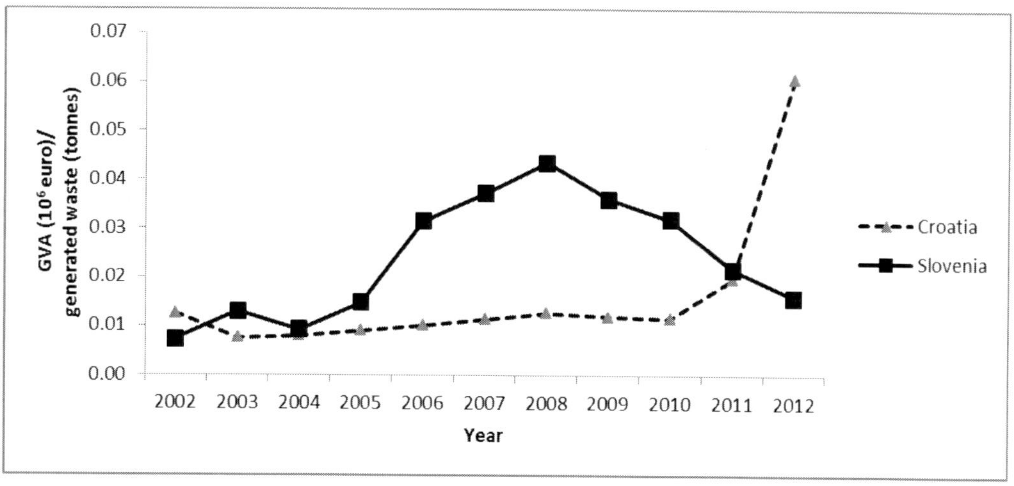

Figure 7. Ratio of the GVA from services sectors and the generation of the waste from services sectors.

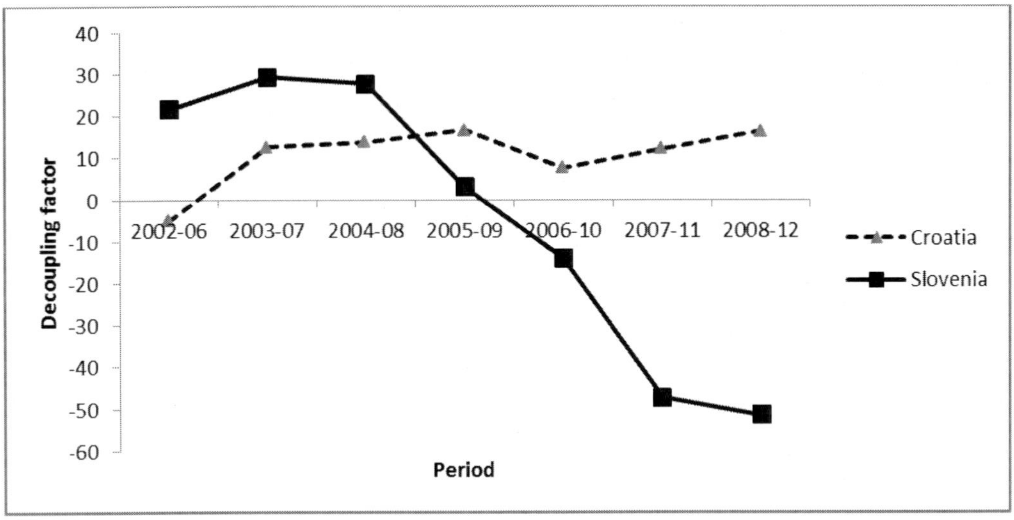

Figure 8. Decoupling factors of the GVA from services sectors and the generation of the waste from services sectors.

Figure 8 shows the results of the dividing of the linear regression slopes of the GVA from services sectors and the generation of the waste from services sectors, during the 5 year period for the period 2002 till 2012 (total of 7 five year periods). The decoupling factor for the Republic of Croatia has slightly increased from the period 2006 till 2010, but varies insignificantly during the whole time series. The decoupling factor for the Republic of Slovenia shows steady decrease, starting with the period 2004 till 2008 all the way to the last period of the time series.

The results for both countries show that in the last years of the observed period GVA from services sectors stagnates without significant changes probably as a result of worldwide recession, while the amounts of generated waste change: in the two last years for Croatia have a significant decrease indicating waste prevention, while for Slovenia in the last four years show significant increase indicating problems with implementation of waste prevention measures in this sectors.

The rate of prevented waste was calculated by projecting the amount of generated waste from services sectors for the year 2012, as the last year in the time series, by using the model of regression analyses of the historical data on the GVA from services sectors and the generation of the waste in those sectors.

The results for the Republic of Croatia show that projected amount for the year 2012 of generated waste from services sector is 467.72% larger than the actual amount of generated waste from those sectors in that year, indicating prevention of waste of 82.38% in relation to the projected amount of generated waste from services sectors for that year.

The results for the Republic of Slovenia show that projected amount for the year 2012 of generated waste from services sectors is 54.53% smaller than the actual amount of generated waste from services sectors in that year, indicating that 119.94% of waste was generated more in the 2012 than the projected amount of generated municipal waste for that year.

Relation between GDP and Domestic Material Consumption

The results of the indicator "Ratio of the GDP and the domestic material consumption" during the time period of 2001 till 2013 are shown in the Figure 9. The ratio has increased for both countries, with the in-between decrease shown for Slovenia in the years 2011 and 2012. The influence of the worldwide recession from year 2008 can be seen for both of the countries, where the amounts of domestic material consumption have more decreased than the GDP.

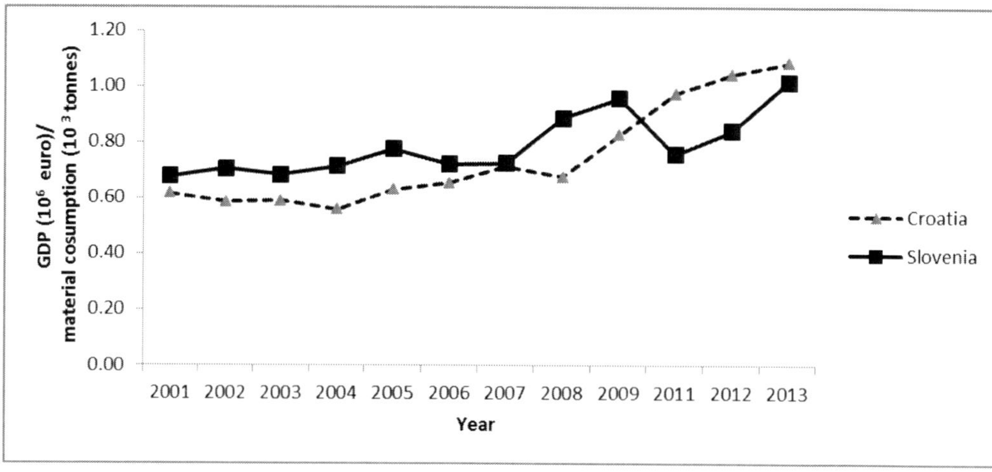

Figure 9. Ratio of the GDP and the domestic material consumption.

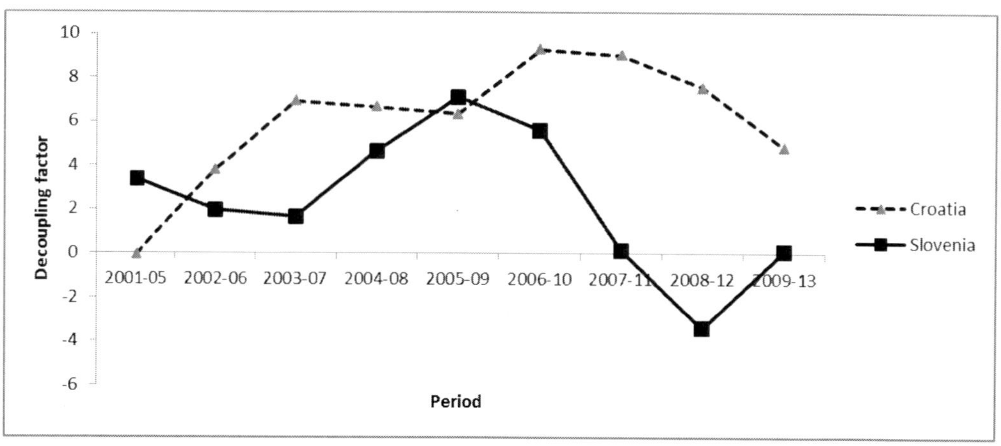

Figure 10. Decoupling factors of the GDP and the domestic material consumption.

Figure 10 shows the results of dividing the linear regression slopes of the GDP and the domestic material consumption for the period 2001 till 2013. Decoupling factors for the Republic of Croatia vary over the observed period with two peaks, one from the first five year period till the period 2003 – 2007 and the second from period 2005 – 2009 to period 2006 – 2010, after which the decoupling factors decrease till the end of the observed period. Decoupling factors for the Republic of Slovenia peaked in the period 2005 – 2009, after which they decrease till the period 2008 – 2012 and then increase in the last observed period.

The reverse decoupling (value below 0) for Slovenia can be seen because in the year 2011 the material consumption increased, although the GDP has decreased.

The rate of prevented waste was calculated by projecting the amount of generated waste from services sectors for the year 2013, as the last year in the time series, by using the model of regression analyses of the historical data on the GDP and the domestic material consumption. The results for the Republic of Croatia show that projected amount for the year 2013 of domestic material consumption is 23.61% larger than the actual amount of domestic material consumption in that year, indicating prevention of waste of 19.10% in relation to the projected amount of domestic material consumption for that year. The results for the Republic of Slovenia show that projected amount for the year 2013 of domestic material consumption is 24.33% larger than the actual amount of domestic material consumption in that year, indicating prevention of waste of 19.57% in relation to the projected amount of domestic material consumption for that year.

6.4. Uncertainties

There is a significant amount of uncertainties regarding this case study, mostly connected with the data. Data is collected from two main sources: national statistical offices (in Croatia also the Environmental Agency) and from Eurostat, with source of data in Eurostat for two countries studied are also those same national statistical offices. The uncertainties arise from different methodologies in collecting the data: ones covering the data on generation of waste and seconds covering the economic variables. The result of this that there is sometimes very little correlation between these two sets of data which is best seen in the calculating the projected amounts of waste which could have been produced by using the linear regression analysis. The coefficients of correlation are especially low between the GVA from services sectors and the amounts of generated waste from those sectors: for Slovenia -0.76 (indicating that with the growth of GVA the amounts of waste decrease) and for Croatia only 0.22. Also, low coefficient of correlation of 0.30 is seen for Croatia between the amounts of GDP and the amounts of domestic material consumption. Furthermore, statistical data on waste is collected by Eurostat every two years so the estimated amounts (in this case by interpolation) do not necessarily mean that this trend really happened that year (e.g. interpolation shows a decrease of amount for the year in-between, a what really happened was in that year was an increase in waste amounts and in the next big decrease in amounts). So the statistical analysis of this data would also show very little connectivity of these two variables. Missing data had to be estimated which further raises the uncertainties. There is also certain amount of uncertainties connected with the transformation of data for missing years from national statistical offices from NACE 2007 classification to NACE 2010 classification which Eurostat uses.

CONCLUSION

Although waste prevention is considered the most desirable option of dealing with waste in the European Union and other European countries, there remain a lot of unsolved issues regarding the measuring of its progress and assessing the impacts which it can have on the environment, economy and society. Without clear measures, prevention activities cannot be chosen, prescribed, implemented or evaluated. It is clear that monitoring approaches of the certain activity (measuring the produced waste before and after the implemented activity) would be the best way to go, this is not easily applicable if the measuring must be done on the macroeconomical level. Mostly used approaches in this case include models for the conducting of life-cycle assessment and econometric models. Application of LCA is one of the mostly used methods for the purpose of detecting and understanding of the influence which the changes in the product and product system would have on the society, environment and economy. For the waste prevention evaluation purposes authors have mostly used their own variations of the model prescribed by ISO 14040 series. The LCA approach is not meant for measuring of waste prevention, rather to make it possible for the decision makers to see the impacts of their decisions regarding implementation of waste prevention. The gaps in data availability authors solve by combining data from national statistics, trade statistics, data from other authors and their own measurements and assumptions, although it raises the level of uncertainties in conducted LCA studies.

Other commonly used approach for the measuring waste prevention is the application of econometric models in which authors connect the generated amounts of generated waste with the economic variables such as spending, consumption and production. These models are used to explain above mentioned relations quantitatively so they can be used to create policy and legislative. One of the models used in the recent time in order to assess the waste prevention are Computable General Equilibrium models which can give the decision makers the predicted results what will happen if some measure is undertook. Mostly used indicator of prevention of waste on the national level is the decoupling of the generation of different waste streams for which data in time series is available (e.g. on municipal waste, manufacturing waste etc.) from the GDP or one of its components. In this paper application of the suggested indicators of waste prevention using the official data of the Republic of Croatia and the Republic of Slovenia has been studied. After the scoping of the available data, four indicators have been chosen using the three calculating methodologies. Results show that waste prevention level in these countries varies depending on the year observed and indicator used, and also varies if different time periods are observed together. Based on the data used, it can be seen that there are periods when the waste prevention is increasing, and there are periods when waste prevention is very low. Different calculating method used on the same indicator yield the similar result, although some may indicate higher realised waste prevention than others. For both of the countries in the last observed time periods the waste prevention makes progress in relation to previous year/period, except for the Republic of Slovenia in the data on the waste from services sector in relation to the GVA from services sector. However, there is a large amount of uncertainties connected with the drawing of conclusions from this study. Namely, the time series of the data is not sufficiently long, data is collected using different methodologies from different sources so there is sometimes very little correlation between the variables studied, which should be the case if economic variable influences the generation of waste. In ordered to apply suggested indicators it is necessary to harmonise the methodologies for collecting the data for the waste and for the economic indicators in order for them to be more closely linked. Also, it is necessary to have larger and higher frequency (e.g. half year, quarterly) time periods available in order to make more accurate conclusions.

REFERENCES

[1] Abrate, G., Ferraris, M. (2010) The Environmental Kuznets Curve in the Municipal Solid Waste Sector, Working Paper 2010/1, Hermes, Moncalieri.

[2] Azapagic, A. (1999) Life cycle assessment and its application to process selection, design and optimisation, *Chemical Engineering Journal* 73, 1 - 21.

[3] Bartoszczuk P., Ma T., Nakamori Y. (2002) Environmental Kuznets Curve for Some Countries–Regression and Ageng-Based Approach. Air Pollution-International Conference, p. 93-102.

[4] BiPRO (2012) Screening of waste management performance of EU Member States. Report submitted under the EC project "Support to Member States in improving waste management based on assessment of Member States' performance". Report prepared for the European Commission DG ENV by Beratungsgesellschaft für integrierte Problemlösungen (BiPRO), Munich.

[5] Böhringer, C., Löschel, A. (2006) Computable general equilibrium models for sustainability impact assessment: Status quo and prospects, *Ecological Economics* 60, 49 – 64.

[6] Bosello, F., Campagnolo, L., Eboli, F., Parrado, R. (2012) Energy from waste: generation potential and mitigation opportunity, *Environmental Economics and Policy Studies* 14(4), 403 – 420.

[7] Bruvoll, A., Fæhn, T., Strøm, B. (2003) Quantifying Central Hypotheses on Environmental Kuznets Curves for a Rich Economy: A Computable General Equilibrium Study, *Discussion Papers Statistics Norway Research Department*, 1 - 34.

[8] Cleary, J. (2009) Life cycle assessments of municipal solid waste management systems: A comparative analysis of selected peer-reviewed literature, *Environment International* 35(8), 1256–1266.

[9] Cleary, J. (2010) The incorporation of waste prevention activities into life cycle assessments of municipal solid waste management systems: methodological issues. *Int. J. Life. Cycle. Assess* 15(6), 579 – 589.

[10] Cleary, J. (2012) Incorporating Waste Prevention Activities into Life Cycle Assessments of Residential Solid Waste Management Systems – A thesis submitted in conformity with the requirements for the degree of Doctor of Philosophy, University of Toronto, Toronto.

[11] Coleman, T., Masoni, P., Dryer, A., McDougall., F. (2003) International expert group on life-cycle assessment for integrated waste management. *Int. J. LCA* 8(3), 175-178.

[12] Creel, M. (2014) Econometrics, Universitat Autònoma de Barcelona, Barcelona.

[13] DECNSW (2004) Producing and consuming efficiently to conserve our resources, Department of Environment and Conservation (NSW), Parramatta NSW, Australia.

[14] DEFRA (2013) Prevention is better than cure - The role of waste prevention in moving to a more resource efficient economy, Department for Environment, *Food & Rural Affairs – DEFRA*, London.

[15] Dinda, S. (2004) Environmental Kuznets Curve Hypothesis: A Survey, *Ecological Economics* 49, 431 – 455.

[16] Directive 2008/98/EC of the European Parliament and of the Council of 19 November 2008 on waste and repealing certain Directives, Official Journal of the European Communities L 312(2008), p. 3 – 30.

[17] EEA (1997) Life Cycle Assessment (LCA) - A guide to approaches, experiences and information sources. European Environmental Agency <http://www.epa.gov/nrmrl/std/lca/pdfs/Issue20report20No206.pdf>. Accessed on 12th of September 2014.

[18] EEA (2014) European Environment Agency – The SOER 2015 thematic fiche on waste. < http://www.eea.europa.eu/soer-2015/europe/waste> . Accessed 25th of April 2014.

[19] Ekvall, T., Assefa, G., Björklund, A., Eriksson, O., Finnveden, G. (2007) What life-cycle assessment does and does not do in assessments of waste management, *Waste Management* 27, 989 – 996.

[20] Eriksson, O., Carlsson Reich, M., Frostell, B., Bjorklund, A., Assefa, G., Sundqvist, J., O.,Granath, J., Baky, A., Thyselius, L. (2005) Municipal solid waste management from a systems perspective, *J. Clean. Prod.* 13, 241–252.

[21] European Commission (2011) Evolution of (Bio-) Waste Generation/Prevention and (Bio-) Waste Prevention Indicators - Final Report. European Commission in association with Bio Intelligence Services, Umweltbundesamt, Arcadis, AEAT & Vito.

[22] Finkbeiner, M., Inaba, A., Tan, R., B., H., Christiansen, K., Klüppel, H., J. (2006) The new international standards for Life Cycle Assessment: ISO 14040 and ISO 14044. *Int. J. Life. Cycle. Assess.* 11, 80 – 85.

[23] FMENCNS (2013) Waste Prevention Programme of the German Government with the Involvement of the Federal Länder, Federal Ministry for the Environment, Nature Conservation and Nuclear Safety (BMU), Bonn, Germany.

[24] Frisch, R. (1933): "Editorial," *Econometrica* (1), 1-4.

[25] Frischknecht, R., Jungbluth, N., Althaus, H-J., Doka, G., Dones, R., Heck, T., Hellweg, S.,Hischier, R., Nemecek, T., Rebitzer, G.,Spielmann, M. (2004) The ecoinvent Database: Overview and Methodological Framework, *Int. J. Life. Cycle. Assess.* (OnlineFirst), 1 – 7.

[26] Gentil, E., C., Damgaard, A., Hauschild, M. Finnveden, G. Eriksson, O., Thorneloe, S., Kaplan, P., O., Barlaz, M., Muller, O., Matsui, Y. Ii, R. Christensen, T., H. (2010) Models for waste Life Cycle Assessment: review of technical assumptions. *Waste Management* 30, 2636 – 2648.

[27] Gentil, E., C., Gallo, D., Christensen, T., H. (2011) Environmental evaluation of municipal waste prevention. *Waste Management* 31(12), 2371–2379.

[28] Giugliano, M., Cernuschi, S., Grosso, M., Rigamonti, L. (2011) Material and energy recovery in integrated waste management systems - An evaluation based on life cycle assessment, *Waste Management* 31, 2092–2101.

[29] Goedkoop, M., Heijungs, R., Huijbregts, M., De Schryver, A., Struijs, J., van Zelm. R. (2009) ReCiPe 2008: A life cycle impact assessment method which comprises harmonised category indicators at the midpoint and the endpoint level. First edition. Report I: Characterisation. Rumte en Milieu. Ministerie van Volkshuisvesting. Ruimtelijke Ordening en Milieubeheer.

[30] Goedkoop, M., Oele, M., Leijting, J., Ponsioen, T., Meijer, E. (2013) Introduction to LCA with SimaPro. PRé, LE Amersfoort, The Netherlands.

[31] Greyson, J. (2007) An economic instrument for zero waste, economic growth and sustainability, *J. Clean. Prod.* 15, 1382 – 1390.

[32] Guinée, J., B. (2004) Handbook on Life Cycle Assessment: Operational Guide to the ISO Standards, Kulwer Academic Publishers, Dordrecht, p. 27.

[33] Hansen, B., E. (2014) Econometrics, University of Wisconsin, Wisconsin.

[34] Hendrickson, C., Horvath, A. (1998) Economic Input-Output Models for Environmental Life-Cycle Assessment, *Policy Analysis* 32(7), 184 – 191.

[35] Hoff, J., L. (2008) Life Cycle Assessment and the LEED Green Building Rating System. RCI 23rd International Convention. Phoenix, p. 121 – 131.

[36] Horne, R., Grant, T., Verghese, K. (2009) Life cycle assessment: Principles, Practice and Prospects, CSIRO Publishing, Collingwood Australia.

[37] IEPA (2014) Irish Environmental Protection Agency - Towards a Resource Efficient Ireland A National Strategy to 2020 incorporating Ireland's National Waste Prevention Programme, Johnstown Castle, Co. Wexford, Ireland.

[38] ISO 14040:2006 Environmental management - Life cycle assessment - Principles and framework (ISO 14040:2006).

[39] ISO 14044:2006 Environmental management - Life cycle assessment - Requirements and guidelines (ISO 14044: 2006).

[40] Jackson, C., Watkins, E. (2012) EU waste law: the challenge of better compliance. Institute for European Environmental Policy, London.

[41] Jeganova, J. (2004) Product Life Cycle Design: Integrating Environmental Aspects into Product Design and Development Process at Alfa Laval – Master's thesis. Lund University, Sweden, p. 9.

[42] Jolliet, O., Margni, M., Charles, R., Humbert, S., Payet, J., Rebitzer, G., Rosenbaum, R. (2003) IMPACT 2002: A New Life Cycle Impact Assessment Methodology. *Int. J. Life. Cycle Assess.* 8(6), 324 – 330.

[43] Katajajuuri, J., M., Hartikainen, H., Jalkanen, L., Koivupuro, H., K., Silvennoinen, K.,Reinikainen, A. (2011) Reduction of food waste in Finnish food production chain as part of life cycle management. MTT Agrifood Research Finland, Helsinki.

[44] Khajuria, A., Matsui, T., Machimura, T.,Morioka, T. (2012) Decoupling and Environmental Kuznets Curve for municipal solid waste generation: Evidence from India. *Int. J. Environ. Sci.* 2(3), 1670 – 1674.

[45] Klöpffer, W. (2012) The critical review of life cycle assessment studies according to ISO 14040 and 14044 - Origin, purpose and practical performance. *Int. J. Life. Cycle Assess.* 17(9), 1087 – 1093. doi: 10.1007/s11367-012-0426-7.

[46] Lindhqvist, T. (2000) Extended Producer Responsibility in Cleaner Production - Policy Principle to Promote Environmental Improvements of Product Systems – Doctorial Dissertation, Lund University, Sweden.

[47] Loesje (2013) Afvalpreventieprogramma Nederland, Loesje International, Arnhem - the Netherlands. (In Dutch).

[48] LRAM (2013) Dėl valstybinės atliekų prevencijos programos patvirtinimo, LIETUVOS RESPUBLIKOS APLINKOS MINISTRO, <http://www3.lrs.lt/pls/inter3/dokpaieska.showdoc_l?p_id=458655&p_tr2=2>. Accessed on 01st of September 2014. (In Lithuanian).

[49] MAAMA (2013) Proyecto de programa estatal de prevencion de residuos, Ministerio de agricultura, alimentacion y medio ambiente, Madrid, Spain. (In Spanish).

[50] Mazzanti, M. (2008) Is waste generation de-linking from economic growth? Empirical evidence for Europe, *Applied Economics Letters* 15(4), 287 – 291.

[51] Mazzanti, M., Zoboli, R. (2008) Waste Generation, Incineration and Landfill Diversion. De-coupling Trends, Socio-Economic Drivers and Policy Effectiveness in the EU, Fondazione Eni Enrico Mattei, Milano.

[52] Moutavtchi, V., Stenis, J., Hogland, W. and Shepeleva, A. (2010) Solid waste management by application of the WAMED model. *Journal of Material Cycles and Waste Management* 12(2), 169 – 183.

[53] MŽRPSR (2014) Program predchádzania vzniku odpadu SR na roky 2014 – 2018, Ministerstvo životného prostredia Slovenskej republiky, <http://www.minzp.sk/files /sekcia-enviromentalneho-hodnotenia-riadenia/odpady-a-obaly/registre-a-zoznamy/ppvo-vlastnymaterial.pdf>. Accessed on 01st of September 2014. (In Slovak).

[54] OECD (2004) Towards waste prevention performance indicators, Organisation for Economic Co-operation and Development. <http://www.oecd.org/officialdocuments/ publicdisplaydocumentpdf/?doclanguage=en&cote=env/epoc/wgwpr/se(2004)1/final>. Accessed on 02th September 2014.

[55] Östblom, G., Söderman, M., L., Sjöström, M. (2010) Analyzing future solid waste generation - Soft linking a model of waste management with a CGE-model for Sweden, working paper of the National Institute of Economic Research (NIER) 118, 1 -51.

[56] Rebitzer, G., Ekvall, T., Frischknecht, R., Hunkelerd, D., Norrise, G., Rydbergf, T., Schmidtg, W., P., Suhh, S., Weidema, B., P., Penningtonf, D., W. (2004) Life cycle assessment - Part 1: Framework, goal and scope definition, inventory analysis, and applications. *Environ. Int.* 30, 701 – 720.

[57] SAIC (2000) Measuring Waste Prevention in New York City, Prepared by: Science Applications International Corporation, Newport – Boston, USA.

[58] Salhofer. S., Obersteiner, G., Schneider, F., Lebersorger, S. (2008) Potentials for the prevention of municipal solid waste. *Waste Management* 28; 245-259.

[59] Sefic Williams, A. (2009) Life Cycle Analysis: A Step by Step Approach. Illinois Sustainable Technology Center, Institute of Natural Resource Sustainability, University of Illinois at Urbana-Champaign.

[60] Sharp, V., Giorgi, S., Wilson, D., C. (2010) Methods to monitor and evaluate household waste prevention, *Waste. Manag. Res.* 28, 269 – 280.

[61] Sjöström, M., Östblom, G. (2010) Decoupling waste generation from economic growth — A CGE analysis of the Swedish case. Ecol. Econ. 69, 1545 –1552.

[62] Skovgaard, M., Moll, S., Andersen, F., M., Larsen, H. (2005) Outlook for waste and material flows Baseline and alternative scenarios, ETC/RWM working paper 2005/1, European Topic Centre on Resource and Waste Management, Copenhagen.

[63] Skovgaard, M., Villanueva, A., Andersen, F., M., Larsen, H. (2007) ETC/RWM working paper 2007/1, European Topic Centre on Resource and Waste Management, Copenhagen.

[64] Šohinger, J. (2002) Mikroekonomska analiza jednosektorskog modela opće ravnoteže u uvjetima diferencijacije proizvoda na strani uvoza i izvoza (eng. Microeconomic analysis of a one sector general equilibrium model with product differentiation on the side of imports and exports), *Ekonomski pregled* 53(9-10), 849 – 863. (In Croatian).

[65] Solano, E. (1999) Integrated solid waste management alternatives in consideration of economic and environmental factors: A mathematical model and development and evaluation - PhD dissertation, North Carolina State University, Raleigh.

[66] Suzuki, H., Dastur, A., Moffat, S., Yabuki, N., Maruyama, H. (2010) Eco^2 Cities – Ecological Cities as Economical Cities, The International Bank for Reconstruction and Development/ The World Bank, Washington DC.

[67] The Scottish Government (2013) Safeguarding Scotland's Resources - Blueprint for a More Resource Efficient and Circular Economy, the Scottish Government. <http://www.scotland.gov.uk/Publications/2013/10/6262/7>. Accessed on 01st of September 2014.

[68] The Welsh Government (2013) Towards Zero Waste One Wales: One Planet - The Waste Prevention Programme for Wales, the Welsh Government. <http://wales.gov.uk /topics/environmentcountryside/epq/waste_recycling/prevention/waste-prevention-programme/?lang=en>. Accessed on 01st of September 2014.

[69] UNEP (2013) Guidelines for national waste management strategies: Moving from challenges to opportunities, United Nations Environment Programme, Nairobi, Kenya.

[70] USEPA (2014) U. S. Environmental Protection Agency. Waste Reduction Model (WARM).< http://epa.gov/epawaste/conserve/tools/warm/index.html>.Accessed 25th of February 2014.

[71] Wang, H., He, J., Kim, Y., Kamata, T. (2011) Municipal Solid Waste Management in Small Towns - An Economic Analysis Conducted in Yunnan China, Policy *Research Working Paper* 5767, The World Bank Group, Washington DC.

[72] Wilts, H. (2012) National waste prevention programs: Indicators on progress and barriers. *Waste. Manag. Res.* 30(9) Sup. 29 – 35, 29 -35.

[73] Winters-Lynch, J. (1993) The Use of Life Cycle Assessment in Environmental Labelling Programs. The United States Environmental Protection Agency, Washington.

[74] Yedla, S. (2005) Modified landfill design for sustainable waste management, *Int. J. Global Energy Issues* 23(1), 93 – 105.

[75] Zorpas, A., A., Lasaridi, K. (2013) Measuring waste prevention, *Waste Management* 33(5), 1047 – 56.

In: Landfills and Recycling Centers
Editor: Carla H. Jackson

ISBN: 978-1-63463-792-3

Chapter 8

FROM DUMPING TO SUSTAINABLE LANDFILLING: THE CONCEPT OF AEROBIC LANDFILLS

Gabriela Kalčíková *[*] ***and Andreja Žgajnar Gotvajn***
University of Ljubljana, Faculty of Chemistry and Chemical Technology, Ljubljana, Slovenia

ABSTRACT

Landfilling has been for decades the most often used method for waste management. However, the growing awareness of the detrimental environmental effects of current waste processing led to important movement towards sustainable waste management practices in Europe. The concept includes clear hierarchy of waste management, where the main priority of waste management is waste reduction, reuse, recycling and composting, energy recovery and the last favorable option is landfilling. However, a reduction of municipal waste has been minimal over past years and landfilling is still the dominating waste management option in many countries of the European Union. Because of that, municipal waste will be still disposed in landfills and therefore, there is still the need for landfilling. Landfilling of municipal waste under aerobic conditions is one of the most promising methods for sustainable municipal waste disposal. Hence, the main aim of our study was to compare different ways of landfilling during past years. Special attention was focused on the concept of aerobic landfilling. As a case study, biodegradation of residual municipal waste in aerobic landfill was investigated. The biodegradation proceeded in a pilot-scale aerobic landfill bioreactor and as a result landfill leachate was generated. Due to the fast and intensive biodegradation, leachate pollution considerably decreased and after several months, all investigated parameters of leachates were below legislative limits. Furthermore, leachate became non-toxic to many organisms and it is expected to cause minimal adverse effect to the environment. Aerobic landfilling thus meet requirements of sustainability concepts and it seems to be an appropriate option for modern landfilling in the framework of the sustainable waste management.

---

[*] Corresponding author: +386 1 479 8568, e-mail: gabriela.kalcikova@fkkt.uni-lj.si

INTRODUCTION

A Brief History of Landfilling

Historically, waste management and landfilling has been inextricably linked with the evolution of human communities, population growth and the emergence and development of commerce because before early humans roamed the Earth, there was no such thing as waste (Letcher, 2011). In the beginning, waste was probably composed of the remains from hunting, gathering, food preparation and human fecal matter. When waste accumulated, nomadic people simply moved to another location (Pichtel, 2005). Those people that settle and stay longer at a location established primitive dump sites (Bilitewski et al., 1994). With development of civilizations the issue connected to waste became significant. Similar primitive "waste management" based on placing of waste away from populated areas into large pits or in rivers occurred in much more developed civilizations as Mesopotamia and Egypt. The cradle of waste management with organized landfilling can be considered ancient Greece. From 5th century BC, Greek municipalities began to establish town dumps that were maintained in a relatively orderly condition (Pichtel, 2005). Their concern about the waste stemmed also from the knowledge of the Greek scholar Hippocrates that suggested an association between personal hygiene, contaminated water, spoiled food and disease outbreak and epidemics (Bilitewski et al., 1994). Roman Empire was another civilization with a relatively developed waste management. During the 1st century CE, many public toilets were already installed in Rome and waste collection and transportation of the refuse to a pit located outside the city gates was well organized. However, with the end of the Roman Empire the knowledge about waste management and basic hygiene were lost. From the Middle Ages, there was no organized waste management and routine procedure was to simply dump waste, including fecal matter, directly out of the window or to discard waste into surface water. The water pollution and absence of hygiene resulted in devastating plaque pandemic in 14th century when 25 million of 80 million of European population died. Although, dumping the garbage into rivers was after the plaque pandemic prohibited garbage from households, animal manure and industrial debris were still dumped into central gutters in the streets (Pichtel, 2005). A breakthrough in waste management occurred during the Industrial Revolution. Except a great increase in waste generation especially in terms of industrial waste, the time of the Industrial revolution brought many important revelations such as that bacteria and viruses cause of diseases and thus spread of diseases could be controlled by the presence or absence of public health measures. Due to the demands for water and sewage treatment plants, cleanness, and for preventive public health measures, engineers and technicians were in the 19th century challenged to develop technologies that would alleviate and solve these problems (Bilitewski et al., 1994). Until the 1960s uncontrolled land dumping without any modern infrastructure or environmental control was the most common method of waste management worldwide (Cossu, 2010). However, during the final years of the 20th century it became apparent that such expanded past economic growth could not be sustained in future without irreparable damage to the environment and landfilling has undergone significant changes (Letcher, 2011). The technical and scientific progress led to modernization of landfills featuring artificial lining system, leachate drainage system, waste compacting, anaerobic process, biogas and leachate collection and treatment and final capping (Cossu, 2010). Such way of landfilling had, however, numerous disadvantages such as long term gas and leachate emissions, possible water pollution and acceptance of waste material that could be potentially reused or recycled (Neczaj et al., 2005). The further development of landfilling, thus, targets on disposal of certain types of waste; only residual waste was aimed to landfilling. Thereby the volume of waste is reduced, production of greenhouse gases is minimized and the environmental impact is lowered (Cossu, 2010). Currently, the most recent trend in landfilling is based on avoiding leaving future generations to manage unacceptable environmental burdens, to protect ecosystems, to sustain a livable environment for people and thus the sustainable landfilling advanced into the forefront.

Sustainable Waste Management

With a growing awareness of the detrimental environmental effects of past and current waste handling, waste management is becoming one of the key problems of the modern world. The first move towards sustainable waste management in Europe was introduced by the application of the Waste Framework Directive in 1975 (Council Directive 75/442/EEC). The Directive pointed out the basic principles of waste management, which are fundamental to the issue of sustainable development defined as 'development that meets the needs of the present without compromising the ability of future generations to meet their needs' (Almorza et al., 2002). The most recent trend in waste management has become Integrated waste management. This strategy refers to the complementary use of a variety of practices to safely and effectively handle municipal solid waste (Almorza et al., 2002) and it has developed into the concept of the hierarchy of waste management given by Council Directive 2008/98/EC (2008). The hierarchy places emphasis on reducing, reusing, and recycling (composting) the majority of wastes, while energy recovery and especially landfilling are the most unfavorable options (Picture 1).

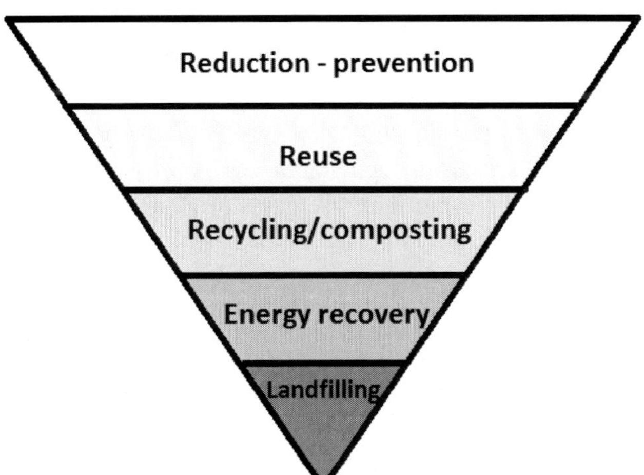

Picture 1. Hierarchy of waste management.

The main priority of waste management is waste reduction that can significantly save raw materials, energy and reduce waste disposal costs (Williams, 2005). However, the reduction of municipal waste in many EU countries has been minimal (e.g. Germany, Hungary) or the generation of municipal waste has even increased (e.g. France, Austria, Switzerland) (Eurostat, 2013). The reason is that the generation of municipal waste is a direct consequence of human activities. In most developed countries the impact is more profound due to rapid economic growth. As a result, the higher purchasing power enables the community to consume various new types of products (Agamuthu and Fauziah, 2011) and thus the economic prosperity of the country and the mean living standard are the major parameters and drivers of municipal waste generation (Lebersorger and Beigel, 2011).

When waste already generated, it should principally be reused. The waste material that is with little or no processing used again for the same purpose or may also include new uses for the item once it has served its original purpose. The best example of well adopted and long-lasting reuse of material is reuse of glass bottles. It involves the collection, cleaning and reuse of these same bottles. However, reuse may not be desirable since the environmental and economic cost of reuse in terms of energy use, cleaning, recovery, transportation, etc., may outweigh its benefits (Williams, 2005).

If waste material cannot be reused it can be often recycled. Recycling of waste is the breaking down of the used item into raw materials which are used to make new items. This process is also well established, e.g. the recycling of aluminium cans (Green, 2007). It may not be appropriate in all cases, for example where the abundance of raw material, energy consumption during collection and re-processing, or the emission of pollutants have a great impact on the environment or it is not cost-effective (Williams, 2005). Some materials can be used for composting. Composting involves conversion of organic waste material into compost that can be used as additive to improve soil quality. The composting of household organic waste has been encouraged for home owners as a direct way of recycling. On the other hand it can also be collected on behalf of municipality and composted together with green wastes from parks and sewage sludge in large scale composters (Rhyner et al., 1995).

Residual waste (non-recyclable and non-usable) can be used for energy recovery. The energy recovery turns waste materials into useable heat, electricity, or fuel through a variety of processes, including combustion, gasification, pyrolization, anaerobic digestion, and landfill gas recovery. This process is often called waste-to-energy (EPA, 2014). However, many combustion facilities represent a high initial capital cost and sophisticated emission control measures are required to clean-up the flue gases (Williams, 2005).

Despite the fact that landfilling is the least favourized option in the hierarchy of waste management it is still used as a primary means of waste treatment in many countries (Eurostat, 2012).

Traditional Landfilling of Municipal Waste

Landfilling of municipal waste still plays a very important role in waste treatment in many countries in Europe. The major advantage associated with landfilling of wastes is the initial low cost of landfills compared to other options and the fact that a wide variety of wastes are suitable for landfill. Many other waste treatment options require the final disposal route for the residues to be landfilled, e.g. incineration bottom and flyashes (Williams, 2005).

However, not only the scarcity of available landfill sites, the long-term deterioration of a landscape and its limited usability, but also the long-term emissions following the disposal of municipal solid waste, make the management of municipal solid waste a major problem (Neczaj et al., 2005).

Traditional old landfills, which are in some cases still in current use or have long been disused, were constructed before the environmental impacts of leachate and landfill gas were recognized (Williams, 2005). In past times, when landfills were not under strict control, legal (or illegal) co-disposal of industrial or hazardous waste in municipal landfills was common (Slack et al., 2005). At present, modern or reconstructed landfills are highly engineered facilities designed to eliminate or minimize the adverse impacts of landfilling on the environment (Wiszniowski et al., 2006).

Degradation Processes in a Traditional Landfill

A municipal landfill is a physically, chemically and biologically heterogeneous system, which behaves like a large-scale anaerobic bioreactor. After landfilling, solid waste undergoes physico-chemical and biological transformations and, consequently, the degradation of the organic fraction of waste proceeds (Bila et al., 2005; Kurniawan et al. 2006) (Table 1). During the placement of waste, hydrolysis – the aerobic degradation stage occurs as a first stage of biodegradation. Aerobic microorganisms start the degradation of organic fractions like cellulose and hemicellulose to produce simpler hydrocarbons, carbon dioxide, water and heat. Temperatures can reach up to 90 °C; however, compacted waste achieves lower temperatures due to the lower availability of oxygen. Water and carbon dioxide are the main products, with carbon dioxide released as gas or absorbed into water to form carbonic acid, which gives acidity to leachate. This stage lasts for only a matter of days or weeks and it depends on the availability of oxygen. Soon aerobic conditions are replaced by anaerobic ones and hydrolysis and fermentation occurs. Different microorganisms – the facultative anaerobes, which can tolerate reduced oxygen conditions, become dominant. Carbohydrates, proteins, and lipids are hydrolyzed and further decomposed to carbon dioxide, hydrogen, ammonia and organic acids. The ammonia is derived largely from the deaminisation of proteins. The organic acids are mainly acetic acids and their formation depends on the initial waste material. The formed organic acids are converted during the next stage – acetogenesis – by acetogenic microorganisms to acetic acids, acetic acid derivates, carbon dioxide and hydrogen under anaerobic condition. Other organisms convert carbohydrates directly to acetic acids in the presence of carbon dioxide and hydrogen. Low hydrogen levels promote the methanogenic microorganisms which generate methane and carbon dioxide from the organic acids. The presence of organic acids generates very acidic conditions with a pH level of 4 or lower. In the next stage – methanogenesis – methanogenic microorganisms generate carbon dioxide and methane from the organic acids created in previous stages. Consequently, the acid concentration becomes depleted, thus, pH rises to about 8. The methanogenesis is very slow and takes many years. The duration of the stage depends on many factors, e.g. moisture level, but it is the longest stage of them all. Oxidation is the final stage of waste degradation. The acids are used up and new aerobic microorganisms slowly replace the anaerobic forms and re-establish aerobic conditions (Kjeldsen et al., 2002; Williams, 2005).

Table 1. Degradation processes in a traditional landfill

Time	From waste placement untill landfill closure				
Process	Hydrolysis	Hydrolysis/ Fermentation	Acetogenesis	Metanogenesis	Oxidation
Conditions	Aerobic	Anaerobic			Aerobic
Products	CO_2, H_2O heat, low polluted leachate	CO_2, H_2 from weakly to strongly acidic leachate		CH_4, CO_2 alkaline leachate	CO_2 low polluted leachate
Duration	Days, weeks	Years		Decades	
Microorganisms	Aerobic	Facultative anaerobic	Acetogenic	Metanogenic	Aerobic

Landfill Gases

Gases occurring in active and closed landfills include methane, carbon dioxide, carbon monoxide, hydrogen, hydrogen sulfide, ammonia, nitrogen and oxygen. The most common gases produced in landfills are methane and carbon dioxide (biogas), with lesser amounts of the other gases occurring, typically in concentrations below 1% (Pichtel, 2005). Biogas formed during the ageing of wastes (mainly during the methanogenic phase) at least has some economic value, because it can and should be harvested and utilized for energy purposes. The total amount of methane recovery for a (final or temporary) covered landfill site can reach even more than 90% (Morcet et al., 2003).

Landfill Leachate

Landfill leachate is one of the major problems connected to landfilling of municipal waste because leachate represents only an economic burden and environmental risk. Leachates occur during landfilling at every site and can be generated for many years after the landfill had been closed. Leachate is defined as heavily polluted liquid, which contains a variety of different pollutants. Leachates present a risk, not only for the environment, but also for human beings (Renou et al., 2008).

Leachate composition has been the subject of numerous research studies in the last decades because the risk connected to water pollution caused by leakage or by release of untreated leachate has been and still remains one of the main problems in landfill management. At the beginning of the 1970s leachate quality studies were focused on a rough characterization of leachate components. Later on, researchers' efforts shifted towards the identification of the main factors influencing the composition of leachate (Christensen et al., 1992).

Pollutants in landfill leachate can be divided into four groups (Kjeldsen et al., 2002; Kulikowska and Klimiuk, 2008; Renou et al., 2008; Comstock et al., 2010):

1. Dissolved organic matter, quantified as Chemical Oxygen Demand (COD) or Total Organic Carbon (TOC), volatile fatty acids and more refractory compounds, such as fulvic and humic compounds.
2. Inorganic macrocomponents as calcium, magnesium, sodium, potassium, ammonium, iron, manganese, chloride, sulfate and hydrogen carbonate.
3. Heavy metals: cadmium, chromium, copper, lead, nickel and zinc.
4. Xenobiotic organic compounds originating from household or industrial chemicals and present in relatively low concentrations (in $\mu g \cdot L^{-1}$). These compounds include, among others, a variety of aromatic hydrocarbons, phenols, chlorinated aliphatics, pesticides and plastizers.

Other compounds may be found in leachate, e.g. borate, sulfite, arsenate, mercury and cobalt. However, in general, these compounds are found in very low concentrations and they are only of secondary importance.

Moreover, the composition and concentrations of pollutants in leachate are very varied, depending on several parameters, including those concerning waste mass and site localization, as well as those deriving from the design and management of the landfill. However, leachate composition is mainly influenced by waste composition, precipitation, seasonal weather variation, and especially by the landfill age (Renou et al., 2008; Tatsi et al., 2003). As a landfill passes through different stages of its life-cycle (Table 1) the leachate composition changes dramatically. Every stage of the landfill's life-cycle generates specific types of leachate (Ziyang et al., 2009).

In the beginning of a landfill's operation, during the aerobic phase, leachate mostly results from the release of moisture during compaction, as well as from precipitation through the buried waste (Kjeldsen et al., 2002; Salem et al., 2008). The amount of leachate leaving the landfill is limited due to the retention of water by waste, until the leachate reaches the collection system and it is drained to the collection basin (Armstrong and Rowe, 1999). This period is relatively unimportant in terms of leachate quality and produces leachate that contains matter, which was simply washed out of the waste (Salem et al., 2008). In the following anaerobic stage the concentration of organic matter starts to increase due to the occurrence of fatty acids generated during hydrolysis of carbohydrates, proteins and lipids. During acetogenesis a high concentration of soluble biodegradable organic compounds is generated. The concentration of organic matter (expressed as COD, BOD_5) significantly increases, as well as the BOD_5/COD ratio, representing potential for biodegradation (Deng and Englehardt, 2006). With the occurrence of acids the pH of leachate becomes acidic on the scale from slight to strong. Timur and Özturk (1999) reported COD of leachate generated after 3.5 years of waste disposal from 16 200–20 000 $mg \cdot L^{-1}$ and BOD_5 from 10 750–11 000 $mg \cdot L^{-1}$. Kjeldsen et al. (2002) presented even more polluted acidic leachate (pH 4.5–7.5), COD was 6000–60 000 $mg \cdot L^{-1}$ and BOD5 400–40 000 $mg \cdot L^{-1}$. Such highly polluted acidic leachate is chemically aggressive and increases the solubility of many compounds, especially metals and thus increases their concentration in landfill leachate. In addition, organic acids, chlorine ions, ammonium nitrogen and phosphate, all in high concentrations in leachate, readily form complexes with metal ions, causing further increases in solubility of metal ions. Hydrogen sulphide may also be produced throughout the anaerobic stage as the sulphate compounds in the waste are reduced to hydrogen sulphide by sulphate-reducing microorganisms. Metal sulphides may be a reaction product of the hydrogen sulphide and metal ions in the solution. Such high-strength leachate typically occurs up to five years after waste disposal. Significant components of leachate in this phase are thus heavy metals and biodegradable organics (Pivato and Gaspari, 2006; Williams, 2005).

After several years, when methanogenic conditions are established, leachate becomes strongly buffered by bicarbonate, while the content of salts in the leachate may still be high. Leachate becomes neutral or slightly alkaline, low-biodegradable, but still polluted by organic matter. COD can still be high at around 3000 $mg \cdot L^{-1}$, but BOD_5/COD is very low – 0.05 (Li et al., 2010; Bila et al., 2005). In the stable methanogenic phase, where methane production decreases, the pH continues to increase and leachate contains mostly recalcitrant compounds – humic and fulvic acids (Armstrong and Rowe, 1999; Kjeldsen et al., 2002; Salem et al., 2008). These organic substances are products of the microbial degradation of waste and they remain in landfills because microorganisms are unable to degrade them further, so they are typical for mature landfill leachate (Batarseh et al., 2010). The concentration of metals is very low, as a consequence of physico-chemical conditions in the body of the methanogenic landfill. Sorption and precipitation are determined as significant mechanisms for metal immobilization and the subsequent low concentrations of metals in the leachate. Wastes containing soils and certain types of organic matter have a significant sorptive capacity, especially at neutral to high pH values prevailing in methanogenic leachate. In addition, the solubility of many metals, sulphides and carbonates at higher pH is low. Sulphides and carbonates are capable of forming precipitates with Cd, Ni, Zn, Cu and Pb. While carbonates are abundant in landfill leachate, the solubility of metal carbonates is generally higher than the solubility of metal sulphides. Cr is an exception because it does not form an insoluble sulphide precipitate, but it forms precipitates with hydroxide. A determination of the sulphur content of landfill waste has shown that the waste does not contain sufficient sulphur to bind all the heavy metals (only 5%) (Williams, 2005; Christensen, 2011; Armstrong and Rowe, 1999). The moisture content of waste is one of the most important factors that affect degradation and the consequential stabilization of a landfill. By recirculating leachate, the refuse moisture content increases from its initial value, which is typically 15–25%, to 40–50%. Leachate recirculation results in a better distribution of nutrients, substrates, and bacteria. It is beneficial for microorganisms inside a landfill to degrade the organic matter and thereby accelerate degradation (Kjeldsen et al., 2002).

The final aerobic stage of waste degradation results in the end of the degradation reactions and the leachate will probably not be significantly changed from the previous stage, except that the salt concentrations may start decreasing as a consequence of flushing. However, such leachates have so far not been investigated further (Christensen, 2011; Armstrong and Rowe, 1999).

Besides pollutants, landfill leachates contain a variety of microorganisms. Municipal waste contains a large microbial population and may be contaminated by pathogenic microorganisms. First microbiological studies carried out on sanitary landfills investigated and emphasized hygienic and sanitary considerations focusing on microorganisms, especially bacteria and viruses of fecal origin. More recent studies have started to closely examine the numbers and physiological activity of microorganisms involved in the stabilization of waste. Various studies observed bacteria populations with varying entities and composition according to landfill age (Christensen et al., 1992). Some bacteria have been found to be partially responsible for the reduction of hexavalent chromium to low-toxic and essential trivalent chromium in landfill leachates (Li et al., 2007). Bacteria can, however, be inhibited in the first stage, because temperatures can reach up to 90 °C, and also in the next stages due to a very low pH. Thus, the concentration of fecal indicator bacteria generally decreases as the landfill age increases and ultimately reaches levels which are no longer detectable. Little information is available in literature on the presence of fungi in leachate. Species observed by various researchers are mainly saprophytes such as *Aspergillus*, *Penicillum* and *Fusarium* (Christensen et al., 1992).

Sustainable Landfilling

The concept of "sustainable landfilling" is relatively young and it occurs as a consequence of sustainable development and sustainable waste management. The term seems to be fairly irrational due to the hierarchy of waste management (Picture 1), where landfilling is the least sustainable option for waste treatment. However, the term "sustainable landfilling" represents the most recent evolution in waste management.

Although, recycling and composting of municipal waste advanced into the forefront during the last decade (Municipal Waste Statistics, 2011) municipal waste or better yet, residual municipal waste (where the recyclable, biodegradable and usable fractions have been removed according to sustainable waste management), will still be (if not incinerated) disposed in landfills and therefore, there is still the need for landfilling. However, the objectives of many traditional landfill operations focus largely on the short term prevention of hazards and risks to the environment that are caused by polluted leachate and landfill gas, as well as on a controlled utilization of the energy provided by extracted landfill gas. Once this option is not feasible or at least not economically beneficial anymore, many landfills will still exhibit a significant emission potential, which tapers off with time to very low levels. At this stage, these traditional landfills are considered to be unsustainable and landfill operations have to be changed to move toward sustainable landfilling (Ritzkowski and Stegmann, 2012).

The aim of sustainable landfilling is to develop landfill sites capable of reducing environmental impacts to a level that no longer significantly affect the quality of surrounding environment, whether air, water or soil, over one generation (30–40 years). A similar condition is habitually termed final storage quality (Huber-Humer and Lechner, 2011; Cossu and Lai, 2012). The main goal is the sustainable conversion of traditional anaerobic landfills into a biologically stabilized state associated with a significantly lowered or nearly eliminated landfill gas emission potential and polluted landfill leachate (Ritzkowski and Stegmann, 2007). As a general rule the landfilled material is made up of different fractions: gasifiable, leachable, stable, and mineralisable. Once the degradables have been degraded and the leachables washed out, stable and mineralized matter – roughly accounting for as much as 50% – represents a potential geological sink. A sustainable landfill design should thus aim to adopt the best strategy suited for dealing with the mobile fraction, posing a potential threat to the environment in the short and long term in a suitable way (Cossu, 2010).

A Case Study

Landfilling of municipal waste under aerobic conditions is one of the most promising methods for sustainable municipal waste disposal. Under aerobic conditions, waste is rapidly stabilized, amount and strength of landfill leachate is reduced and landfill gas emission is significantly lowered or nearly eliminated (Ritzkowski and Stegmann, 2012). So far, researchers have mostly focused on the investigation of gas and leachate emissions during aerobic stabilization of unsorted or synthetic municipal waste (Giannis et al., 2008; Xiaoli et al., 2013) and some of them highlighted prevalence of anaerobic conditions in the initial phases of landfill operation in spite of intensive aeration. Such occurrence of anaerobic conditions may be directly linked to consumption of all available oxygen during the massive degradation of organic matter in unsorted municipal waste. However, less research attention was given to the current composition of municipal waste that has been significantly changed over last past years due to implementation of European legislation preventing the organic fraction of municipal waste from being landfilled (Municipal Waste Statistics, 2011). In particular, there have been no studies that investigated degradation of residual municipal waste (fraction which is not-recyclable and cannot be used for composting) under aerobic conditions. In this context, the objective of this case study was to investigate characteristics of leachates generated during aerobic degradation of residual municipal waste in a pilot-scale landfill exposed to natural climatic conditions. It is hypothesized that if only residual uncompacted waste is landfilled, the aerobic conditions within the body of landfill can be achieved only by introduction of ambient air into the landfill through its surface and therefore in this study pilot-scale landfill was not additionally aerated.

Materials and Methods

A pilot-scale landfill was located outside at the field of Center of Waste Management (Vrhnika, Slovenia, 45°58'30.3"N 14°18'38.6"E) and exposed to natural climatic conditions. A high resistant high density polyethylene container (HDPE, ZIEGLER d.o.o., Picture 2) with a volume 650 dm^3 was installed at the site, washed out by water and one pipe was screwed in the front part of the container for collecting of leachates. Gravel was placed on the bottom of the container as drainage (approximately 20 cm) and one layer of nonwoven fabric was used as membrane to isolate gravel from the waste. Municipal waste was obtained from daily municipal waste collection truck from a location where waste separation practice among inhabitants is well adopted. The municipal waste was roughly shredded by shredding machine and additionally sorted to remove minor fractions of waste which is not supposed to be landfilled (e.g. pharmaceuticals, batteries, remains of fruits and vegetables). Such sorted waste can be regarded as representative of residual non-recyclable and non-usable waste that is expected to be (if not incinerated) always disposed on landfills. Waste was filled into the container and manually compacted in the reactor. The top was gently covered by soil to prevent loosing of waste if strong wind occurs. Air was not injected into the bioreactor as described in some studies (e.g. Giannis et al., 2008; Erses et al., 2008) but aerobic conditions were achieved only by introduction of ambient air into the landfill through its surface. The temperature was measured using a digital thermometer with sensors installed in the middle of the container and on the top of disposed waste. Temperature (inside and outside of landfill) and meteorological conditions were monitored daily. Leachate was collected in HDPE container (10 L). Stabilization of municipal waste in pilot-scale aerobic landfill started in March, 15th 2012 and it was monitored for 326 days.

The quality of landfill leachate was evaluated by following physico-chemical parameters: pH, redox potential, BOD_5 (Biochemical Oxygen Demand) (ISO 5815–1, 2003), COD (Chemical Oxygen Demand) (ISO 6060, 1989), DOC (Dissolved Organic Carbon) (ISO 8245, 1999), concentration of ammonium nitrogen (ISO 7150–1, 1984), concentration of nitrite and nitrate nitrogen, alkalinity (expressed as $CaCO_3$), orthophosphates, and chlorides (Standard Methods, 2012). Each analysis was performed twice and the results are given as mean values. The list of ISO standards is given in Kalčíková et al. 2014.

For determination of metals, samples in three parallels were decomposed by $HNO_3:H_2O_2 = 1:1$ in teflon cups (561B, Savillex, Minnesota, USA) at 200 °C for 2 hours. Samples were vaporized and quantitatively transported to polypropylene tubes for centrifugation and diluted to 15 ml by Mili-Q water. The concentration of metals was determined by ICP-MS (Agilent 7500ce). Blank samples were taken in account.

Picture 2. HDPE container used for pilot scale landfill project.

A battery of biassays containing different organisms was chosen for evaluation of the acute toxicity of leachate (Table 2).

Bacterium *Vibrio fischeri* brine shrimp *Artemia salina* and white mustard *Sinapis alba* were chosen due to their widespread applications in toxicity studies and sensitivity to landfill leachates.

For evaluation of leachate's toxicity, an effective concentration (EC_{50}) was used. The EC_{50} is the concentration of added toxicant/wastewater that in given time under given conditions brings about a specified effect in 50% of the test population (Duffus and Worth,

2006). The IC_{50} is employed instead of EC_{50} to express the concentration that inhibits/reduces the response observed in the control by 50%. In this case, the measurement is not quantal but quantitative. It was used in the test with white mustard *Sinapis alba*; the measure is expressed as IC value because it focuses on changes in root growth and not in the number of organisms that show effects (Domínguez, 2008).

Table 2. Description of bioassays used for evaluation of landfill leachate toxicity

Species	Ecosystem	Endpoint	Test duration	Evaluation	Type of the test
Luminescent bacterium *Vibrio fischeri*	Aquatic, marine	Luminescence inhibition	30 min	EC_{50}	Acute
White mustard *Sinapis alba*	Terrestrial	Root growth inhibition	72 hours	IC_{50}	Acute
Brine shrimp *Artemia salina*	Aquatic, saline	Immobilization	24 hours	EC_{50}	Acute

The acute test with brine shrimp *Artemia salina* is based on the determination of the inhibition of mobility of testing organisms.

The test was performed according to the ARC-test procedure (Persoone et al., 1984). Standard artificial seawater contained 24.70 $g{\cdot}L^{-1}$ NaCl, 0.54 $g{\cdot}L^{-1}$ KCl, 4.59 $g{\cdot}L^{-1}$ $MgCl_2{\cdot}6H_2O$, 6.29 $g{\cdot}L^{-1}$ $MgSO_4{\cdot}7H_2O$, 0.15 $g{\cdot}L^{-1}$ $CaCl_2$, and 0.04 $g{\cdot}L^{-1}$ $NaHCO_3$ (Bagshaw et al., 1986).

The artificial seawater was used for hatching, as well as a medium for preparing different concentrations of tested leachates. Cysts of *Artemia salina* were incubated in a beaker with 300 mL artificial seawater and gently aerated for 24 hours. The hatched larvae of the first instar were transported to a 300 mL beaker with artificial seawater and after another 24 hours the second and third instar nauplii of *Artemia salina* were harvested. Standard artificial seawater was used for the preparation of different concentrations of leachate and the inhibition of mobility of tested organisms was evaluated after 24 hours of exposition in the dark at 20 ± 1 °C.

The test using freeze-dried luminescent bacterium *Vibrio fischeri* is based on the fact that the bioluminescence is affected by the toxicity of the sample. The pH of samples was corrected to 7.0 ± 0.2 and subsequently the samples were diluted by 2% NaCl solution to obtain different test concentrations.

The inhibition of bioluminescence was measured prior and after incubation (30 min) by a LUMIStox luminometer (Dr. Lange, Germany). Bioassay was performed according to standard procedure (ISO 11348–3, 2007).

Twenty seeds of the white mustard *Sinapis alba* were germinated in each Petri dish with a filtration paper on the bottom. Each tested leachate was diluted by growth medium to obtain different test concentrations. The length of the roots was measured after three days of incubation in the dark, at constant temperature 20 ± 1 °C (MŽP, 2007). Seeds were provided by Semenarna Ljubljana, Slovenia and they were not previously treated.

Results and Discussions

Considering the previous studies, traditional landfilling is not an option for sustainable waste management.

It is not essential only to change the composition of waste and disposed non-usable and non-biodegradable residual waste, but also the landfills' operation. Therefore, in this study, long-term biodegradation of residual municipal solid waste in the pilot-scale aerobic landfill bioreactor was monitored as a function of time during 326 days of operation. The pilot-scale landfill was set up with waste on 15th March 2012.

Degradation of waste was most intensive during April 2012 (Figure 1, IV.), resulting in a significant temperature increase in the landfill body (inside temperature) where temperatures reached 32.2 °C in comparison to the outside temperature being -1.2 °C (Figure 1). This phase of the aerobic landfill cycle is comparable to the first phase of the traditional landfill cycle, where aerobic degradation occurs only for a short period of time resulting in an increase of temperature (Table 1 – hydrolysis).

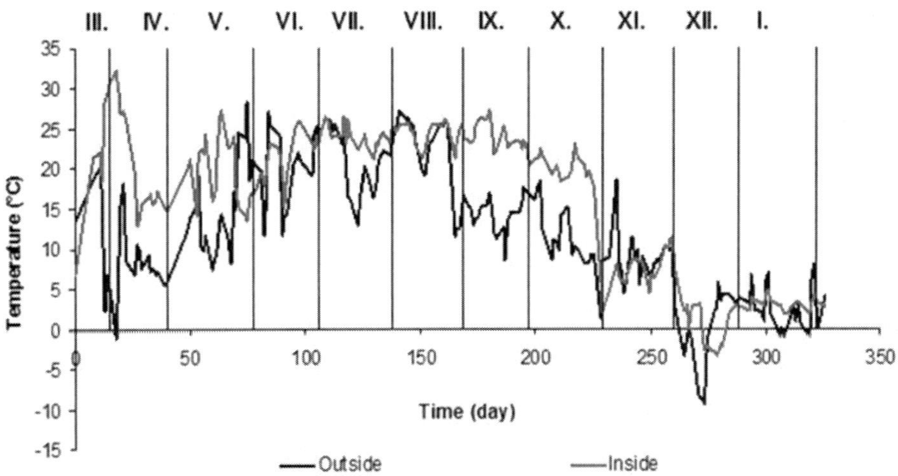

Figure 1. Temporal variations of inside temperature (in the landfill body) and outside temperature (air temperature) during monitoring periods (Roman numerals express months).

However, the highest temperature measured within the aerobic landfill containing only residual waste is considerably lower in comparison to landfills filled by waste with high biodegradable fractions where temperature can reach from 55 °C (Giannis et al., 2008) up to 90°C (Williams, 2008).

The first leachate occurred at the beginning of April 2012 (Figure 2, IV.), 22 days after the pilot-scale landfill was set up, and the generation of leachate increased almost linearly (Figure 2).

The volume of leachate was monitored daily also the occurrence of rain or snow was noted daily.

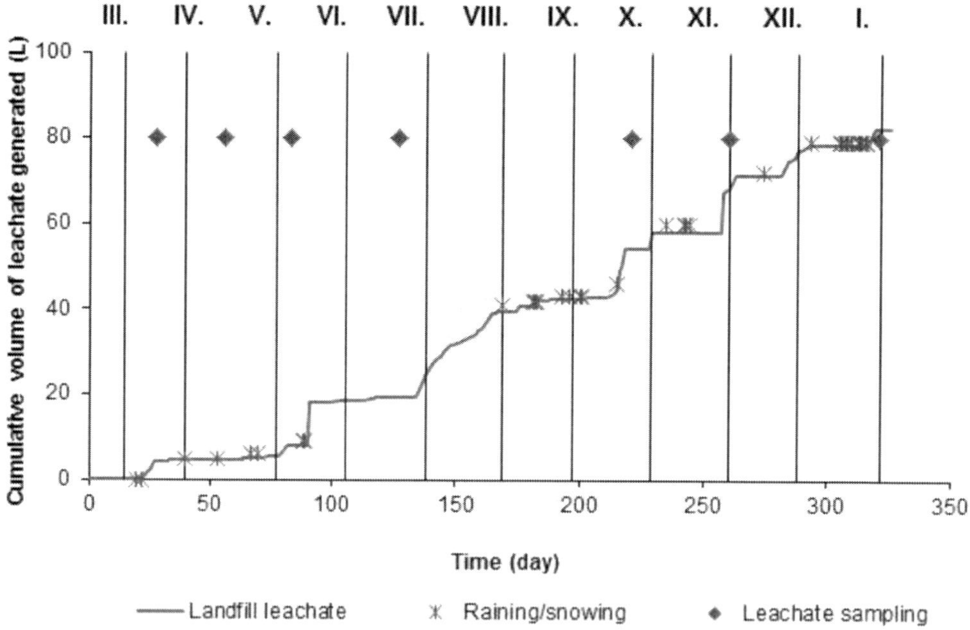

Figure 2. Monitoring of leachate volume generated in the pilot-scale landfill, monitoring of rain/snow during monitoring periods and time of leachate sampling for analyses (Roman numerals express months).

After each precipitative event the landfill needed up to 4 days for saturation and afterwards leachate occurred.

Quality of leachate was monitored in time and leachates were sampled after 28 days (Sample No. 1), 56 days (Sample No. 2), 83 days (Sample No. 3), 127 days (Sample No. 4), 221 days (Sample No. 5), 260 days (Sample No. 6), and 321 days (Sample No. 7) of pilot-scale landfill operation, when a minimal volume needed for analyses (1000 mL) was collected. COD, DOC, BOD_5, pH, oxidation reduction potential (ORP), concentrations of metals, ammonium, nitrate and nitrite nitrogen, chlorides, alkalinity, orthophosphates, toxicity to white mustard *Sinapis alba*, brine shrimp *Artemia salina* and bacterium *Vibrio fischeri* were determined.

The oxidation reduction potential (ORP) was always in the positive range, signalizing aerobic conditions in the landfill (Figure 3).

ORP increased from 50 mV to 150 mV in 250 days, indicating the presence of more oxidized components in the leachate.

The pH of leachates was always above neutral values and it did not drop below pH = 7, which would indicate the formation of organic acids during anaerobic decomposition of organic matter in waste.

Therefore, the introduction of ambient air into the landfill only through its surface was sufficient to maintain aerobic conditions. Our measurements are fairly consistent with the study by Borglin et al. (2004) and Giannis et al. (2008), where a pH decrease and ORP increase were observed in leachates during the aerobic degradation of municipal waste.

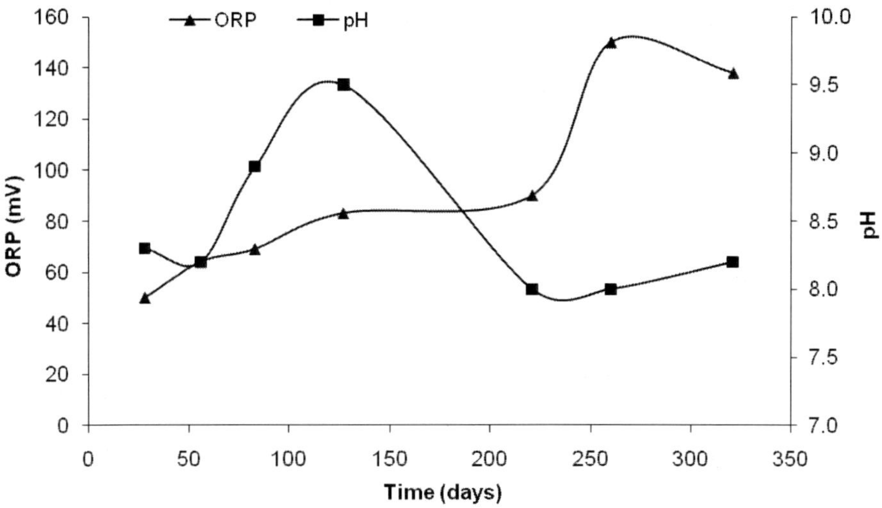

Figure 3. Monitoring of leachate pHs and oxidation reduction potential (ORP).

Concentrations of organic matter (COD, DOC, BOD_5) and their changes in time are presented in Figure 4. In the early stage of waste decomposition leachate contained high concentrations of organic matter; the COD and DOC in the leachate increased, reaching a maximum of 7103 $mg \cdot L^{-1}$ and 3290 $mg \cdot L^{-1}$, respectively. However, the concentration of biodegradable organics expressed as BOD_5 was very low in leachate and, thus, the BOD_5/COD ratio, which reached a maximum of 0.08, was almost 10 times lower than is expected in young leachate from traditional landfills. The concentration of organics (COD, DOC) was stable during the first 50 days. BOD_5 ratio decreased rapidly during this period and it was a consequence of aerobic biological degradation in the landfill body, indicated by increased temperature during this time (Figure 1).

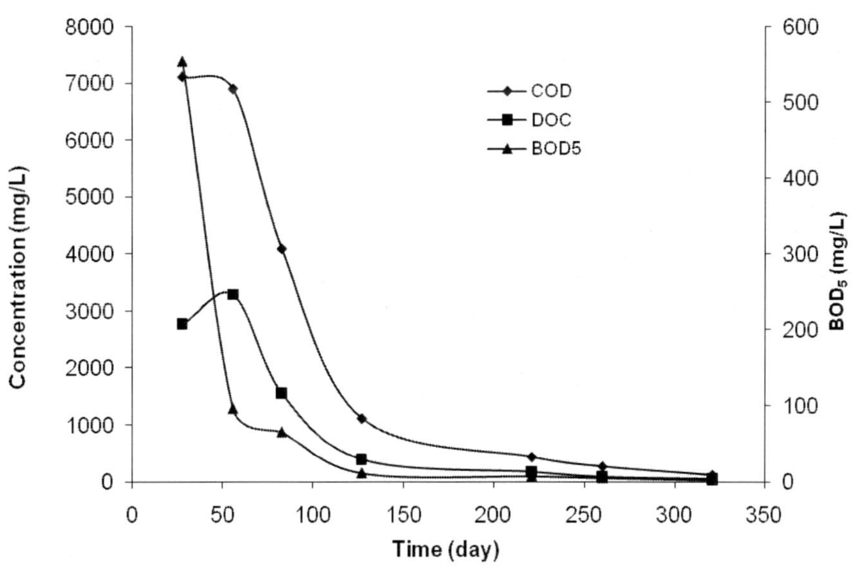

Figure 4. Concentrations of organic matter (COD, DOC, BOD_5) during monitoring period.

Afterwards, the concentration of organic matter (COD, DOC) started to decrease and continued to decrease until the end of the experiment. Most of the organic matter was decomposed very quickly, in a relatively short period of aerobic landfill operations. The removal efficiency of COD and BOD_5 by the end was 98% and 99%, respectively. The decrease of organic matter content was visible also in the decolorization of leachates with time (Picture 3).

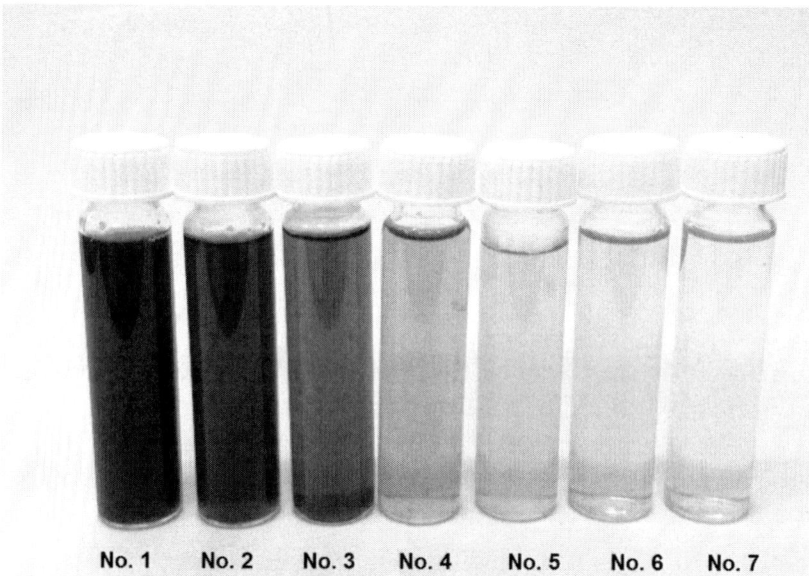

Picture 3. Decolorization of leachate with time.

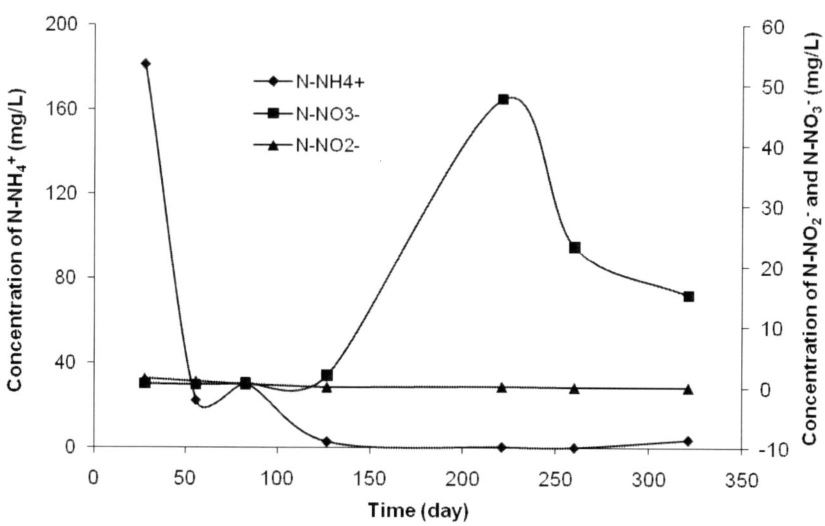

Figure 5. Concentrations of nitrogen (ammonium nitrogen $N\text{-}NH_4^+$, nitrite nitrogen $N\text{-}NO_2^-$ and nitrate nitrogen $N\text{-}NO_3^-$) during monitoring period.

In aerobic landfill biological nitrification proceeds and thus ammonium nitrogen released during degradation of protein is oxidized, resulting in nitrate. The changes in ammonium nitrogen $N\text{-}NH_4^+$, nitrite nitrogen $N\text{-}NO_2^-$ and nitrate nitrogen $N\text{-}NO_3^-$ are displayed in Figure 5.

The concentration of $N\text{-}NH_4^+$ started decreasing immediately and by the end of the experiment 98% was removed. The concentration of $N\text{-}NO_3^-$ was stable and after 100 days started to increase and after 200 days reached a maximum of 48 $mg{\cdot}L^{-1}$. The concentration of $N\text{-}NO_2^-$ was fairly stable during the monitored period. The decline in ammonium nitrogen concentration was coupled with an increase in nitrate concentration and, thus, ammonium nitrogen removal can be attributed to nitrification. A stechiometric increase in nitrate concentrations was not observed because denitrification was expected to occur. Part of the ammonium nitrogen was expected to be utilized by the heterotrophs for biomass synthesis and the remainder was oxidized to nitrate by nitrifying bacteria (Giannis et al., 2008).

The concentration of $P\text{-}PO_4^{3-}$ was 26 $mg{\cdot}L^{-1}$ and it decreased linearly to 2 $mg{\cdot}L^{-1}$. The decline in $P\text{-}PO_4^{3-}$ concentration may have been the results of phosphorous assimilation by microorganisms. The concentration of chlorides and alkalinity also rapidly decreased in the first 100 days of landfill operations (Figure 6). A reduction in alkalinity may have been connected to decreased ammonium nitrogen concentration in leachate, because significant amounts of bicarbonate are consumed in the conversion of ammonia to nitrite (Henze et al., 2001). A decrease in chloride is linked to the dilution and washout effect and it is not expected to have been affected by other factors (Erses et al., 2008).

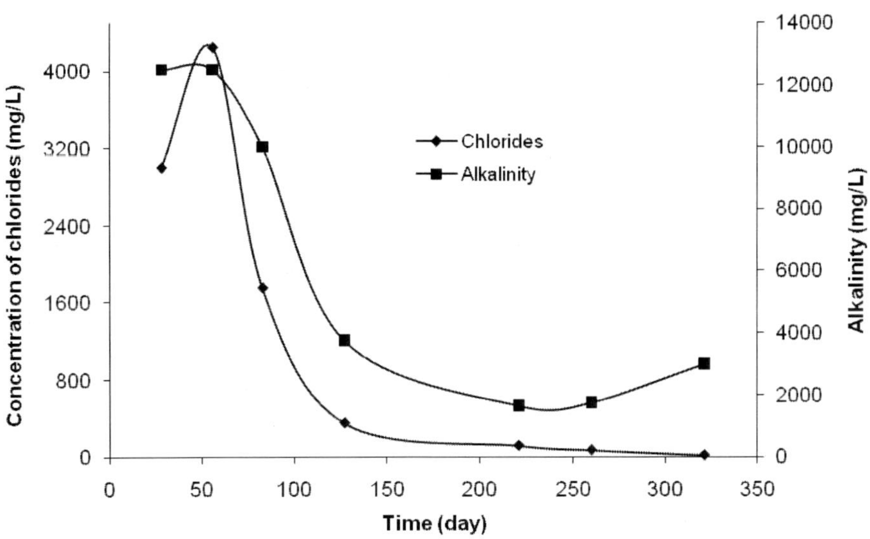

Figure 6. Concentrations of chlorides and alkalinity during monitoring period.

In sustainable landfills concentrations of metals should be minimal due to the separation of metals from municipal waste before landfilling. The pH of the leachate from the pilot aerobic landfill was from 8.0–9.5 (Figure 3) and thus if some metals were present in municipal waste they should be retained inside the landfill body.

Table 3 showed concentrations of metals in four leachates (No. 1, No. 3, No. 5, No. 7) sampled during operation.

Table 3. Concentrations of metals in four leachates (No. 1, No. 3, No. 5, No. 7). Parameters are given as mean values with standard deviations (±). Values exceeding legislative limits (Official Gazette, 2008) are in bold

Metals ($mg{\cdot}L^{-1}$)	Leachates – No. of sample			
	1	3	5	7
Al	3.7 ± 0.1	2.0 ± 0.03	0.248 ± 0.003	0.073 ± 0.005
V	0.069 ± 0.001	0.054 ± 0.001	0.0091 ± 0.0002	0.0043 ± 0.0002
Cr	**0.68** ± 0.01	**0.53** ± 0.01	0.039 ± 0.001	0.0178 ± 0.0004
Mn	3.2 ± 0.05	1.49 ± 0.02	0.0574 ± 0.001	0.0128 ± 0.0004
Fe	15.9 ± 0.2	6.8 ± 0.2	0.94 ± 0.03	0.34 ± 0.02
Co	0.365 ± 0.005	0.117 ± 0.001	0.021 ± 0.001	0.0073 ± 0.0001
Ni	0.292 ± 0.003	0.191 ± 0.002	0.065 ± 0.001	0.0225 ± 0.0004
Cu	**0.63** ± 0.02	0.299 ± 0.003	0.187 ± 0.003	0.081 ± 0.001
Zn	**2.00** ± 0.04	0.851 ± 0.004	0.36 ± 0.01	0.165 ± 0.005
As	0.0310 ± 0.0002	0.0164 ± 0.0003	0.0026 ± 0.000	< 0.002
Mo	0.090 ± 0.001	0.0111 ± 0.0002	0.0316 ± 0.0005	0.0092 ± 0.0004
Ag	0.0084 ± 0.0000	0.0028 ± 0.0003	< 0.002	< 0.002
Cd	0.0058 ± 0.0001	< 0.002	< 0.002	< 0.002
Sb	0.0915 ± 0.0027	0.0406 ± 0.0001	0.0209 ± 0.0004	0.0074 ± 0.0001
Ba	0.52 ± 0.01	0.341 ± 0.004	0.024 ± 0.001	0.006 ± 0.001
Pb	0.105 ± 0.001	0.0228 ± 0.0002	< 0.002	< 0.002

As can be seen from Table 3, metals appeared in the leachate only at the initial stage of the experiment and only Cr, Zn and Cu exceeded legislative limits for discharge into the environment (0.5 $mg{\cdot}L^{-1}$, 2.0 $mg{\cdot}L^{-1}$ and 0.5 $mg{\cdot}L^{-1}$, respectively; Official Gazette, 2008). The reason could be the association of different metals with organic matter (Zupančič et al., 2009), because high concentrations of organic matter were leached out from the landfill at the beginning of the landfill operation (Figure 4).

After 83 days of landfill operation (Sample No. 3) the leachate contained only a higher amount of Cr, and after 221 days (Sample No. 5) the leachate contained very low concentrations of metals.

Similarly to our study, Bilgili et al. (2007) reported Cr concentrations of up to 0.8 $mg{\cdot}L^{-1}$ in the initial stage of the aerobic landfill operation, while after 50 days the concentration of Cr dropped below 0.2 $mg{\cdot}L^{-1}$. At the end of the experiments the concentrations of Cd, Cu, Pb were also below 0.2 $mg{\cdot}L^{-1}$. Therefore, the metal content of landfill leachate should not be a major concern for aerobic landfills.

The toxic effect of leachates was determined by three bioassays using brine shrimp *Artemia salina*, white mustard *Sinapis alba* and bacterium *Vibrio fischeri*. Toxicity of four leachate samples generated after 28 days (No. 1), 83 days (No. 3), 221 days (No. 5) and 321 days (No. 7) of experimental time were determined (Table 4).

The toxicity of the leachate generated in the first days of landfill operation (28 and 83 days) showed some toxic effect to all the applied organisms due to high pollution of the leachate. At the first sampling (day 28) the concentration of COD, ammonium nitrogen, chlorides and alkalinity were 7103 $mg{\cdot}L^{-1}$, 181 $mg{\cdot}L^{-1}$, 300 $mg{\cdot}L^{-1}$, and 12 500 $mg{\cdot}L^{-1}$, respectively. Similarly, the next sampled leachate (day 83) contained 6892 $mg{\cdot}L^{-1}$, 22 $mg{\cdot}L^{-1}$, 4250 $mg{\cdot}L^{-1}$, and 12 500 $mg{\cdot}L^{-1}$, respectively.

Table 4. Toxicity of four leachates sampled in the aerobic landfill to three organisms

No. of sample	t * (days)	EC_{50} or IC_{50} (% v/v)		
		Artemia salina	*Sinapis alba*	*Vibrio fischeri*
1	28	39.7 ± 3.2	23.1 ± 0.1	29.5 ± 8.9
3	83	74.5 ± 7.4	32.9 ± 0.5	39.6 ± 4.3
5	221	> 100	> 100	> 100
7	321	> 100	> 100	> 100

* – Days after pilot-scale landfill was set up.

The concentration of COD only decreased slightly and the only significant decrease was in terms of ammonium nitrogen removal resulting in a decrease in toxicity to *Artemia salina*. Furthermore, the fast stabilization of organic matter and a decrease of inorganic components led to the leachate's detoxification to all organisms after 221 days of the aerobic landfill operation. The results (Table 4) were compared to the results of our previous study (Kalčíková et al., 2011), where toxicity of young leachate generated in two years old traditional anaerobic landfill was investigated. The young leachate from the traditional anaerobic landfill caused toxic effects to *Artemia salina* $24hEC_{50}$ = 52.6% v/v, to white mustard *Sinapis alba* $72hIC_{50}$ = 8.5% v/v and bacterium *Vibrio fischeri* $30minEC_{50}$ = 22.2% v/v (recalculated from Toxic Units). It was noticed that even the most polluted leachate produced after 28 days in the aerobic landfill is less toxic (except for *Artemia salina*) than the young leachate generated after 2 years of traditional anaerobic landfill operations.

Comparison of Traditional Anaerobic and Aerobic Landfills

Traditional landfilling and processes occurring under anaerobic degradation in the landfill body are well recognized, as well as the changes of the landfill leachate composition during ageing of the landfill. In traditional anaerobic and in sustainable aerobic landfills, only the first stage of the landfill life-cycle is comparable. The first stage is aerobic where mainly hydrolysis takes place. In the traditional landfill this stage is relatively short and afterwards anaerobic conditions are established. Under those conditions, cellulose and hemicellulose, which are important biodegradable organic components in municipal solid waste (Lamborn, 2009), are sources of carbon for methanogenic bacteria, which convert organics into biogas. Cellulose contributes to 90% of total methane production, but is not easily biodegradable under anaerobic conditions and its half-time is about 15 years, resulting in slow stabilization of waste (Erses et al., 2008). Similarly, proteins are deaminated under anaerobic conditions to ammonium nitrogen and its concentration remains stable for years. During anaerobic degradation the most important by-products are biogas and polluted landfill leachate. On the other hand, landfills operating under aerobic conditions can be compared to composting. Processes in aerobic landfills are probably slightly different in comparison to composting because the organic matter content is significantly lower. In any case, cellulose, hemicellulose, proteins and other components are hydrolyzed under aerobic conditions. Hydrolysis of solid waste components is the rate-limiting step during degradation, while water soluble organics as starch, carboxylic acids, amino acids, and phenols are rapidly degraded. Furthermore, solid and water soluble organics both are decomposed (at a faster or slower rate) under aerobic conditions into carbon dioxide and water, with stabilized humic material remaining (Castaldi et al., 2005; Komilis, 2006). Nitrogen, mainly in the form of ammonium is oxidized into nitrate and partially assimilated in biomass. Waste decomposition in aerobic landfills takes place within a few years, instead of several decades, and it results in the generation of a slightly polluted leachate and low amounts of harmless gases (Cossu et al., 2003; Erses et al., 2008, Giannis et al., 2008).

CONCLUSION

Traditional landfilling has been used as a waste treatment method for decades. Such waste management results in the generation of hazardous landfill leachate and landfill gases. The stabilization of waste under anaerobic conditions takes many years and even after decades landfill leachate is still polluted and toxic. The treatment of leachate is a complicated process and not even combinations of different methods leads to fulfilling of legislative limits. Accordingly, it is clear that traditional landfilling is definitely not a sustainable option for a current waste management. If a real sustainable approach is employed, whereby no environmental problems should be left to future generations, it represents the main goal to be achieved by modern selective waste collection and corresponding landfill strategies. The most recent approach is the disposal of residual waste to a landfill operating under aerobic conditions. Biodegradation of organic matter is fast, as is the elimination of inorganic components. Detoxification of leachate proceeded in a short period of time. Overall, the results of our study showed that the degradation of residual municipal waste in an aerobic landfill is a viable option for sustainable waste management.

REFERENCES

Agamuthu, P. & Fauziah, S.H. (2011). Challenges and issues in moving towards sustainable landfilling in a transitory country – Malaysia. *Waste Management and Research*, 29, 13-19.

Almorza, D., Brebbia, C.A., Sales, D. & Popov V. (2002). Waste management and the environment. Southampton, WIT Press.

Armstrong, M.D. & Rowe R.K. (1999). Effect of landfill operations on the quality of municipal solid waste leachate. In: Seventh International Waste Management and Landfill Symposium, S. Margherita di Pula (Cagliari). Proceedings Sardinia 99, 81-88.

Bagshaw, J.C. Rafiee, P., Matthews, C.O. & MacRae T.H. (1986). Cadmium and zinc reversibly arrest development of Artemia larvae. *Bulletin of Environmental Contamination and Toxicology*, 37, 289-296.

Batarseh, E.S. Reinhart, D.R. & Berge N.D. (2010). Sustainable disposal of municipal solid waste: Post bioreactor landfill polishing. *Waste Management*, 30, 2170-2176.

Bila, M.D. Montalvão, F.A. Silva, A.C. & Dezotti, M. (2005). Ozonation of a landfill leachate: evaluation of toxicity removal and biodegradability improvement. *Journal of Hazardous Materials*, 117, 235-242.

Bilitewski, B. Härdtle, G. & Marek K. (1994) Waste Management. Berlin, Springer.

Bilgili, M.S. Demir, A. Ince, M. & Ozkaya, B. (2007). Metal concentrations of simulated aerobic and anaerobic pilot scale landfill reactor. *Journal of Hazardous Materials*, 145, 186-194.

Borglin, S.E. Hazen, T.C. Oldenburg, C.M. & Zawislanski, P.T. (2004). Comparison of aerobic and anaerobic biotreatment of municipal solid waste. *Journal of the Air & Waste Management Association,* 54, 815-822.

Castaldi, P. Alberti, G. Merella, R. & Melis P. (2005). Study of the organic matter evolution during municipal solid waste composting aimed at identifying suitable parameters for the evaluation of compost maturity. *Waste Management*, 25, 209-213.

Christensen, T. Cossu, R. & Stegmann, R. (1992). Landfilling of waste: Leachate. London, E & FN SPON.

Christensen, T. (2011). Solid waste technology and Management. Chichester, John Wiley & Sons.

Comstock, S.E.H. Boyer, T.H. Graf, K.C. & Townsend T.G. (2010). Effect of landfill characteristics on leachate organic matter properties and coagulation treatability. *Chemosphere*, 81, 976-983.

Cossu, R. (2010). Technical evolution of landfilling. *Waste Management*, 30, 947-948.

Cossu, R. & Lai, T. (2012). Washing of waste prior to landfilling. *Waste Management*, 32, 869-878.

Cossu, R. Raga, R. & Rossetti, D. (2003). The PAF model: an integrated approach for landfill sustainability. *Waste Management*, 23, 37-44.

Council Directive 75/442/EEC, 1975. Available online (21.1.2013): http://eur-lex.europa.eu/LexUriServ/LexUriServ.do?uri=CELEX:31975L0442:EN:HTML

Council Directive 2008/98/EC, 2008. Available online (27.5.2013): http://eur-lex.europa.eu /LexUriServ/LexUriServ.do?uri=OJ:L:2008:312:0003:0030:en:HTML

Deng, Y. & Englehardt, J.D. (2006). Treatment of landfill leachate by Fenton process. *Water Research,* 40, 3683-3694.

Domínguez, J.B. (2008). Soil contamination research trends. New York, Nova Publisher.

Duffus, J.H. & Worth H.G.J. (2006). Fundamental toxicology. Cambridge, The Royal Society and Chemistry.

EPA, 2014. Energy Recovery from Waste. Available on-line (22.9.2014) http://www.epa.gov /osw/nonhaz/municipal/wte/

Erses, A. S. Onay, T.T. & Yenigun O. (2008). Comparison of aerobic and anaerobic degradation of municipal solid waste in bioreactor landfills. *Bioresource Technology*, 99, 5418-5426.

Eurostat, 2012. Available online (4.6.2012): http://epp.eurostat.ec.europa.eu/cache/ ITY_PUBLIC/8-27032012-AP/EN/8-27032012-AP-EN.PDF

Eurostat, 2013. Available online (3.1.2013): http://epp.eurostat.ec.europa.eu/tgm/graph.do ?tab=graph&plugin=1&pcode=tsdpc240&language=en&toolbox=data

Giannis, A. Makripodis, G. Simantiraki, F. Somara, M. & Gidarakos W. (2008). Monitoring operational and leachate characteristics of an aerobic simulated landfill bioreactor. *Waste Management,* 28, 1346-1354.

Green, J.A.S. (2007). Aluminum recycling and processing for energy conservation and sustainability. ASM International.

Henze, M. Harremoes, P. Jansen, J. & Arvin E. (2001). Wastewater treatment: Biological and chemical processes. Berlin, Springer.

Huber-Humer, M. & Lechner, P. (2011). Sustainable landfilling or sustainable society without landfilling? *Waste Management*, 31, 1427-1428.

ISO 11348–3 (2007). Water Quality – Determination of the inhibitory effect of water samples on the light emission of *Vibrio fischeri* (Luminescent bacteria test), Part 3: Method using freezed-dried bacteria, Edition 2, Geneva.

Kalčíková, G. Babič, J. Pavko, A. & Žgajnar Gotvajn, A. (2014). Fungal and enzymatic treatment of mature municipal landfill leachate. *Waste Management*, 34, 798-803.

Kalčíková, G. Vávrová, M. Zagorc-Koncan, J. & Žgajnar Gotvajn, A. (2011). Evaluation of the hazardous impact of landfill leachates by toxicity and biodegradability tests. *Environmental Technology*, 32, 1345-1353.

Kjeldsen, P. Barlaz, M.A. Rooker, A.P. Baun, A. Ledin, A. & Christensen T.H. (2002). Present and long-term composition of MSW landfill leachate: A Review. *Critical Reviews in Environmental Science and Technology*, 32, 297-336.

Komilis, D.P. (2006). A kinetic analysis of solid waste composting at optimal conditions. *Waste Management*, 26, 82-91.

Kulikowska, D. & Klimiuk, E. (2008). The effect of landfill age on municipal leachate composition. *Bioresource Technology*, 99, 5981-5985.

Kurniawan, T.A. Lo, W.H. & Chan G.Y.S. (2006). Physico-chemical treatments for removal of recalcitrant contaminants from landfill leachate. *Journal of Hazardous Materials*, 129, 80-100.

Lamborn, M. (2009). Characterisation of municipal solid waste composition into model inputs. In: Third international workshop "Hydro-Physico-Mechanics of landfills" Braunschweig, Germany, 10-13 March.

Lebersorger, S. & Beigl, P. (2011). Municipal solid waste generation in municipalities: Quantifying impacts of household structure, commercial waste and domestic fuel. *Waste Management*, 31, 1907-1915.

Letcher, T.M. & Vallero, D.A. (2011). Waste: A Handbook for managers. Burlington, Academic Press.

Li, W. Hua, T. Zhou, Q. Zhang, S. & Li, F. (2010). Treatment of stabilized landfill leachate by the combined process of coagulation/flocculation and powder activated carbon adsorption. *Desalination*, 264, 56-62.

Li, Y. Low, G.K-C. Scott, J.A. & Amal, R. (2007). Microbial reduction of hexavalent chromium by landfill leachate. *Journal of Hazardous Materials*, 142, 153-159.

Morcet, M. Aran, C. Bogner, J. Chanton, J. Spokas, K. & Hebe, I. (2003). Methane mass balance: A review of field results from three French landfill case study. In: Ninth international waste management and landfill symposium, S. Margherita di Pula (Cagliari). Proceedings Sardinia 2003.

Municipal waste statistics (2011): Available online (22.1.2013): http://epp.eurostat.ec.europa.eu/statistics_explained/index.php/Municipal_waste_statistics

MŽP. (2007). The Guideline of the Department of waste management to determine the ecotoxicity of waste. Ministry of the Environment of the Czech Republic, Prague, (In Czech).

Neczaj, E. Okoniewska, E. & Kacprzak M. (2005). Treatment of landfill leachate by sequencing batch reactor. *Desalination*, 185, 357-362.

Official Gazette of Republic of Slovenia. (2008). Decree on the emission of substance in the discharge of landfill leachate, ULRS 62/2008, 8219–8221, (In Slovene).

Persoone, G. Jasperse, E. & Claus, C. (1984). Ecotoxicological testing for the marine environment. Bredene, State university Ghent and institute for marine scientific research.

Pichtel, J. (2005). Waste management practice. Municipal, hazardous and industrial. Boca Raton, Taylor and Francis.

Pivato, A. & Gaspari L. (2006). Acute toxicity test of leachates from traditional and sustainable landfills using luminescent bacteria. *Waste Management*, 26, 1148-1155.

Renou, S., Givaudan, J.G., Poulain, S., Dirassouyan, F. & Moulin P. (2008). Landfill leachate treatment: Review and opportunity. *Journal of Hazardous Materials*, 150, 468-493.

Rhyner, C.R. Schwartz, L.J. Wenger, R.B. & Kohrell, M.G. (1995). Waste management and resource recovery. Florida, CRC Press.

Salem, Z. Hamouri, K. Djemaa, R. & Allia K. (2008). Evaluation of landfill leachate pollution and treatment. *Desalination*, 220, 108-114.

Ritzkowski, M. & Stegmann, R. (2007). Controlling greenhouse gas emissions through landfill in situ aeration. *International Journal of Greenhouse Gas Control*, 1, 281-288.

Ritzkowski, M. & Stegmann, R. (2012). Landfill aeration worldwide: Concepts, indications and findings. *Waste Management*, 32, 1411-1419.

Slack, R.J. Gronow, J.R. & Voulvoulis, N. (2005). Household hazardous waste in municipal landfills: contaminants in leachate. *Science of the Total Environment*, 337, 119-137.

Standard methods for the examination of water and wastewater. (2012) American water works association/American public works association/Water environment, Edition 22, Washington.

Tatsi, A.A. Zouboulis, A.I. Matis, K.A. & Samaras, P. (2003). Coagulation–flocculation pretreatment of sanitary landfill leachates. *Chemosphere*, 53, 737-744.

Williams, P.T. (2005). Waste treatment and disposal. Chichester, John Wiley & Sons.

Wiszniowski, J. Robert, D. Surmacz-Gorska, J. Miksch, K. & Webere J.V. (2006). Landfill treatment methods: A review. *Environmental Chemistry Letters*, 4, 51-6.

Timur, H. & Özturk I. (1999). Anaerobic sequencing batch reactor treatment of landfill leachate. Water Research, 33, 3225-3230.

Xiaoli, C. Yongxia, H. Guixiang, L. Xin, Z. & Youcai, Z. (2013). Spectroscopic studies of the effect of aerobic conditions on the chemical characteristics of humic acid in landfill leachate and its implication for the environment. *Chemosphere*, 91, 1058-1063.

Ziyang, L. Youcai, Z. Tao, Y. Yu, S. Huili, C. Nanwen, Z. & Renhua H. (2009). Natural attenuation and characterization of contaminants composition in landfill leachate under different disposing ages. *Science of the Total Environment*, 407, 3385-3391.

Zupančič, M. Zupančič Justin, M. Bukovac, P. & Šelih, V.S. (2009). Chromium in soil layers and plants on closed landfill site after landfill leachate application. *Waste Management*, 29, 1860-1869.

INDEX

#

20th century, 33, 113, 200

A

access, vii, viii, 25, 27, 49
accounting, 177, 207
acetic acid, 72, 102, 104, 203
acid, ix, 6, 56, 59, 70, 71, 72, 77, 92, 98, 102, 103, 104, 105, 106, 107, 108, 109, 121, 203
acidic, 57, 82, 83, 85, 96, 100, 203, 204, 205
acidity, 108, 119, 203
activated carbon, ix, 69, 70, 72, 73, 74, 75, 76, 79, 81, 82, 83, 85, 90, 91, 92, 93, 100, 101, 102, 103, 104, 106, 107, 108, 220
activation energy, 79
additives, 77, 105
adhesives, 10
adjustment, 30
adsorption, ix, 2, 10, 59, 60, 70, 72, 73, 74, 75, 76, 78, 79, 81, 82, 83, 85, 91, 92, 100, 102, 104, 106, 107, 108, 150
adsorption isotherms, 75, 81, 85
adults, 10
adverse effects, 118, 170
aerobic conditions, x, 117, 199, 203, 207, 208, 212, 217, 218, 221
aerobic landfill bioreactor, xi, 199, 211
aerogels, 106
aesthetic(s), 122, 131
affirming, 176
age, 8, 10, 19, 56, 66, 71, 115, 123, 152, 205, 206
agricultural soils, viii, 53, 56, 60, 62
agriculture, 61, 72, 183
air emissions, 182
Air Force, 128, 145
air pollutants, 152
air temperature, 43, 211
airborne thermal mapping, viii, 26
Alaska, 141
aldehydes, 74
alfalfa, 127, 129
alkalinity, 54, 56, 102, 209, 212
aluminium, 202
aluminum oxide, 59
ambient air, x, 147, 150, 151, 155, 156, 158, 159, 163, 208, 212
amino, 102, 116
amino acid(s), 116
ammonia, 55, 61, 62, 65, 66, 102, 116, 153, 203, 204
ammonium, 57, 60, 65, 102, 205, 209, 212, 214, 215
anaerobic digestion, 54, 103, 202
aniline, 142
anions, vii, 1, 2, 6, 7, 9, 10, 11, 12, 59, 87, 101, 108
ANOVA, 7
anti-pest compounds, ix, 69, 70
aqueous solutions, 70, 80, 96, 101, 103, 108, 109
aqueous suspension, 74, 106
aquifers, 27, 44, 49, 51, 52
ARC, 210
Argentina, 116, 135
aromatic compounds, 81, 88
aromatic hydrocarbons, 205
aromatic rings, 81, 96
Arrhenius equation, 79
Artemia, 209, 210, 212
aseptic, 175
Asia, 164
assessment, vii, viii, x, 1, 2, 5, 13, 22, 25, 26, 27, 29, 47, 48, 102, 135, 137, 151, 152, 167, 170, 172, 173, 174, 175, 176, 192, 193, 194, 195, 196
assets, 176
atmosphere, 2, 60, 61, 117, 120, 124, 127, 136, 142, 148, 151, 152, 162
atoms, 59, 96
Austria, 122, 179, 202

authorities, 122
avoidance, 182
awareness, x, 26, 114, 170, 199, 201

B

bacteria, 10, 21, 83, 84, 85, 92, 100, 108, 117, 125, 139, 140, 200, 206
bacterium, 210, 212
band gap, 74, 79
banks, 176
barriers, 29, 119, 125, 138, 197
base, 33, 59, 86, 106
batteries, 114, 169, 208
beer, 174
Belgium, 116, 117, 179
beneficial effect, 83
benefits, ix, 10, 111, 121, 125, 131, 171, 174, 175, 202
benzene, x, 74, 106, 147, 148, 152, 153, 155, 156, 157, 159, 160, 161, 163
bias, 180
bicarbonate, 56, 87, 101, 108, 206
biodegradability, 80
biodegradation, ix, x, 65, 73, 83, 103, 106, 111, 199, 203, 205, 211
biodiversity, vii, 2, 3, 131
bioenergy, 120
biogas, 36, 38, 142, 200, 204
biological processes, ix, 58, 59, 60, 69, 70
bioluminescence, 210
biomarkers, 5
biomass, 56, 62, 63, 66, 83, 127, 131, 134, 139, 141
biomedical applications, 104
bioremediation, 73
birds, vii, 1, 10, 114
bisphenol, ix, 70, 71, 75, 89, 102, 103, 104, 105, 107
Bisphenol A (BPA), 70, 105
blue baby, 55
body weight, 152
boreholes, viii, 26, 30, 31, 32, 38, 44, 45, 46, 47, 49
branching, 59
Brazil, 53, 54, 55, 57, 64, 65, 66
breeding, 143
bursa, 16
by-products, 80, 95, 121

C

Ca^{2+}, 56, 79
cadmium, 56, 127, 205
calcium, 56, 119, 144, 205
calcium carbonate, 119
canals, 148
cancer, 152, 176
capillary, 125, 128, 129, 130, 140, 145
carbohydrates, 171, 203, 205
carbon, ix, 54, 55, 56, 57, 60, 61, 70, 72, 73, 74, 75, 76, 77, 79, 81, 83, 84, 85, 90, 91, 92, 100, 101, 104, 106, 108, 116, 117, 118, 148, 150, 175, 203, 204
carbon dioxide, 56, 83, 116, 117, 118, 148, 203, 204
carbon materials, 74
carbon monoxide, 204
carbon nanotubes, 74, 106
carbonyl groups, 82
carboxyl, 71, 82, 91, 92, 101
carboxylic acid(s), 74, 80
carboxylic groups, 82
carcinogen, 152
carcinogenesis, 72
carcinogenicity, 103, 147, 151, 152, 163
case study(s), x, 20, 22, 35, 47, 51, 52, 134, 142, 183, 191, 199, 208
catalyst, 74, 76, 77, 105
catalytic activity, 106, 108
cation, 58, 59, 83
cattle, 65
causality, 177
CEC, 59
cellulose, 56, 203
CFI, 120
challenges, 23, 55, 197
charge density, 82, 83
chelates, 59
chemical(s), viii, ix, 2, 3, 6, 7, 12, 13, 19, 21, 22, 30, 31, 36, 38, 42, 47, 49, 53, 54, 56, 57, 58, 59, 60, 61, 69, 70, 71, 72, 73, 74, 76, 83, 91, 94, 95, 96, 97, 102, 103, 104, 118, 119, 125, 126, 132, 136, 152, 203, 206
chemical characteristics, 57
chemical properties, 2, 118, 136
chemical reactions, 59
children, 10, 55, 71, 103
China, 50, 145, 164, 197
chlorinated aliphatics, 205
chlorine, 205
chromatography, 6, 7, 143, 149, 150
chromium, 56, 205, 206
CIA, 176
circulation, 28, 30, 39
city(s), 54, 175, 196
citizens, 138, 169
civilization, 200
classes, 7, 13, 14, 15

classification, 75, 107, 108, 134, 141, 184, 191
clay caps, ix, 111, 112, 119, 123, 125, 126, 129, 132, 133, 144
cleaning, 138, 171, 202
cleanup, 139
climate(s), 19, 20, 55, 62, 65, 120, 126, 130, 134, 135, 156, 176
climate change, 65, 120, 176
closure, viii, ix, 6, 26, 29, 34, 111, 114, 128, 131, 132, 134, 136, 141, 204
CO_2, 56, 73, 76, 92, 98, 204
coal, 42
cobalt, 205
collaboration, 182
colonisation, 144
colonization, 3, 19
color, 56
coma, 10
combustion, 174, 175, 202
commerce, 200
commercial(s), 58, 79, 114, 171, 180
community(s), vii, ix, x, 2, 5, 13, 14, 15, 19, 20, 23, 26, 111, 114, 118, 147, 148, 151, 155, 156, 157, 158, 159, 160, 161, 163, 184, 200, 202
community service, 184
compaction, 56, 114, 117, 205
comparative analysis, 193
competition, 75, 95, 101, 126
complement, 26, 27, 45
complexity, viii, 22, 26, 54, 175
compliance, 195
composites, 106
composition, ix, 5, 8, 13, 19, 34, 55, 56, 66, 69, 70, 115, 116, 118, 126, 133, 174, 182, 204, 205, 206, 208, 211
compost, 66, 140, 182, 202
composting, x, 54, 169, 174, 175, 182, 199, 201, 202, 207, 208
compounds, vii, ix, x, 1, 10, 12, 15, 20, 22, 56, 57, 59, 60, 69, 70, 73, 74, 75, 78, 94, 95, 96, 98, 101, 102, 107, 109, 116, 117, 137, 147, 148, 152, 153, 155, 157, 158, 159, 164, 205, 206
computer, 137
computer simulations, 137
conceptual model, 37, 118
condensation, 112
conditioning, 48
conductivity, 28, 34, 36, 38, 42, 50, 59, 61, 114, 119, 122, 123, 125, 126, 135, 140, 141, 142, 143, 153
configuration, 42
conformity, 193
Congress, 143
connectivity, 191
consensus, 173
conservation, 31
conserving, 27
constituents, 2, 115
construction, 6, 44, 47, 112, 119, 123, 132, 134, 180, 183
consumers, 172, 182
consumption, x, 72, 148, 167, 171, 174, 176, 177, 178, 179, 180, 181, 185, 190, 191, 192, 208
contaminant, 13, 45, 47, 56
contaminated sites, 141
contaminated soil(s), 48, 139, 141
contaminated water, 71, 200
contamination, vii, viii, 5, 25, 26, 27, 28, 29, 31, 32, 33, 34, 44, 45, 46, 47, 49, 50, 51, 55, 58, 60, 61, 63, 64, 102, 112, 113, 115, 116, 134, 135, 138, 148, 155, 182
contamination plume, viii, 25, 26, 28, 34, 44, 45, 46, 47, 49
contour, 45
control measures, 202
controversial, 15
controversies, 61
COOH, 74
cooking, 12
copper, 56, 205
correlation(s), 43, 47, 153, 161, 163, 170, 177, 181, 191, 192
cost, ix, 54, 70, 72, 111, 123, 125, 126, 132, 134, 202
covering, 7, 47, 169, 191
cracks, 123, 137
creep, 14, 127
critical value, 159
Croatia, 22, 102, 167, 182, 183, 184, 185, 186, 187, 188, 189, 190, 191, 192
crop(s), 3, 6, 10, 61, 62, 63, 65, 70, 72, 122, 142
crop production, 122
crystalline, 28
crystallinity, 79
cultivation, 6
culture, 135, 139
cure, 193
cyanide, 137
cycles, 123
cycling, 141
cyclodextrins, 105

D

damages, 63, 139, 153, 162
data analysis, 33
data availability, 173, 192
data processing, viii, 25, 49

data set, 42, 176
database, 176, 184
decay, 108
decision makers, x, 167, 192
decomposition, ix, 12, 34, 54, 56, 69, 70, 73, 80, 86, 94, 102, 138, 212, 213
decoupling, 170, 177, 178, 179, 185, 187, 188, 189, 190, 192
deficiencies, 3
DEFRA, 180, 193
degradation mechanism, 107
degradation process, 101, 120, 123
degradation rate, 73, 75, 85, 86, 88, 97
denitrification, 60, 61
Denmark, 51, 145
Department of Energy, 123
deposition, 3, 29, 30, 31, 40, 46, 117, 119
deposits, 30, 31, 54, 115, 136
depth, 26, 27, 28, 30, 33, 36, 38, 41, 42, 44, 45, 46, 48, 56, 64, 78, 126, 127, 129, 131, 132, 133
derivatives, 72, 74, 106
desiccation, 123, 132, 134
designers, 171
detectable, 207
detection, 21, 22, 28, 52
detection system, 52
detergents, 183
detoxification, 102, 104
developed countries, 2, 120, 148, 202
developing countries, viii, 53, 54, 122, 132, 148, 178
diffusion, 58, 79, 119
digestion, 6
direct measure, 30, 155, 163
disability, 176
diseases, 72, 176, 200
dispersion, x, 58, 61, 62, 63, 81, 83, 100, 147, 155, 156, 157, 158
displacement, 45, 47
distress, 13
distribution, ix, 20, 21, 36, 39, 40, 41, 43, 70, 71, 83, 135, 140, 147, 153, 163, 172, 206
diversity, 3, 13, 15, 54, 176
DNA, x, 147, 152, 153, 161, 162, 163, 165
DNA damage, x, 147, 152, 153, 161, 162, 163
DOC, 209, 212, 213, 214
DOI, 51
dominance, 131
dosage, 77
drainage, 5, 128, 137, 138, 143, 200, 208
drawing, 41, 192
drinking water, 70, 71, 72
drought, 126, 127
drying, 118, 123
dumping, 3, 19, 113, 148, 151, 162, 169, 200
durability, 69, 70, 76, 123
dyes, 105

E

ecological restoration, 19
economic activity, 170, 178, 180
economic growth, 178, 195, 196, 200, 202
economic indicator, 177, 192
economic theory, 177
ecosystem(s), vii, 1, 2, 3, 4, 5, 10, 13, 15, 19, 20, 23, 47, 66, 126, 141, 142, 176, 201
ecotoxicological, 3, 5, 22
ecotoxicology, 13
EEA, 171, 183, 194
effluent(s), viii, 53, 54, 60, 63, 64, 134
Egypt, 69, 200
elaboration, 62
electrical conductivity, 4, 10, 37, 38, 39, 46, 54, 56, 61
electricity, 120, 132, 202
electrodes, 33
electrolysis, 80
electrolyte, 76
electromagnetic, viii, 26, 46, 52
electron(s), 21, 59, 73, 74, 79, 81, 82, 85, 95, 96, 101, 105, 108, 109
electron state, 85
elongation, 123, 133
emission, ix, 6, 66, 111, 113, 117, 120, 122, 124, 132, 133, 134, 136, 140, 142, 143, 145, 147, 148, 149, 150, 151, 152, 153, 154, 155, 157, 158, 162, 163, 174, 175, 182, 202, 207, 208
employees, 180
encephalopathy, 10
endocrine, 70, 71, 72, 76, 102, 103, 104
endothermic, 100
energy, x, 40, 49, 74, 75, 80, 81, 87, 93, 100, 116, 139, 168, 169, 170, 171, 172, 173, 174, 175, 177, 178, 181, 194, 199, 201, 202, 204, 207
energy consumption, 80, 202
energy recovery, x, 168, 172, 174, 178, 194, 199, 201, 202
engineering, viii, 26, 33, 36, 49, 50, 123, 144
England, 117, 142, 180
environment(s), viii, ix, x, xi, 2, 5, 7, 10, 13, 21, 22, 26, 27, 28, 29, 36, 49, 54, 63, 69, 70, 73, 102, 112, 113, 114, 118, 121, 130, 132, 136, 137, 139, 143, 148, 167, 168, 169, 170, 171, 172, 173, 174, 177, 182, 183, 192, 199, 200, 202, 203, 204, 207
environmental characteristics, 60
environmental conditions, 13

environmental contamination, 56
environmental control, 200
environmental degradation, 168
environmental effects, ix, x, 69, 70, 138, 199, 201
environmental factors, 127, 197
environmental impact, vii, viii, 1, 3, 5, 12, 25, 26, 27, 29, 54, 61, 63, 113, 114, 115, 118, 120, 121, 123, 133, 137, 168, 171, 201, 203, 207
environmental issues, 113, 171
environmental laws, ix, 111
environmental performance, ix, 111
environmental protection, 106, 171, 178
Environmental Protection Agency, 23, 71, 72, 113, 120, 138, 164, 174, 175, 195, 197
enzyme, 66
EPA, 6, 7, 10, 23, 112, 113, 116, 119, 120, 121, 122, 123, 125, 127, 128, 129, 131, 132, 138, 151, 164, 175, 202
epoxy resins, 71
equilibrium, 75, 79, 87, 97, 177, 193, 196
equipment, viii, 25, 180
erosion, 127
erosive processes, vii, 1
ester, 103, 104
ethanol, 79
ethylene, 119
Europe, x, 70, 127, 178, 196, 199, 201, 202
European Commission, 169, 178, 179, 193, 194
European Community, 71
European Parliament, 194
European policy, 2
European Union (EU), x, 6, 71, 167, 169, 170, 178, 179, 191, 193, 195, 196, 199, 202
eutrophication, viii, 53, 55
evaporation, 112, 121, 124, 126, 151
evapotranspiration, 55, 64, 125, 130, 134, 141, 142
Evapotranspiration covers, 141, 143
evidence, 33, 70, 121, 170, 178, 196
evolution, vii, 2, 3, 26, 44, 45, 46, 49, 116, 120, 200, 207
excavations, 37
expenditures, 185, 186
experimental condition, 77, 82, 83, 87, 89
exploitation, 49, 169, 171, 172
exports, 196
exposure, x, 79, 95, 127, 147, 152, 153, 159, 160, 162, 163
extraction, 46, 103, 176
extracts, 2

F

factories, 71
families, 145
fatty acids, 116, 205
fauna, 3
fermentation, 203
fertility, 63
fertilizers, 61, 117
fibrate, 107
field surveys, viii, 25
filters, 134, 142
filtration, 210
financial, 65, 180
financial support, 65
Finland, 51, 116, 140, 195
fire hazard, 122
fires, 42, 118
flexibility, 123, 176
flight, 42
flocculation, 60, 103, 104
flora, 114, 131
flora and fauna, 114, 131
flue gas, 202
fluid, viii, 25, 27, 33, 39, 40, 58
fluid flow direction, viii, 25
fluidized bed, 104
fluorine, 11
food, 12, 61, 169, 176, 195, 200
food production, 195
food web, 61
forage crops, 140
forbs, 14, 128, 131
force, 179, 185
forest management, 72
formation, viii, 50, 53, 54, 55, 56, 57, 59, 60, 61, 73, 80, 83, 87, 100, 123, 203, 212
France, 118, 202
free radicals, 74
freezing, 123
freshwater, 104
fruits, 208
fungi, 139, 207

G

gamma radiation, ix, 70, 73, 75, 80, 94, 95, 98, 99, 107
garbage, 200
gasification, 202
GDP, x, 167, 170, 177, 178, 179, 180, 181, 183, 184, 185, 190, 191, 192
gel, 102
genus, 117
geochemical assessment, vii, 1
geology, viii, 26, 29, 30, 31, 33, 49, 56

geometry, viii, 25, 50
geophysical methods, vii, viii, 25, 26, 27, 28, 29, 30, 35, 36, 44, 45, 50, 51
Georgia, 130
Germany, 117, 118, 179, 194, 202, 210
GHG, 174, 175
global warming, 112, 117
graph, 95
grass(s), 5, 14, 15, 70, 72, 117, 121, 123, 126, 127, 128, 129, 131, 136, 144
grasslands, 14, 15
gravitational force, 56
gravity, viii, 26, 40, 41, 51, 52
grazing, 6, 19
Greece, 23, 52, 57, 67, 116, 200
green alga, 73, 104
greenhouse, 60, 61, 116, 134, 142, 145, 148, 182, 201
greenhouse gas(s), 60, 134, 142, 145, 148, 182, 201
greenhouse gas emissions, 145
groundwater, vii, viii, ix, 1, 19, 21, 23, 25, 26, 27, 28, 29, 30, 31, 33, 34, 38, 44, 45, 46, 47, 48, 49, 54, 55, 57, 58, 59, 60, 61, 63, 64, 69, 70, 102, 112, 113, 115, 116, 120, 135, 136, 138, 139, 140, 142
growth, 2, 15, 55, 62, 73, 108, 117, 125, 126, 127, 128, 131, 133, 170, 178, 191, 210
growth rate, 127
Guangzhou, 164
guidelines, 195

H

habitat(s), 13, 176
hazardous substances, 55, 61, 169, 176
hazardous waste(s), 23, 71, 102, 103, 114, 145, 148, 180, 203
hazards, 2, 120, 132, 207
HDPE, ix, 111, 114, 119, 122, 123, 132, 208, 209
health, vii, 1, 2, 5, 7, 10, 20, 21, 22, 103, 148, 151, 159, 161, 176, 200
health effects, vii, 10, 151, 159
heavy metals, vii, viii, 1, 2, 3, 7, 8, 10, 12, 20, 21, 53, 56, 57, 59, 61, 64, 115, 127, 206
height, 122, 151
hemicellulose, 203
hemisphere, 127
herbicide(s), ix, 69, 70, 72, 75, 104, 106, 107, 118
herbivores, vii, 1, 19
heterogeneity, 19, 27, 43
heterotrophic microorganisms, 60
high density polyethylene, 208
historical data, 181, 185, 188, 189, 191
history, vii, viii, 20, 25, 26, 52
human, 2, 54, 55, 56, 72, 148, 152, 169, 176, 182, 200, 202, 204
human development, 54
human exposure, 182
human health, 2, 54, 55, 56, 169, 176
humidity, 123
Hungary, 202
hunting, 200
hybrid, 127, 129, 131, 139
hydrocarbons, 6, 203
hydrochemical data, viii, 26, 45
hydrogen, 87, 95, 101, 105, 108, 116, 203, 204, 205
hydrogen atoms, 108
hydrogen peroxide, 105
hydrogen sulfide, 204
hydrogeology, viii, 26, 29, 30, 31, 33
hydrolysis, 203, 205, 211
hydrophobicity, 81, 85, 100
hydrosphere, 61
hydroxide, 206
hydroxyl, 73, 76, 78, 95, 101, 105, 108
hydroxyl groups, 78
hygiene, 200
hypothesis, 88, 97

I

ideal, 5, 54
identification, 27, 28, 34, 39, 49, 133, 171, 174, 204
IEA, 142
illumination, 77, 105
image(s), 26, 43, 44, 84, 113, 153
imagery, 42
immersion, 83
immobilization, 206
impact assessment, 175, 176, 193, 194, 195
imports, 196
improvements, 49, 171, 173, 174
in vivo, 165
income, 177, 178, 180, 182
independent variable, 80
India, 116, 140, 144, 195
indirect effect, 83
industrial chemicals, 205
industrial waste(s), vii, viii, 1, 2, 3, 6, 10, 11, 25, 36, 40, 58, 65, 200
industry, 103, 176, 180, 182
infrastructure, 26, 174, 182, 200
ingestion, 10
inhibition, 78, 92, 99, 100, 210
insects, 10, 122
integration, 47

integrity, 13, 28, 31, 118, 132, 139
interface, 74, 108
interference, 44
Internal structure, 40
International Bank for Reconstruction and Development, 197
international standards, 176, 194
international trade, 103
inversion, 37, 51
ions, 54, 59, 64, 66, 73, 79, 83, 85, 101, 105, 205
Iowa, 141
Iran, 116, 135
Ireland, 140, 179, 195
iris, 164
iron, 56, 59, 137, 205
irradiation, 77, 80, 94, 95, 96, 98, 99, 100, 105, 106, 107
irrigation, 20, 121, 136
Islam, 101
isolation, 149, 152
isotherms, 42, 81, 83, 84
issues, 3, 26, 131, 137, 138, 144, 169, 176, 191, 193
Italy, 50, 122, 135, 138, 140, 142, 143, 145, 179

J

Japan, 103, 135

K

K^+, 56
Kenya, 197
kill, 10
kinetic model, 94, 107
kinetic parameters, 95
kinetics, 79, 83, 86, 94, 95, 97, 101, 107
Kuznets Curve, 178, 193, 195

L

land disposal, 112
Landfill capping, ix, 111, 120, 121, 133, 138, 139
landscape(s), 122, 135, 203
larvae, 210
Latin America, 54, 55
laws, ix, 111, 169
layering, 34
LCA, x, 167, 170, 171, 173, 174, 175, 176, 177, 192, 193, 194
leachates, vii, viii, ix, xi, 1, 2, 3, 5, 7, 8, 10, 19, 20, 21, 22, 34, 35, 36, 37, 38, 53, 54, 55, 57, 60, 61, 62, 63, 64, 66, 69, 70, 75, 101, 102, 103, 107, 143, 199, 206, 208, 209, 210, 212, 214, 215, 216, 217, 220, 221
leaching, ix, 59, 61, 62, 64, 69, 70
lead, 5, 26, 46, 56, 58, 59, 61, 108, 115, 118, 168, 178, 180, 205
leakage, viii, 25, 28, 33, 204
leaks, 33, 49
learning, 171
LEED, 195
legislation, 169, 208
life cycle, 170, 171, 172, 173, 175, 176, 193, 194, 195
life-cycle assessment, x, 167, 170, 174, 192, 193, 194
lifetime, 54, 55, 73, 152, 159, 160
light, 79, 80, 103, 105, 151
lignin, 56
limestone, 6, 11
lipids, 203, 205
liquid interfaces, 108
liquid phase, 59
liquids, 76, 105, 143
lithology, 28
Lithuania, 180
liver, 10, 72
liver damage, 72
livestock, 65, 66
local community, 114
local conditions, 27, 125, 126
localization, 205
logging, 28, 29
logical reasoning, 175
low risk, 118, 159
lying, 112

M

magazines, 12
magnesium, 56, 205
magnitude, 45, 117, 174
majority, 155, 201
man, 26
management, ix, x, 10, 111, 113, 114, 131, 133, 136, 143, 144, 168, 169, 170, 171, 174, 193, 194, 195, 197, 199, 200, 201, 203, 204, 205, 207
manganese, 56, 205
manipulation, 139
manufacturing, 171, 183, 184, 185, 186, 187, 188, 192
manure, 66, 200
mapping, viii, 26, 29, 40, 41, 50
marshy areas, vii, 1, 19
mass, 7, 22, 56, 90, 93, 144, 175, 179, 205

mass spectrometry, 7
material resources, 171
materials, vii, ix, 1, 6, 10, 34, 37, 42, 43, 56, 69, 70, 74, 76, 108, 112, 121, 125, 168, 169, 171, 172, 173, 174, 175, 177, 180, 181, 202
mathematics, 177
matrix, 106
matter, 27, 56, 59, 60, 76, 83, 101, 200, 203, 205, 206, 207, 213, 214
measurement(s), x, 41, 42, 43, 128, 137, 143, 151, 155, 167, 192, 210, 212
meat, 174
media, 27, 33, 34, 96, 144
median, 155
Mediterranean, v, vii, 1, 3, 7, 11, 14, 15, 19, 21, 22, 23, 67
Mediterranean climate, 7, 19, 23, 67
Mediterranean countries, 11
mercury, 7, 22, 205
Mesopotamia, 200
metabolism, 137
metal ion(s), 205
metals, 9, 10, 12, 22, 29, 56, 59, 205, 206, 209, 212
meter, 151
methane, ix, 55, 56, 60, 111, 112, 115, 116, 117, 118, 119, 120, 121, 122, 123, 124, 125, 128, 131, 132, 133, 135, 136, 137, 139, 140, 142, 143, 144, 145, 147, 148, 149, 151, 153, 154, 155, 203, 204, 206, 217
methane emission rate, x, 147, 153
methemoglobinemia, 55
methodology, 47, 107, 136, 137, 173, 174, 175, 182
Mexico, 69, 128, 130, 137
Mg^{2+}, 56, 79
mice, 10, 103
microbial cells, 60
microbial community, 60
microgravity, 50
micronutrients, vii, 1, 13, 20, 22
microorganism(s), 10, 60, 73, 83, 84, 85, 100, 117, 124, 153, 203, 205, 206
micropollutant, ix, 70, 107
migration, 50, 58, 71, 103, 116, 136, 153, 162, 163
mine soil, 136
mineralization, 34, 60, 73, 77, 80, 98, 100
Minneapolis, 50
mixing, 49
modelling, 129, 138, 140
models, x, 32, 37, 79, 127, 129, 130, 131, 136, 167, 170, 171, 177, 183, 192, 193
modernization, 200
moisture, 14, 15, 54, 55, 56, 125, 131, 136, 143, 145, 153, 203, 205, 206
moisture content, 136, 153, 206
molecular weight, 57, 73, 98, 116
molecules, 57, 59, 75, 82, 83, 95, 96, 100
Montana, 128
mosaic, 15
Moscow, 117
MSW, 5, 8, 66, 102, 112, 113, 114, 116, 122, 128, 135, 142
multimedia, 51
municipal solid waste, 5, 20, 21, 22, 23, 54, 67, 102, 117, 136, 138, 143, 148, 163, 174, 193, 195, 196, 201, 203, 211
municipal waste disposal, x, 199, 207
mutagen, 71

N

Na^+, 56, 87
NaCl, 76, 210
nanoparticles, 105
Nanostructures, 105
National Institute for Occupational Safety and Health, 71
National Strategy, 179, 195
native species, 111, 127, 131, 145
natural depressions, vii, 1, 19
natural disaster(s), 126
natural resources, 168, 176
nematode, 23
Netherlands, 22, 179, 194, 195
neutral, 136, 206, 212
New South Wales, 181
New Zealand, 1
nickel, 205
Nigeria, 20
nightmares, 135
nitrates, 6, 11, 13
nitrification, 60, 61, 140, 215
nitrite, 60, 64, 209, 212, 214, 215
nitrogen, viii, 6, 20, 53, 54, 55, 56, 57, 59, 60, 61, 65, 66, 117, 204, 205, 209, 212, 214, 215
nitrogen fixation, 60
nitrous oxide, 60, 61
N-N, 57, 214, 215
North America, 127, 143
Norway, 116, 117, 193
Norway spruce, 117
nuclear magnetic resonance, 135
nuclei, 152
nuisance, 114
nutrient(s), viii, 3, 7, 9, 10, 11, 14, 15, 53, 55, 56, 60, 61, 62, 126, 127, 142, 206
nutrient concentrations, 10

nutrition, 61, 63
nutritional imbalance, 61

O

obesity, 72
OECD, 112, 169, 170, 196
OH, 74, 86, 87, 98, 108
oil, 114
oilseed, 140
one dimension, 144
operations, 27, 31, 33, 34, 47, 118, 172, 207, 214
opportunities, 171, 197
organic chemicals, viii, 53, 61
organic compounds, 7, 8, 10, 11, 12, 20, 21, 56, 73, 86, 95, 102, 105, 116, 148, 205
organic matter, 3, 10, 12, 54, 56, 57, 58, 59, 60, 65, 73, 77, 98, 100, 101, 103, 104, 116, 126, 205, 206, 208, 212, 213, 214
organic solvents, 115
OSHA, 159
osmosis, 119, 141
oxidation, 72, 73, 74, 75, 76, 77, 85, 91, 92, 100, 101, 102, 103, 104, 105, 106, 117, 118, 119, 120, 123, 125, 131, 132, 135, 136, 137, 140, 145, 212, 213
oxidation rate, 117, 118
oxygen, 56, 74, 86, 117, 118, 119, 123, 132, 203, 204, 208
ozonation, 73, 77, 104
ozone, 91, 92, 108

P

Pacific, 135
parallel, 75
participants, 182
pasture(s), 5, 6, 15, 72
pathogens, viii, 53, 56, 61, 64
pathways, 39
PCBs, 12
percentile, 155, 157
percolation, ix, 58, 69, 70, 111, 122, 124, 125, 126, 127, 128, 129, 130, 132
performance indicator, 196
permeability, 36, 56, 123
permission, 49
permit, 81
personal hygiene, 200
pesticide(s), 10, 21, 56, 69, 70, 72, 75, 105, 107, 108, 205
PET, 175
Petroleum, 49
pH, 4, 6, 7, 8, 9, 10, 11, 38, 47, 54, 56, 57, 58, 59, 72, 73, 75, 76, 77, 79, 80, 81, 82, 83, 84, 85, 86, 87, 88, 89, 90, 93, 94, 96, 97, 98, 99, 100, 106, 109, 115, 153, 203, 205, 206, 207, 209, 210, 212
pharmaceutical(s), 75, 107, 208
phenol, 82, 91, 92
Philadelphia, 141
phosphate(s), 6, 205
phospholipids, 85
phosphorus, viii, 15, 53, 55
photocatalysis, ix, 69, 74, 78, 91, 92, 106
photocatalysts, 74, 79, 106
photocatalytic degradation, ix, 70, 74, 78, 91, 105, 106, 107
photochemical degradation, 78
photodegradation, 73, 74, 77, 78, 80, 85, 86, 88, 90, 92, 93, 101, 107
photo-degradation, ix, 70
photolysis, 73, 80, 85, 101, 104
Photolysis, ix, 69
photooxidation, 73, 75, 77, 78
phthalates, vii, 2, 10, 23, 70, 71, 100
phthalic acid, ix, 70, 103, 104, 107
physical characteristics, 54
physical properties, 27, 41, 61, 127
physico-chemical parameters, 209
Phytocapping, v, ix, 111, 123, 124, 125, 131, 134, 144, 145
phytoremediation, 19, 127, 138
pioneer species, 13
pitch, 151
plant growth, viii, 21, 53, 61, 101, 116, 124, 126, 131
plants, vii, ix, 1, 5, 7, 10, 19, 20, 22, 60, 61, 71, 111, 117, 118, 124, 125, 126, 127, 129, 131, 133, 136, 139, 143, 148, 152, 153, 164, 200
plaque, 200
plasticizer, 10, 123, 133
plastics, 71, 114, 176
playing, 27
policy, 168, 170, 172, 192
policy makers, 172
politics, 177
pollutants, vii, ix, 2, 3, 5, 6, 8, 10, 12, 13, 19, 23, 54, 55, 59, 61, 69, 70, 71, 72, 74, 75, 76, 80, 85, 94, 95, 96, 101, 102, 108, 127, 148, 178, 202, 204, 205, 206
pollution, xi, 2, 3, 5, 6, 10, 13, 15, 19, 22, 23, 54, 59, 61, 101, 115, 116, 135, 140, 141, 178, 199, 200, 204
polycarbonate(s), 71, 103, 104
polycarbonate plastics, 71

polychlorinated biphenyl, 6, 21
polycyclic aromatic hydrocarbon, 22
polyesters, 71, 104
polymer, 71, 119
polymer structure, 71
polymers, 69, 70
polypropylene, 209
polysaccharides, 56
polyurethane, 77
polyurethane foam, 77
population, 2, 5, 48, 112, 151, 174, 177, 178, 181, 200, 206, 209
population density, 178
population growth, 177, 200
porosity, 36, 48, 61, 126
Portugal, 23, 25, 29, 42, 49, 50, 51, 52
potassium, viii, 53, 55, 56, 205
potato, 102
poultry, 66
precipitation, 8, 10, 57, 59, 60, 112, 125, 130, 131, 144, 151, 205, 206
preparation, 62, 200, 210
prevention, x, 27, 167, 168, 169, 170, 171, 174, 175, 176, 178, 179, 180, 181, 182, 183, 185, 188, 189, 191, 192, 193, 194, 196, 197, 207
principles, x, 144, 167, 171, 173, 201
processing systems, vii
procurement, 171
producers, 175
profit, 27
prognosis, 5
project, 32, 35, 44, 49, 75, 123, 128, 129, 176, 178, 193, 209
propagation, 45
prosperity, 202
protection, 5, 33, 48, 140, 170, 171, 181
proteins, 203, 205
prototype, 141, 142
public health, 113, 200
public opinion, 26, 171
pumps, ix, 111, 124
purchasing power, 202
PVC, 103, 119, 122, 123, 132, 141, 183

Q

quality of life, 176
quantification, 171
Queensland, 112, 113, 114, 122, 131, 134, 138

R

radar, viii, 26, 50
radiation, v, ix, 69, 73, 74, 75, 80, 85, 92, 93, 94, 95, 96, 97, 101, 105, 106
radical formation, 108
radical mechanism, 104
radical reactions, 96
radicals, ix, 70, 73, 74, 76, 78, 80, 85, 86, 87, 88, 92, 94, 95, 96, 97, 100, 101, 105, 107, 108
radioactive waste, 28, 140
rainfall, 3, 54, 55, 111, 121, 123, 126, 128, 129, 131, 132, 133, 136, 144
rainy season, x, 147, 153, 155, 161
rangeland, 72
raw materials, 181, 202
reaction rate, 73, 85, 88, 95, 98, 100, 101
reaction rate constants, 95, 101
reaction temperature, 86
reaction time, 79
reactions, 10, 59, 60, 74, 85, 87, 95, 108, 109, 206
reactivity, 76, 88, 105
real time, 49
receptors, 151
recession, 179, 188, 189, 190
recognition, 171
recombination, 74, 79, 95, 97
recovery, vii, viii, 2, 15, 25, 26, 47, 113, 120, 121, 145, 168, 170, 174, 202, 204
recovery plan, 120
recovery process(s), vii, viii, 2, 25
recycling, vii, x, 3, 54, 113, 114, 117, 138, 168, 169, 172, 174, 175, 178, 182, 197, 199, 201, 202, 207
reductive species, ix, 70
reform, 177
refuse zoning, viii, 25, 49
regenerate, 126
regeneration, 51
Registry, 71, 103
regression, 178, 179, 181, 185, 187, 188, 189, 190, 191
regression analysis, 185, 191
regulations, 113, 122, 148, 174
regulatory bodies, 174
regulatory requirements, 126
rehabilitation, 22, 135, 138
reliability, 130, 176
remediation, ix, 21, 29, 49, 71, 102, 111, 114, 119, 121, 132, 134, 144
remediation procedures, 49
remote sensing, 42, 50
repellent, 10
requirements, xi, 113, 171, 193, 199

researchers, 7, 71, 120, 123, 131, 204, 207, 208
Residential, 148, 193
residues, viii, 19, 22, 53, 54, 56, 61, 63, 202
resistance, 58, 142
resolution, 27, 43, 44, 64, 152
resource availability, 176
resources, ix, 69, 70, 135, 170, 173, 178, 181, 193
response, 107, 210
restaurants, 184
restoration, 116
retail, 176, 184
reusability, 76
revegetation, vii, 2
rice field, 148
risk(s), vii, viii, x, 1, 5, 6, 7, 10, 19, 20, 21, 22, 27, 53, 54, 55, 56, 58, 60, 61, 62, 63, 64, 72, 135, 147, 148, 151, 152, 159, 160, 161, 204, 207
risk factors, 10
risk management, 135
rodents, 114, 122
room temperature, 6
root(s), 64, 123, 124, 125, 126, 127, 131, 139, 152, 210
root growth, 126, 210
root system, 64
rotifers, 10
roughness, 151, 155
Royal Society, 137
rules, 173, 176
runoff, 3, 66, 112, 125, 130
Russia, 140

S

safety, 54
salination, 21
salinity, vii, 1, 8, 10, 13, 66, 126, 127, 135
salt concentration, 3, 61, 206
salts, 7, 8, 20, 22, 56, 115, 206
saturation, 83, 120, 212
savings, 182
scarcity, 203
scavengers, 85, 100, 101
science, 65, 108, 177
scientific progress, 200
scope, 28, 148, 173, 177, 196
Scots pine, 117
sea level, 36
seasonal changes, 123
seasonal flu, 19
sediments, 31, 43, 50, 72, 107, 117
seed, 3
segregation, 113
selectivity, 70, 72
semiconductor, 74
sensing, 42, 50
sensitivity, 13, 128, 209
sensors, 42, 208
serum, 5
services, 66, 174, 183, 184, 185, 188, 189, 191, 192
sewage, 54, 55, 60, 66, 102, 117, 200, 202
shape, vii, viii, 25, 47, 172
sheep, 6, 10, 15
shelter, 131
shoot, 152
showing, 38, 62, 85, 155
shrimp, 209, 210, 212
shrubs, 123, 126, 127, 128, 131
signs, 15, 33
silica, 80, 102
simulation(s), 127, 138, 140, 151, 177
Singapore, 144
SiO_2, 105
skin, 10
Slovakia, 179, 180
sludge, 54, 65, 66, 72, 103, 119, 135, 175, 202
SO_4^{2-}, 4, 7, 56, 101, 115
social indicator, 183
social structure, 182
society, x, 167, 174, 192
sodium, 56, 108, 205
software, viii, 7, 25, 37, 48, 49, 129, 139, 144, 175, 176
soil contamination, vii, viii, 25, 33
soil erosion, 126
soil particles, 58, 59
soil pollution, 12, 59
soil seed bank, 126
soil type, 58, 62, 117
sol-gel, 79
solid phase, 54, 59
solid waste, vii, ix, 1, 7, 41, 54, 55, 56, 112, 116, 120, 134, 135, 137, 143, 144, 147, 148, 149, 151, 153, 155, 159, 161, 163, 194, 196, 197, 203
solubility, 10, 59, 60, 118, 205, 206
solution, 46, 54, 59, 64, 72, 73, 75, 76, 78, 79, 80, 82, 83, 85, 86, 87, 92, 93, 98, 100, 102, 103, 104, 105, 107, 108, 206, 210
solvents, 10
sorption, 81
South Africa, 116, 143
SP, 28, 29, 35, 36, 39
Spain, 1, 4, 5, 20, 21, 22, 23, 195
speciation, 108
species, ix, 3, 7, 13, 14, 15, 16, 17, 19, 20, 21, 61, 69, 70, 73, 75, 81, 83, 92, 94, 95, 96, 97, 98, 100,

101, 111, 117, 126, 127, 128, 131, 132, 133, 139, 141, 152, 153, 162, 176
species richness, 13, 15
specific heat, 50
specific surface, 79
spectrophotometry, 6
speech, 10
spending, 177, 178, 183, 192
SS, 115
stability, 74, 76, 101, 104, 131
stabilization, 8, 19, 56, 60, 206, 208
stabilizers, 69, 70
standard deviation, 9, 10, 11
state(s), 2, 15, 36, 59, 71, 73, 117, 120, 130, 138, 142, 151, 170, 179, 181, 183, 185, 207
statistics, 176, 177, 179, 183, 192
steel, 149
storage, ix, 111, 124, 125, 130, 148, 207
storms, 126
stream waters, viii, 53
stress, 23, 119
striatum, 16
structural changes, 180
structure, viii, 5, 25, 40, 41, 50, 52, 58, 66, 78, 94, 98, 140, 145
styrene, 161
subjectivity, 173
substitution, 61, 175, 176
substrate(s), 6, 11, 12, 73, 124, 206
succession, 13, 20, 56
sugarcane, 66, 70, 72
sulfate, 56, 73, 108, 205
sulfur, 60
sulphur, 206
Sun, 103, 104
suppliers, 171
surface area, 3, 81, 83, 91
surface layer, 130
surface modification, 108
surface properties, 108
survival, 125, 132
suspensions, 78, 105
sustainability, xi, 121, 125, 126, 127, 132, 133, 193, 195, 199
sustainable development, 201, 207
sustainable waste management, x, 197, 199, 201, 207, 211, 218
Sweden, 10, 116, 134, 178, 179, 195, 196
swelling, 141
Switzerland, 202
syndrome, 13, 55
synergistic effect, 74, 91

T

target, x, 167
taxa, 15
taxes, 174
TCE, 136
techniques, 6, 20, 26, 33, 49, 50, 52, 54, 119, 123
technological advances, 54
technology(s), viii, ix, 53, 54, 69, 71, 77, 107, 114, 116, 118, 119, 120, 121, 123, 132, 133, 137, 200
teflon, 209
temperature, 41, 42, 43, 56, 77, 79, 86, 88, 93, 121, 129, 136, 151, 155, 156, 208, 210, 211, 213
temporal variation, 130
terrestrial ecosystems, 23
test procedure, 210
testing, 22, 141, 142, 143, 210
texture, 58, 83, 106, 125, 145
Thailand, v, ix, 147, 148, 163, 164
thermodynamics, 79
thin films, 106
threats, ix, 69, 70
tides, 171
time periods, 192
time series, 177, 178, 184, 185, 188, 189, 191, 192
tissue, 152
toluene, 152, 153, 155, 161
total energy, 85
toxic byproducts, ix, 69, 70
toxic effect, 103
toxic products, 169
toxic substances, 2
toxicity, ix, x, 5, 8, 10, 69, 70, 73, 92, 93, 98, 100, 101, 102, 104, 147, 152, 159, 161, 163, 209, 210, 212
toys, 71, 103
trade, 182, 184, 192
transference, 59
transformation(s), 59, 60, 66, 92, 109, 191, 203
transparency, 48
transpiration, 111, 112, 126, 127, 130, 131, 133
transport, 3, 58, 59, 101, 103, 142, 144, 148, 156, 159, 175
transportation, 151, 152, 171, 173, 200, 202
treatment, viii, 51, 53, 54, 55, 57, 58, 60, 66, 72, 73, 75, 76, 78, 85, 89, 90, 91, 92, 98, 100, 101, 102, 103, 104, 105, 107, 108, 114, 117, 119, 121, 134, 135, 136, 143, 169, 174, 200, 202
tremor, 10
trial, 117, 123, 126, 129, 134, 137
trophic networks, vii, 1, 3, 12
turnover, 136

U

ultrasonic frequency, 77
unification, 177
unions, 175
United, 65, 197
United Kingdom (UK), 21, 22, 23, 50, 52, 65, 117, 136, 142, 144
United Nations, 197
United States, 65, 197
updating, 175
urban, vii, 1, 2, 3, 12, 14, 21, 22, 112
urban population, 112
urine, 5
USA, 20, 23, 50, 52, 135, 138, 139, 140, 141, 143, 144, 145, 148, 164, 196, 209
USSR, 142
UV, ix, 70, 73, 74, 75, 77, 78, 79, 80, 86, 87, 88, 90, 91, 92, 93, 100, 101, 103, 104, 105, 106, 107
UV irradiation, 78
UV light, 100
UV radiation, 75, 80, 87

V

validation, 142
valuation, 176
variables, x, 3, 5, 6, 7, 8, 10, 58, 80, 167, 177, 178, 183, 191, 192
variations, 28, 34, 42, 98, 176, 192, 211
vector, 176
vegetables, 208
vegetation, 6, 7, 14, 22, 43, 55, 72, 115, 118, 124, 125, 126, 130, 133, 134, 138, 139, 142
vehicles, 3, 49, 174, 178
velocity, 30, 32, 33, 35, 47
viruses, 200, 206
vision, 29
visualization, 48
VOCs, ix, 147, 148, 149, 150, 151, 153, 155, 159, 161, 162, 163
volatile organic compounds, ix, 147, 148, 161
volatility, 10
volatilization, 61, 62, 66

W

Wales, 180, 197
Washington, 108, 130, 135, 138, 197
waste activities, x, 167, 170
waste disposal, vii, x, 1, 2, 5, 6, 10, 12, 19, 23, 29, 51, 52, 54, 139, 147, 148, 153, 161, 162, 163, 170, 199, 202, 205, 207
waste disposal sites, vii, 1, 2, 5, 12, 19
waste incineration, 116
waste management, ix, x, 69, 70, 169, 170, 171, 172, 173, 174, 175, 177, 178, 179, 182, 183, 193, 194, 196, 197, 199, 200, 201, 202, 207, 211
waste treatment, 174, 202, 207
waste water, 116
wastewater, 65, 66, 70, 71, 72, 98, 99, 100, 101, 134, 139, 209
water quality, 50, 101
water resources, 2
wavelengths, 74
wealth, 179
welfare, 103
wells, 19, 32, 36, 38, 45, 49, 64
Western Australia, 116, 137
wetlands, vii, 1, 5, 10, 11, 141
wetting, 123
wild animals, 3, 10
wildlife, ix, 69, 70
wind speed, 151, 156
windows, 153
Wisconsin, 140, 195
wood, 14, 175
woodland, 137
workers, x, 147, 148, 151, 157, 159, 160, 161, 163
World Bank, 197
worldwide, viii, 53, 70, 127, 145, 188, 189, 190, 200

Y

yield, 61, 95, 96, 97, 100, 143, 171, 174, 192

Z

zinc, 205
ZnO, 74, 105